Smart Grids 2017

MATEC Web of Conferences Volume 141 (2017)

Tomsk, Russia
9 - 13 October 2017

Editors:

Alla A. Kozyreva
Alena O. Zhdanova
Geniy V. Kuznetsov

ISBN: 978-1-5108-5352-2

Printed from e-media with permission by:

Curran Associates, Inc.
57 Morehouse Lane
Red Hook, NY 12571

Some format issues inherent in the e-media version may also appear in this print version.

Printed by Curran Associates, Inc. (2018)

For additional information, please contact EDP Sciences – Web of Conferences at the address below.

EDP Sciences – Web of Conferences
17, Avenue du Hoggar
Parc d'Activité de Courtabœuf
BP 112
F-91944 Les Ulis Cedex A
France

Phone: +33 (0) 1 69 18 75 75
Fax: +33 (0) 1 69 28 84 91

contact-edps@webofconferences.org

Additional copies of this publication are available from:

Curran Associates, Inc.
57 Morehouse Lane
Red Hook, NY 12571 USA
Phone: 845-758-0400
Fax: 845-758-2633
Email: curran@proceedings.com
Web: www.proceedings.com

TABLE OF CONTENTS

Statement of Peer review

In submitting conference proceedings to MATEC Web of Conferences, I certify to the Publisher that I adhere to the **Policy on Publishing Integrity** of the journal in order to safeguard good scientific practice in publishing.

1. All articles have been subjected to peer review administered by the proceedings editors.
2. Reviews have been conducted by expert referees, who have been requested to provide unbiased and constructive comments aimed, whenever possible, at improving the work.
3. Proceedings editors have taken all reasonable steps to ensure the quality of the materials they publish and their decision to accept or reject a paper for publication has been based only on the merits of the work and the relevance to the journal.

Title, date and place of the conference

The fifth international youth forum
Smart grids 2017
October 9-13, 2017
Tomsk, Russia

Proceedings editor(s):

Alla A. Kozyreva
Alena O. Zhdanova
Geniy V. Kuznetsov

Date and editor's signature

November 22, 2017

Convection influence on ice formation process on the evaporator heat pump pipes

Vyacheslav Maksimov[1,*], and *Saloum* Amer[1]

[1]National Research Tomsk Polytechnic University, 634050 Tomsk, Russia

Abstract. The purpose of this research is to identify the effect of convection on the ice growth process on the evaporator of heat pump, and how this growth can affect the heat exchange, and the evaporator performance. An experimental analyse is done on the heat pump station which depends on water as heat source. A 2D model is built in Comsol environment depending on Brinkman and heat transfer equations. The experimental and mathematical simulation results confirm that the ice formation has a negative effect on the evaporator heat exchange efficiency, and the convection has high effect on ice growth rate and this effect increases when the pipe is closer to the heat exchanger base. The effect of block size on the convection and ice growth rate must be studied in future.

1 Introduction

The ice formation on a surface is a special case of classical Stefan problem [1], which aims to describe the temperature gradient in medium undergoing phase change.

A lot of studies are available for freezing problem in pipe flow [2, 3], but there are no studies for the effect of convection on ice formation on a cylindrical pipe, the heat transfer analysis undergoing convection and phase change is needed for the applications which include the use of pipes with temperature under zero in water.

Freezing problem in pipe flow case was studied by Jong Suk Lee from Iowa state university of USA, and they concluded that the ice layer profile at the final steady state increases linearly along the test section. Stefan number, and thermal conductivity affect the freezing rate and freezing time, but do not change the steady state ice layer profile, and the pressure drop along horizontal pipe has significant effect on ice growth rate [1].

2 Experimental technique

An experimental study is done to analyze the ice formation process and heat exchange in the heat pump evaporator (fig. 1). The experimental method is explained in [4].

* Corresponding author: elf@tpu.ru

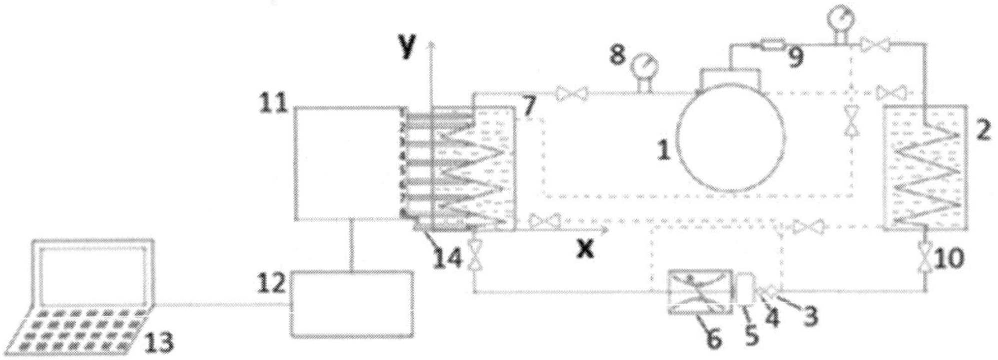

Fig. 1. Scheme of the experiment: 1 – compressor; 2 – the condenser; 3 – filter; 4 – sight window; 5 – drier; 6 – capillary tube; 7 – Evaporator; 8 – manometer; 9 – flowmeter; 10 – valve; 11 – digital thermometer; 12 – converter; 13 – computer; 14 – thermocouple.

If the effect of convection is ignored, the next equations can be used to describe the ice formation process:

$$\delta = \frac{\dfrac{\Delta T_W}{\ln \dfrac{r_W}{r_P}} \cdot 2 \cdot \pi \cdot K_W}{\left(r_W \cdot C_{P,W} \cdot \dfrac{dT_W}{dt} + \dfrac{I_i \cdot \rho_i}{dy} \right) \cdot 2 \cdot \pi \cdot r_P} \tag{1}$$

$$I_i \cdot \rho_i \cdot 2 \cdot \pi \cdot r_i \cdot \frac{dr_i}{dt} = \frac{\Delta T_i}{\ln \dfrac{r_i}{r_p}} \cdot 2 \cdot \pi \cdot K_i - \frac{\Delta T_W}{\ln \dfrac{r_W}{r_i}} \cdot 2 \cdot \pi \cdot K_W \tag{2}$$

where δ – initial ice thickness; T_w – water temperature; T_i – ice temperature; r_w –the distance between the pipe center and the temperature sensor; K_w – water thermal conductivity; K_i – ice thermal conductivity; r_p – pipe radius; r_i – ice radius; $C_{p,w}$ – water heat capacity; ρ_i – ice density; ρ_w – water density; I_i–latent heat of freezing.

Ice growth Stops when the following condition is available:

$$\frac{r_{Ice(Maximum)}}{r_{pipe}} = e^{\frac{2}{Su \cdot Nu}} \tag{3}$$

where Nu – Nusselt number.

$$Su = \frac{k_W \cdot (T_0 - T_W)}{k_I \cdot (T_I - T_0)} \tag{4}$$

where Su – number of superheating.

The convection effect on ice growth rate can be seen in Fig.2, It is noticed that this effect is higher, when the pipe height is less (when the pipe is closer to the base of heat exchanger) (c, d), and the difference between theoretical and experimental curve is bigger when the convection effect is higher which can be concluded by comparing (a), (b) with (c), (d) in fig.2.

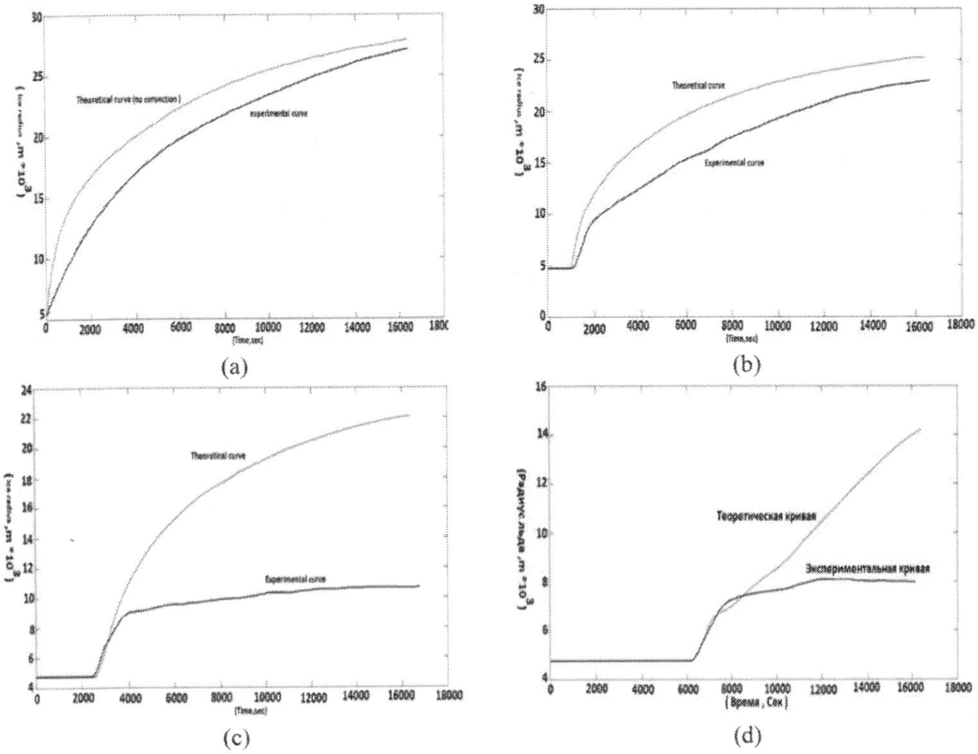

Fig. 2. Theoretical (without convection) and experimental ice radius versus time at different heights (22, 17.5, 15, 12.5) cm with (a, b, c, d) consequently.

3 Simulation study

2D model is built the block dimensions are (10×10) cm and the pipe in the center on the block with diameter 9.54 mm. The following assumptions are taken into consideration: the ice layer is static and nondeformable, the average ice permeability 10^{-12} m^2, transition interval between water and ice is 1 K. Brinkman, phase change, and heat transfer in fluid equations are used for the calculations in the model [5]. The initial water temperature is 280 K, the boundary temperature for the pipe wall is 263 K. The walls of water container are full isolated. The model was built by coupling the heat transfer with flow equations. As the water turns into ice the permeability in the freezing zone is decreased to a very small value. The temperature gradient with height in different moments can be seen in Fig.3.

And the modeling results agree with the experimental results which were published in [4], which confirm the fact that when the temperature is more than 277.15 K the temperature increase with height, but when temperature is less than 277.15, it decreases with height. The velocity diagrams are shown in fig.4. The water flow moves downwards in the first 420 sec and after that the flow direction changes to upward because of the changes in water density with temperature.

In Fig.5. (a) The heat flux around the pipe changes with time, line number 3 represent the case where the phase change is ignored. Fig.5. (b) represents the ice growth with time, line number 1 represent the case of ice growth if the effect of convection is ignored. It is noticed that when the ice thickness increases, the heat exchange rate decreases, that means that the ice growth has a bad effect on heat exchange, in spite of the fact that the coefficient

DOI: 10.1051/matecconf/201714101001

of ice thermal conductivity (2.16 W/m·K) is higher than the liquid water thermal conductivity (0.54 W/m·K).

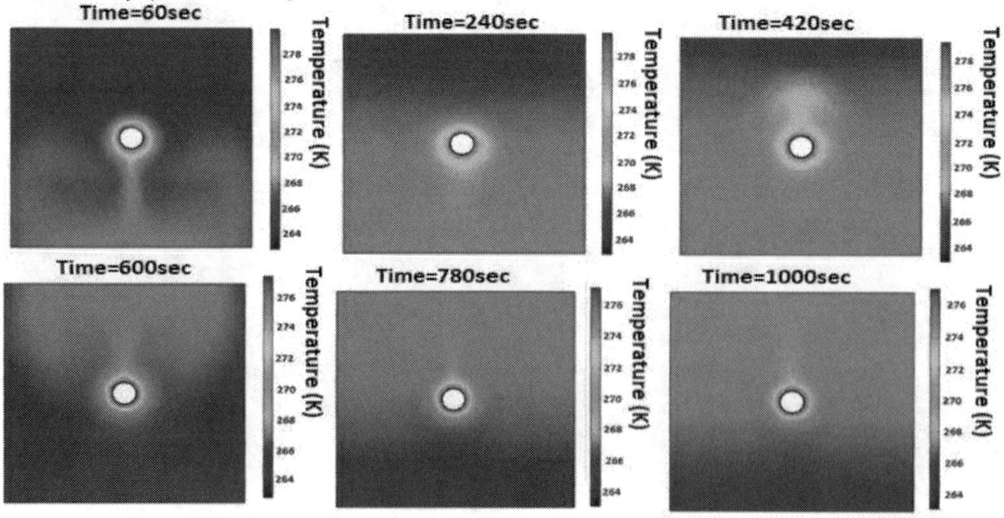

Fig. 3. Temperature diagrams.

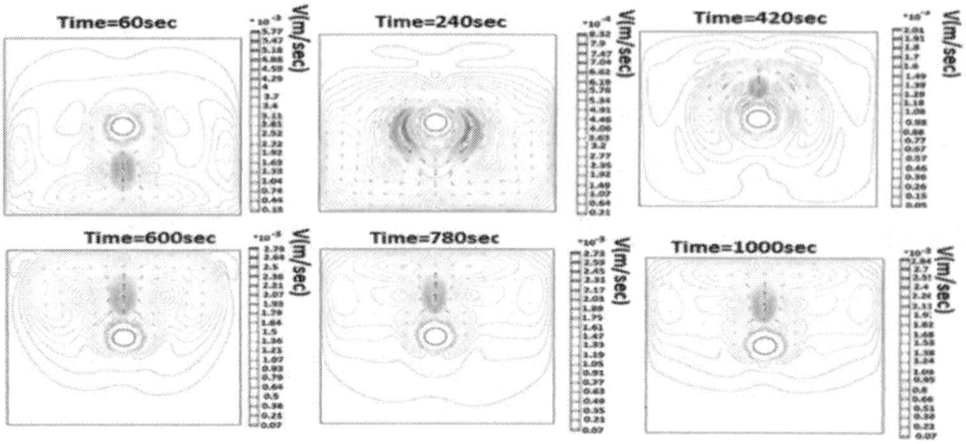

Fig. 4. Velocity diagrams.

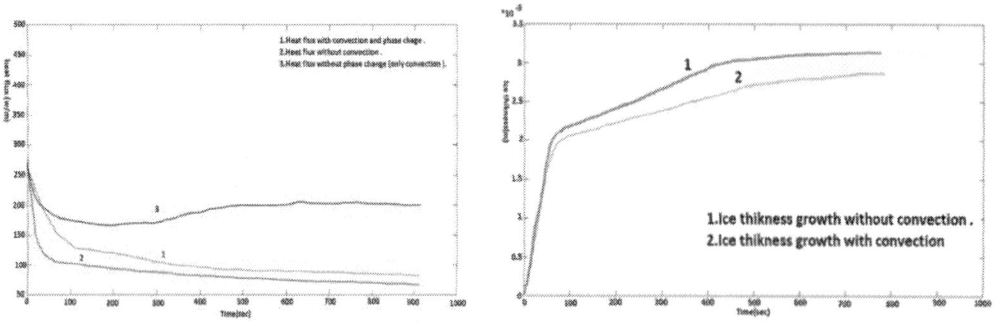

(a) (b)

Fig. 5. Heat flux with time (a), Ice thickness change with time (b).

4 Conclusion

Temperature values in the main nodes of the heat pump plant are registered, the temperature sensors are placed on the surface of the heat evaporator pipes, and at different heights in the evaporator container, the thickness of the ice on the surface of the evaporator pipe is measured. With the use of energy equation, the ice thickness change with time is also calculated.

It is concluded that the difference between the ice thickness values which are obtained analytically and experimentally can exceed 50%, depending on the height of the pipe in the evaporator container. Mathematical model is built for heat exchange case including convection and phase change, A comparing is done between heat flux values in the case of convection and without it. It can be concluded that, convection currents are formed around the pipe and these currents have a significant effect on ice formation, It is found that in case of neglecting convection in the model the ice thickness is less by 16 %.

The article results can be used to assess the heat pump performance in the case of ice formation on the evaporator pipes in water source heat pump systems «autumn, winter, and spring», which will help to improve the efficiency calculating methods for water -heat pump systems.

The work was supported by the Russian President's grant (Scientific School project - 7538.2016.8).

References

1. J.S. Lee, *Freezing problem in pipe flows* (Iowa State University, Ames, 1993)
2. T.A. Kwaleweski, A. Cublski, Int. Conf. on Heat Transfer with Change of Phase, **61**, 2 (1996)
3. V.I. Maksimov, T.A. Nagornova, V.S. Chernyshev, MATEC Web of Conferences, **23**, 01051 (2015), DOI:10.1051/matecconf/20152301051
4. V.I. Maksimov, A. Saloum, MATEC Web of Conferences, **110**, 01074 (2017), DOI:10.1051/matecconf/201711001074
5. R. Hu, Q. Liu. Excerpt from the Proceedings of the COMSOL Conference in Munich, (2016)

The influence of insulation of walls of industrial objects on thermal regime at the heating system of gas infrared radiators

Tatiana Nagornova[1,*], and *Vladimir* Emelenchuk[1]

[1]National Research Tomsk Polytechnic University, 634050 Tomsk, Russia

Abstract. The results of a numerical study of the process of heat transfer from the gas infrared emitters in the heated accommodation are represented. Simulation was conducted taking into account the heat withdrawal in the enclosing constructions and of heat exchange with the environment. The estimation of the average values of temperatures of air indoors in the dependence on the different intensity of heat withdrawal into the vertical walls is carried out (when the layer of insulation is present, and without it).

1 Introduction

The contemporary tendencies of development of science and technology are directed toward strengthening of the economic characteristics of resource use [1, 2]. The savings of the energy and thermal resources has great significance [3-6]. The minimization of expenditures for the heating of large production accommodations will make it possible to reduce capital investments and prime cost of production. At present, widely adapt alternative water of heating system, for example, gas infrared emitters (GIE) [7]. Besides the replacement of heating system possible to use different design methods of an improvement in the thermal insulation properties of premises – the internal isolation of the lateral enclosing constructions. The joint application of GIE and vertical enclosing constructions, covered with thermal insulation materials, is urgent task. Reduction in the intensity of heat withdrawal through the walls of accommodation can lead to substantial changes in the thermal condition. Similar studies earlier were not conducted, but they are sufficiently interesting. The purpose of the work – quantitative assessment is the influence of the heat insulation of walls on the thermal condition of premises with the heating system on the base of gas infrared emitters.

2 Formulation of the problem

To describe the conjugate heat transfer in a room of an industrial facility heated by GIE, a rectangular region (fig. 1) with a source of radiant heating was chosen. As a basic system of dimensionless differential equations describing the process of natural convection, a mathematical model similar to [8] is adopted. The equations of thermal conductivity for

* Corresponding author: tania@tpu.ru

each of the elements of the enclosing constructions were solved. For a gas, the differential Navier-Stokes and energy equations were considered (in the transformed variables "the function of current – vorticity – temperature"). On the external contour of the solution region, boundary conditions were adopted, taking into account heat exchange with the surrounding medium. The boundary conditions of the first kind were set on the surface of the radiant-heating source. At the lower boundary of the medium (the lower concrete base), the heat flux coming from the GIE was taken into account.

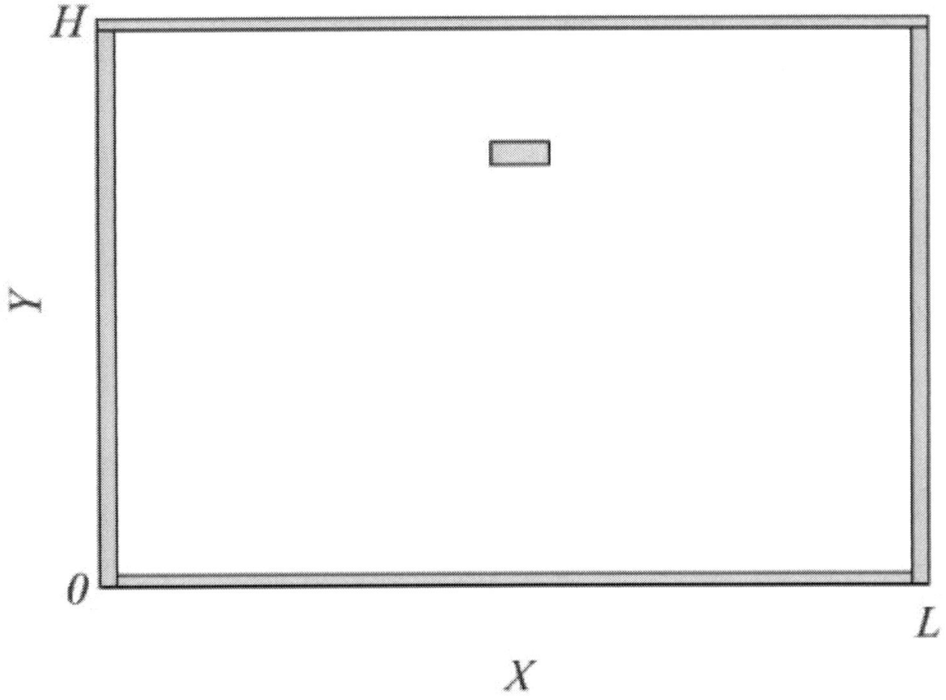

Fig. 1. Area of solution of the problem: L, H - transverse size and height of the solution region, respectively.

Equations with the appropriate initial and boundary conditions were solved by the finite-difference method on the uniform grid. To solve the system of difference equations, the sweep method and the variable direction scheme are used [9]. Was used algorithm [8,10], developed for the numerical solution of the problems of natural convection in closed rectangular regions with local energy sources.

3 Results and discussion

In Fig. 2.3 shows the dynamics of the change in temperature fields and streamlines at different time values for two different heat dissipation options in enclosing structures. The conditions for heat insulation on the vertical walls and its absence were examined.

DOI: 10.1051/matecconf/201714101002

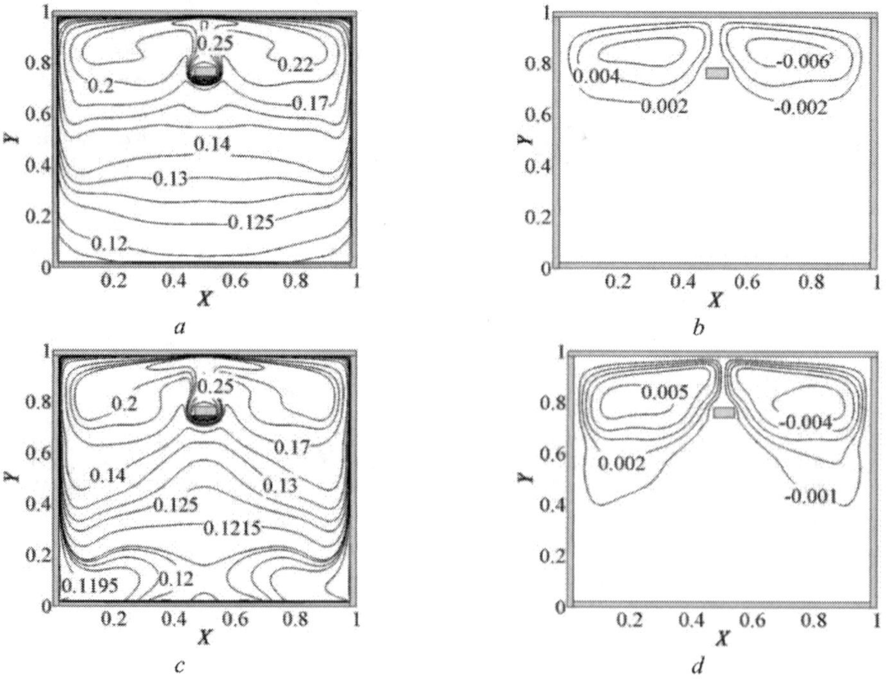

Fig. 2. Temperature fields and isolines of the current function at $\tau = 120 \cdot 10^3$ in the case of insulation (*a*, *b*) of vertical enclosing structures and without it (*c*, *d*).

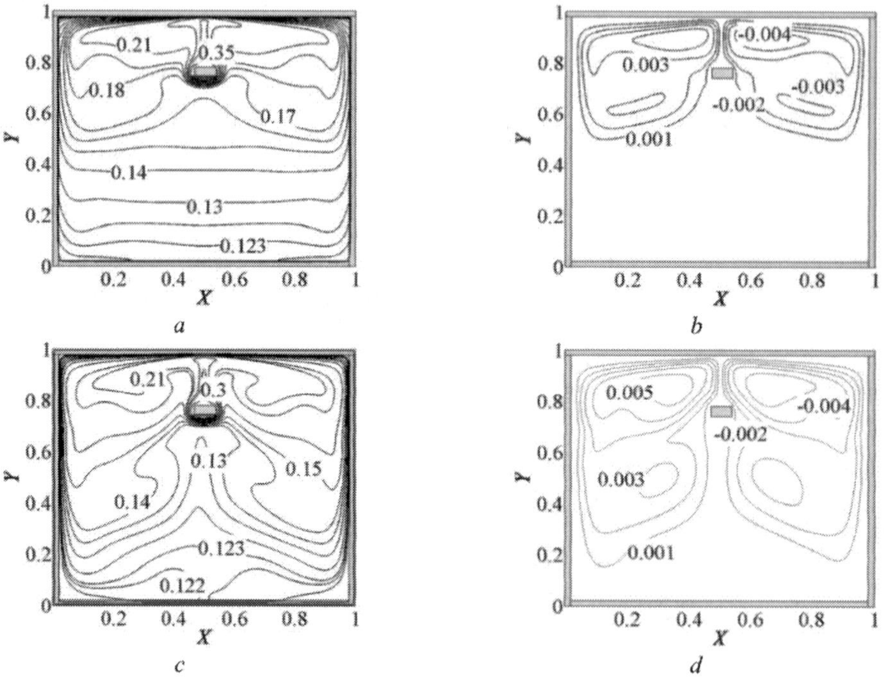

Fig. 3. Temperature fields and isolines of the current function at $\tau = 240 \cdot 10^3$ in the case of insulation (*a*, *b*) of vertical enclosing structures and without it (*c*, *d*).

Analysis of the results Fig. 2 and 3 it shows that the introduction of changes in the construction of the enclosing constructions has an essential effect on the structure of convective flows in heated of premises and the temperature fields (Fig. 4). In the case of wall insulation, there is no formation of vortex structures under the source of radiant heating and, accordingly, there is no intensive mixing of air masses. The stratification of isotherms near the lower "air-floor" interface has been established (Figs. 2, 3).

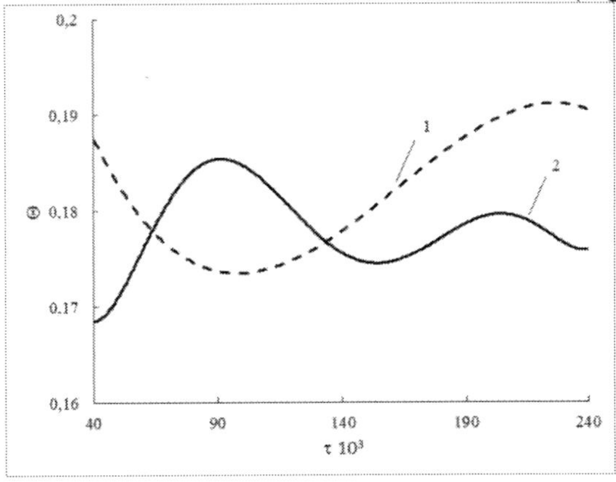

Fig. 4. Dependence of the average temperature of the air environment on the time when the GIE is operating with the use of insulation (1) and in the case of its absence (2).

The results of the numerical simulation (Figure 4) also show a variety of changes in the mean temperatures over time. In the time period $70 \cdot 10^3$ and $130 \cdot 10^3$, the values of Θ for the two variants coincide. Nonmonotonic changes in mean temperatures with time are caused by the processes of heat withdrawal to the enclosing constructions. In the initial period of time, the structure of the flow of heated gas flows upwards. But with an increase τ in the value of mean temperatures in the conditions for the heat insulation of vertical boundaries it becomes more, increases difference numerical values Θ for two versions in question.

4 Conclusion

The results of theoretical studies of heat transfer processes in a building heated by a gas infrared emitter allow us to say that insulating the vertical enclosing structures leads to a significant change in the structure of the flows and temperature distributions of heated air in the room. In this case, in the course of time the influence of heat withdrawal becomes ever more essential. The change in the structure of the flow in time leads to an uneven increase in the average air temperature.

The work was supported by the Russian President's grant (Scientific School project - 7538.2016.8).

References

1. Federal Law from 23.11, N 261-FZ (Red. 03.07.2016) "On Energy Saving and Improvement of Energy Efficiency" (2009) [in russian]
2. European Parliament and the Council of European Parliament, Directive 2010/31/EU of the Parliament and of the Council of 19 May 2010 on the energy performance of building (recast), Off. J. Eur. Union (2010)

3. Xiaojing Meng, Yi Wang, Tiening Liu, Xiao Xing, Yingxue Cao, Jiangping Zhao, Appl. Therm. Eng., **96**, 473-480 (2016), DOI: 10.1016/j.applthermaleng.2015.11.105
4. P. Rohdin, B. Moshfegh, Build. Environ., **46**, 2365-2374 (2011), DOI: 10.1016/j.buildenv.2011.05.019
5. X. Ouyang, B. Lin, Energy and Build., **109**, 316-327 (2015), DOI: 10.1016/j.enbuild.2015.09.068
6. E. Rebollo, C.A. Platero, F.R. Blanquez, F. Blazquez, J.A. Gaonab, Procedia Comput. Sci., **83**, 886-893 (2016), DOI: 10.1016/j.procs.2016.04.180
7. K.-N. Rhee, B.W. Olesen, K.W. Kim, Build. Environ., **112**, 367-381 (2017), DOI: 10.1016/j.buildenv.2016.11.030
8. G.V. Kuznetsov, N.I. Kurilenko, V.I. Maksimov, G.Ya. Mamontov, T.A. Nagornova, J. Eng. Phys. Thermophys., **86**, 3 (2013), DOI: 10.1007/s10891-013-0863-6
9. T. Cebeci, *Analysis of Turbulent Flows with Computer Programs* (The Boulevard, Langford Lane, Kidlington, Oxford, 2013)
10. M.A. Sheremet, Thermophys. Aeromechanics, **18**, 1 (2011), DOI: 10.1134/S0869864311010124

Numerical simulation of heat and mass transfer in a closed two – phase geothermal thermosyphon

Lily Valieva[1,*], and *Alexander* Nee[1]

[1]National Research Tomsk Polytechnic University, 634050 Tomsk, Russia

Abstract. Mathematical model of the heat and mass transfer process in a two-phase thermosyphon with allowance for the coolant phase transformations is presented in this study. The formulated boundary-value problem was solved by the finite differences method. The numerical experiment results are presented and the temperature fields are obtained for different operation times of the geothermal thermosyphon.

1 Introduction

The unconventional energy sources are attracting increasing attention in the conditions of continuous growth of energy consumption and fossil fuels depletion [1-4]. The active solar radiation and wind use to generate electricity and heat is hampered by huge capital and operating costs. Moreover, such energy sources are not stable. Their power depends significantly on the climatic conditions of the terrain and time.

Geothermal energy has the greatest prospects [5]. U-shaped heat exchangers are used to extract energy [6]. However, the use of such heat transfer devices is accompanied by difficulties in mounting. Geothermal closed two-phase thermosyphon (GTS) can be an alternative substitute [7].

Only two studies have been published in the open press [7, 8] devoted to the operation analysis of closed two-phase thermosyphon under geothermal energy transfer conditions.

The various parameters influence on the characteristics of a geothermal thermosyphon was studied by Frank Hartmann et. al.[7] Heat and mass transfer two-dimensional numerical model was resolved and a physical experiment was carried out with different wetting coefficients, tube materials, radii and thermal conductivities of the head well.It is proved that the reduction in the well diameter leads to an increase in the heat flux. The well and soil thermal conductivity are important influencing parameters. The wetting effect, soil material and thermosyphon on the heat transfer coefficient are modeled based on the quasi-three-dimensional thermal equation conductivity in [8]. It is proved that the wetting factor and the GTS body material affect to the heat fluxes transmitted. It is established that polyamide materials can serve as an alternative substitute for the steel body of the heat exchanger.

* Corresponding author: lil.Valiewa@yandex.ru

It is assumed that heat in the system is transmitted only by conduction. Moreover, the issue of heat transfer in the direction of the vertical coordinate was not considered. It is very important in analyzing the geothermal thermosyphon efficiency. Therefore, the present work purpose of is mathematical modeling of heat and mass transfer in a closed two-phase geothermal thermosyphon.

2 Mathematical model and numeral method

The geometric problem formulation is shown in fig. 1. The vertically oriented TS consists of a sealed copper body (4) filled with working fluid (water). Heat is supplied to the thermosyphon bottom cover (1). The intensive evaporation process begins at the boundary (5). Because of forced convection, the vapor moves to the top cover (3). It condenses on the surface (6). The condensate formed due to gravity flows downward. The heat is removed from the top boundary (3) according to the Newton-Richmann law.

It was assumed that the thermophysical characteristics of the refrigerant and the body are temperature independent [9].

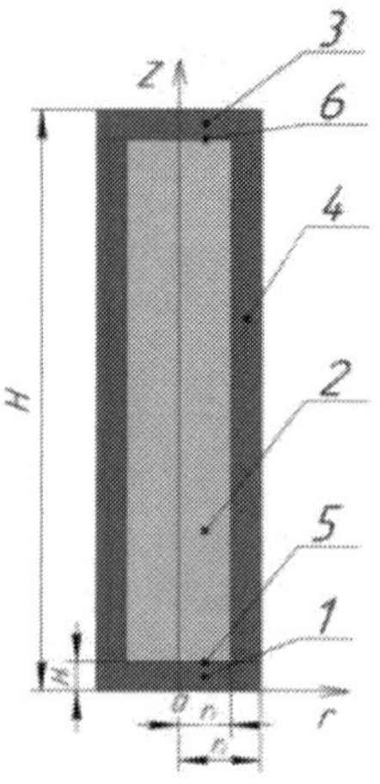

Fig. 1. Solution domain: 1 – bottom cover; 2 – vapor channel; 3 - top cover; 4 – enclosures; 5 – evaporation section; 6 - condensing section.

The heat and mass transfer process in a two-phase thermosyphon is described by the following equations:

- Thermal conductivity equation for the enclosures:

$$c_1\rho_1 \frac{\partial T_1}{\partial t} = \lambda_1 \left(\frac{1}{r} \frac{\partial}{\partial r} \left(r \frac{\partial T_1}{\partial r} \right) + \frac{\partial^2 T_1}{\partial z^2} \right); \qquad (1)$$

- Thermal conductivity equation for the vapor channel:

$$c_2\rho_2 \left(\frac{\partial T_2}{\partial t} + W \frac{\partial T_2}{\partial z} \right) = \lambda_2 \left(\frac{1}{r} \frac{\partial}{\partial r} \left(r \frac{\partial T_2}{\partial r} \right) + \frac{\partial^2 T_2}{\partial z^2} \right); \qquad (2)$$

where $W = \frac{Wev}{\rho}$; ρ - density, kg/m^3; $W_{ev/co}$ – specific mass evaporation/condensation rate, $kg/(m^2 \times s)$.

Rate of evaporation/ condensation is calculated according to the formula Hertz-Knudsen [1]:

$$W_{ev/co} = \frac{A(P_n - P^*)}{\sqrt{\frac{2 \cdot \pi \cdot R \cdot T}{M}}}; \quad (3)$$

where A – accommodation coefficient; P_n – saturation pressure, Pa; P^* – partial pressure, Pa; R – universal gas constant, $J/(mole \times K)$; T – temperature, K; M – molecular weight, $kg/mole$.

- At the initial time:

$$T_1(r, Z, 0) = T_2(r, Z, 0) = T_0,$$

where T_0 – temperature at t=0 s, K.

- at the boundary $Z = 0, 0 \le r \le r_2$;

$$T = T_h;$$

where T_h – temperature of a geothermal source, K.

- at the boundary $Z = H, 0 \le r \le r_2$;

$$-\lambda_1 \frac{\partial T_1}{\partial t} = \alpha(T_c - T_e), \quad (4)$$

where T_c – surface temperature, K; T_e – ambient temperature, K; α – heat transfer coefficient, $Wt/(m^2 \cdot K)$.

- at the boundary $Z = H_1, 0 \le r \le r_1$;

$$\begin{cases} -\lambda_1 \cdot \frac{\partial T_1}{\partial r} = -\lambda_2 \cdot \frac{\partial T_2}{\partial r} - W_{ev}; \\ T_1 = T_2. \end{cases}$$

- at the boundary $Z = H, 0 \le r \le r_1$;

$$\begin{cases} -\lambda_1 \cdot \frac{\partial T_1}{\partial r} = -\lambda_2 \cdot \frac{\partial T_2}{\partial r} + W_{co}; \\ T_1 = T_2. \end{cases}$$

- at the boundary $r = 0, 0 \le z \le H + 2 \cdot H_1$;

$$\frac{\partial T}{\partial r} = 0 \text{ – the symmetry condition.}$$

- at the boundary $r = r_2, 0 \le z \le H + 2 \cdot H_1$;

1) In the first series of mathematical modeling:

$$\frac{\partial T}{\partial r} = 0;$$

2) In the second series of mathematical modeling

$$T = T_{pr},$$

where T_{pr} – ground temperature, K.

at the boundary conditions of the fourth kind:

$$\begin{cases} -\lambda_1 \cdot \dfrac{\partial T_1}{\partial n} = -\lambda_2 \cdot \dfrac{\partial T_2}{\partial n}; \\ T_1 = T_2. \end{cases}$$

The problem was solved by means of the finite difference method [10]. The convective term in the energy equation was discredited based on the locally one-dimensional Samarskogo`s scheme and monotonic A.A. Samarskogo`s approximations [11].In order to verify the mathematical model, algorithm used and solution method were tested on a model problem of heat transfer [11].

3 Results and discussion

The energy transfer investigation in a closed two-phase thermosyphon was carried out with the following parameters $T_h = 373\,K$, $H = 50\,m$, $10 \le t \le \infty s$, $\alpha = 23\,Wt/(m^2 \cdot K)$. The temperature closed two-phase thermosyphon fields in the output to the stationary mode of operation is shown in fig. 2.

Thus, the temperature of the top cover of the gas turbine at a running time of 10 s is the distance over which the pairs overcome $S = 13m$; for $t = 50\,s - T = 316\,K, S = 28m$; $t = 100\,s - T = 324\,K, S = 35m$; $t = 250\,s - T = 343\,K, S = 50m$; $t = \infty\,s - T = 360\,K, S = 50\,m$; the upper cover temperature increases with the two-phase thermosyphon operating time. The temperature difference between the GTS top and bottom cover decreases from too due to an increase in the vaporization process intensity and the vapors velocity.

The temperature fields in the geothermal thermosyphon with a known temperature distribution in the soil are presented. In fig. 3. thus, the upper cover temperature of the gas turbine grows with increasing operating time from 312 to 315 K due to an additional heat exchange between the steam and the vertical walls. The temperature difference in the steam channel also decreases, as in the first experiments series.

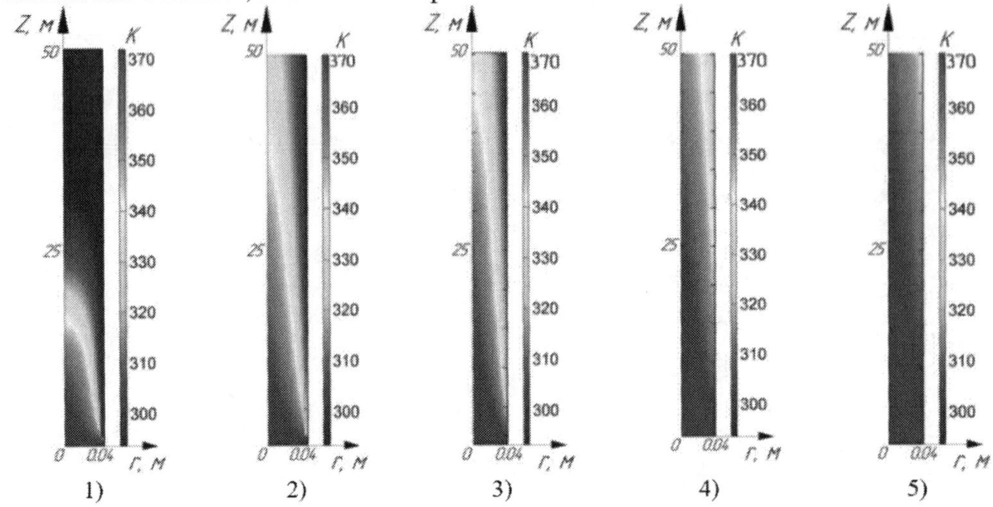

Fig. 2. The steady state temperature fields for: 1) t=10 s; 2) t=50 s; 3) t=100 s; 4) t=250s; 5) t=∞ s.

According to the results of mathematical modeling, the main findings are as follows:

1. The mathematical model of heat and mass transfer in a closed two-phase thermosyphon with allowance for phase transformations of the refrigerant is formulated.

2. The temperature difference between the top and bottom cover changes $293 \le T \le 360\,K$, with increasing operating time $100 \le t \le \infty\ s$ of the geothermal thermosyphon.

3. The upper cover temperature also growths $313 \le T \le 315^0 C$ with increasing operating time taking into account the temperatures distribution in the ground and on the vertical wall of the gas turbine. The temperature is less than in the first approximation. The temperature difference varies within $50 \le \Delta T \le 60\,K$.

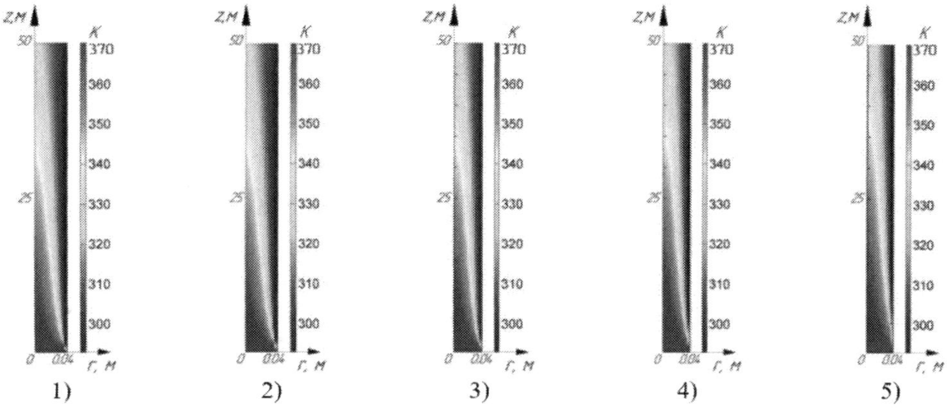

Fig. 3. The steady state temperature fields for: 1) t=100 s; 2) t=500 s; 3) t=1000 s; 4) t=5000 s; 5) t=∞ s.

The reported research was supported by Russian Federation President Grant for state support of the Russian Federation leading scientific schools SS-7538.2016.

References

1. S. Yankovskiy, A. Matveev, A. Tolokolnikov, A. Zenkov, MATEC Web of Conferences, **92**, 01036 (2016), DOI: 10.1051/matecconf/20179201036
2. S. Yankovsky, A. Tolokolnikov, A. Matveev, Y.Marysheva, MATEC Web of Conferences, **110**, 01089 (2017), DOI: 10.1051/matecconf/201711001089
3. K.B. Larionov, I.V. Mishakov, A.A. Gromov, A.V. Zenkov, MATEC Web of Conferences, **91**, 01007 (2017), DOI: 10.1051/matecconf/20179101007
4. A.V. Zenkov, K.B. Larionov, V.Gubin, MATEC Web of Conferences, **110**, 01091 (2017), DOI: 10.1051/matecconf/201711001091
5. S. Carlino, R. Somma, A. Troiano, M.G. Di Giuseppe, C. Troise, G. De Natale, J. Ren. En., **62**, 177-196 (2014), DOI: 10.1016/j.renene.2013.06.052
6. A. Franco, M. Vaccaro, J. Appl. Th. Eng., **59**, 189-199 (2013), DOI: 10.1016/j.applthermaleng.2013.05.024
7. A. Hantsch, U. Gross, J. Geoth., **47**, 97-103 (2013), DOI: 10.1016/j.geothermics.2013.03.003
8. F. Hartmann, R. Behrend, A. Hantsch, T. Grab, U. Gross, J. Geoth., **55**, 99-107 (2015), DOI: 10.1016/j.geothermics.2015.01.014
9. G. V. Kuznetsov, M. A. Sheremet, Therm. and Aerom., **16**, 123-133 (2009), DOI: 10.1007/s11510-009-0012-z
10. V.M. Paskonov, V.I. Polezhaev, and L.A. Chudov, *Numerical Modeling of heat and mass Exchange Processes* (Nauka, Moscow, 1984)
11. Saedi S.M., Khodadadi J.M., J. Heat and Mass Transf., **49**, 1896-1906 (2006), DOI: 10.1016/j.ijheatmasstransfer.2005.10.033

MATEC Web of Conferences **141**, 01004 (2017) DOI: 10.1051/matecconf/201714101004
Smart Grids 2017

Influence of strouhal number on the temperature fields in the locked space with the working source of the radiant heating

Alime Dudak[1], *Vyacheslav* Maksimov[1,*], and *Tatiana* A. Nagornova[1]

[1]National Research Tomsk Polytechnic University, 634050 Tomsk, Russia

Abstract. Is carried out a numerical study of the heat transfer process in the closed domain with the source of radiant heating. Is solved the system of differential equations in the dimensionless setting, which describes free convection in the airspace and thermal conductivities in the enclosing constructions. Examined the influence of the dimensionless parameter, which characterizes the unsteady motion of liquids or gases (Strouhal number), on the formation of temperature fields both near the heater and in the entire region. Is carried out the estimation of the average temperature, depending on the Sh.

1 Introduction

Recently increasingly more attention is given to the prognostication of thermal comfort in residential and industrial areas [1-4]. Used as commercial software packages [3] and developed by the authors [1, 2] mathematical models. In this case are used or standard balance models [2] or systems of nonstationary differential equations [1] into the partial derivatives, which consider the influence of different parameters, entering the dimensionless criteria, such, for example, as numbers of Prandtl, Reynolds and others. The number of studies is dedicated to the study of the influence of the criterion of eddy formation (Strouhal number) on the structure of the flow of gas or liquid after passage by them the obstacle [5-9]. Strouhal number is correlated with this phenomenon as the instability of the flow of working medium. The influence of this dimensionless parameter on the thermal characteristics of the heated region is not thus far studied. By the purpose of the work is the study of the influence of Strouhal number on the structure of airflow with the numerical simulation of the process of heating the source of the radiant heating of closed domain.

2 Formulation of the problem

In formulating the problem of heat transfer in a room with a working source of radiant heat was considered a rectangular area shown in Fig. 1.

* Corresponding author: elf@tpu.ru

DOI: 10.1051/matecconf/201714101004

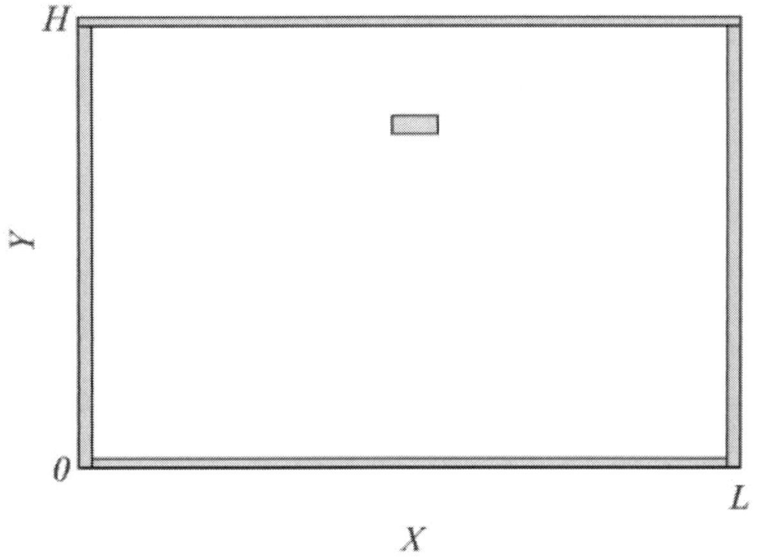

Fig. 1. Study region: L, H - the width and height of the region.

The heat sink to the enclosing heat-conducting walls was taken into account similarly to [1]. On the lower interface of media was assigned the value of heat flux q (coming from emitter), which varied depending on coordinate x. On the outer boundaries of the enclosing walls the nonlinear boundary conditions of heat exchange with the environment were assigned (caused by the convection and the emission). It was assumed that air of pure and it is possible to consider transparent for emission.

With the mathematical formulation of the problem of equation, that describe the process of free convection in the region with regime of conjugate heat transfer, were recorded by analogy [1] in the dimensionless variables. System of equations in the partial derivatives is solved by the finite-difference method with the use of an algorithm [1, 10], developed for the numerical solution of the conjugate natural convection in closed rectangular areas with local energy sources.

3 Results and discussion

In Fig. 2-4 are represented the fundamental characteristics of heat transfer in the locked space with the working source of radiant heating with different numerical values of Strouhal number.

Values Sh were selected in the range change, that corresponds to the conditions for the process from the quasi-stationary state of system in question "GIE – air – the enclosing constructions" to the substantially unsteady.

With the small numbers Sh in the upper part of the region (Fig. being investigated. 2 *a*, *b*) it is formed 2 small of vortex. It is noticeable (Fig. 2), that with an increase in Strouhal number the turbulence above the source of radiant heating decreases, their speeds are reduced (Fig. 2 *d*) and then disappear completely (Fig. 2 *f*). For example, with Sh = 10 are observed only main flows, distributed along the entire upper interface of media (Fig. 2 *e*, *f*).

A change in the structure of the movement of air near the source of heating leads to a significant change in temperatures in this region (to 20 % (Fig. 3)), (to their value practically they coincide near $Y = H/2$) the lower interface of media.

 DOI: 10.1051/matecconf/201714101004

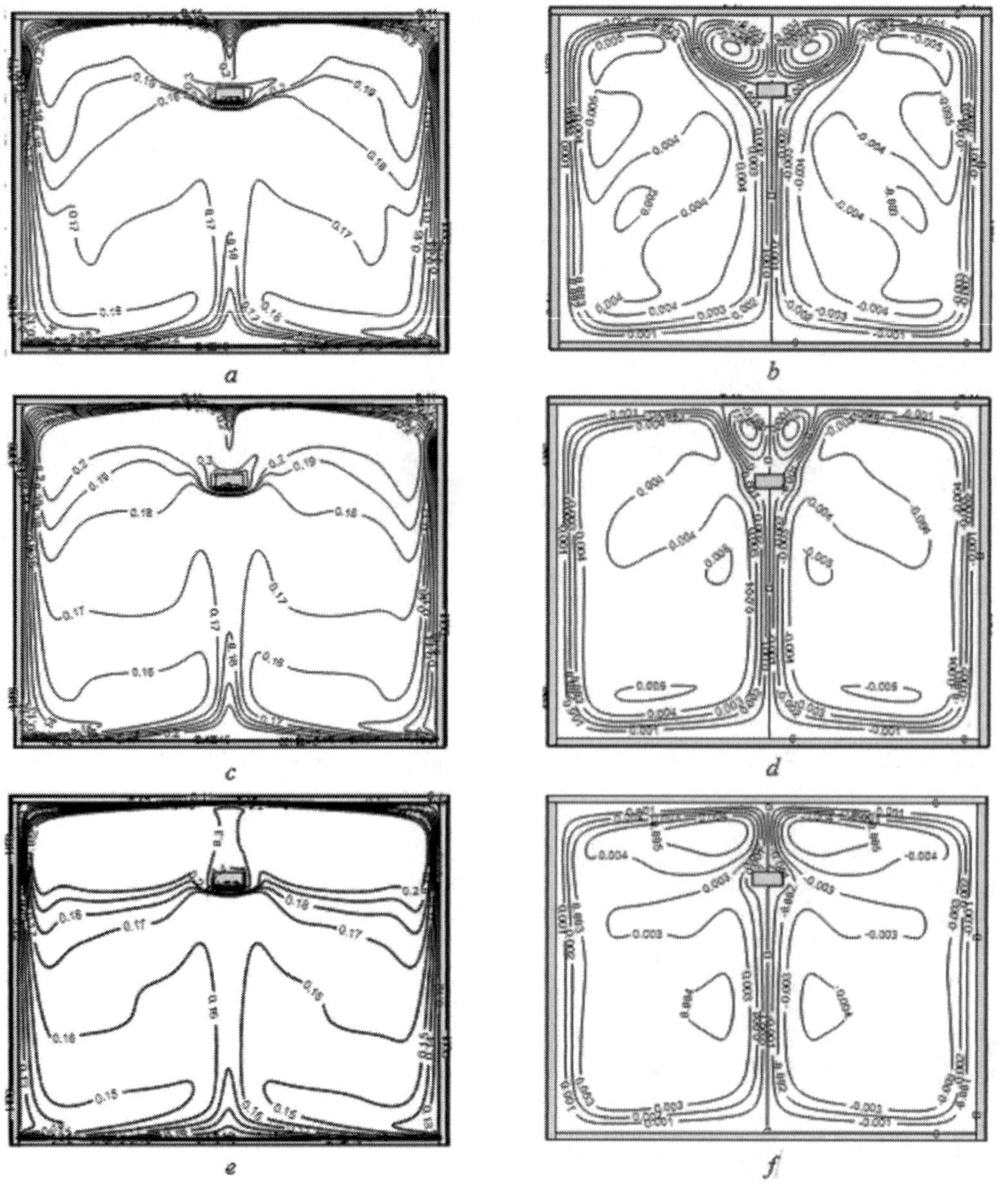

Fig. 2. Area of solving the problem: *L*, *H* - the height and length of solutions area respectively. The temperature field (*a*, *c*, *e*) and the isolines of the current function (*b*, *d*, f) at the instant $\tau = 24{\cdot}10^4$ for different values of the Strouhal number: Sh = 0.1 (a, b); Sh = 1 (c, d); Sh = 10 (d, e).

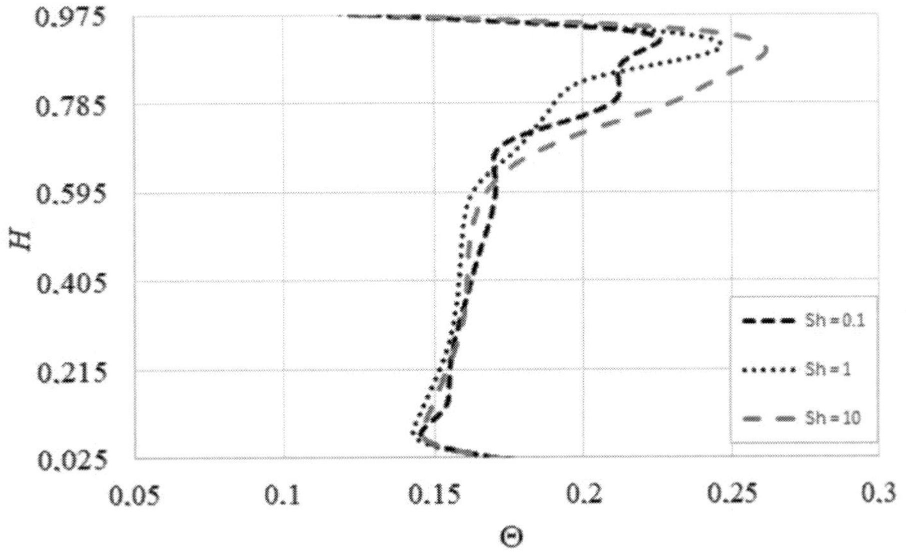

Fig. 3. Temperature distribution along the height of the investigated region in the cross section $X = L/4$ at time $\tau = 24{\cdot}10^4$.

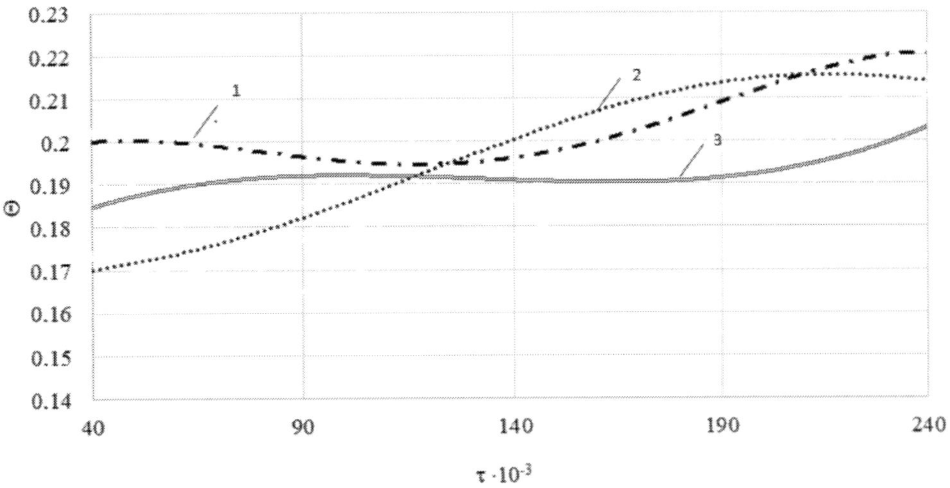

Fig. 4. Dependences of the average temperature of the air medium on time for different values of the Strouhal number: 1) Sh = 0.1; 2) Sh = 1; 3) Sh = 10.

Reduction in the speeds (Fig. 2 *b*, *d*, *f*) and the intensity of mixing respectively reduce average air temperatures (Fig. 4). With Sh = 10 temperature are minimum (Fig. 4), but with greater τ the difference in the values Θ becomes less 10 %. Interesting is the agreement of the values of average dimensionless temperature with $\tau = 12{\cdot}10^4$ (1 the hour of physical time), calculated with the different Sh.

4 Conclusion

Is carried out the numerical simulation of the process of heat transfer in the closed domain, heated by the source of radiant heating. The results of numerical studies illustrate a change in the structure of the flow of heated air with an increase in the number Sh only near the source of the radiant heating, which has high temperature. In the environment of the lower interface of media changes in temperatures are insignificant. It is established that an increase Sh from 0.1 to 10 leads to an unessential change in the structure of the flows of heated air and temperature fields in the closed domains (to 20 %) and practically it does not influence the temperature of the surface of the enclosing constructions.

The work was supported by the Russian President's grant (Scientific School project - 7538.2016.8).

References

1. G.V. Kuznetsov, N.I. Kurilenko, V.I. Maksimov, G.Ya. Mamontov, T.A. Nagornova, J. Eng. Phys. Thermophys., **86**, 3 (2013), DOI: 10.1007/s10891-013-0863-6
2. D.A. Garcia, Build. Environ., **106**, 378-388 (2016), DOI: 10.1016/j.buildenv.2016.07.013
3. Jong Bum Kim, WoonSeong Jeong, Mark J. Clayton, Jeff S. Haberl, Wei Yan, Autom. Constr., **50**, 16-28 (2015), DOI: 10.1016/j.autcon.2014.10.011
4. Xiaojing Meng, Yi Wang, Tiening Liu, Xiao Xing, Yingxue Cao, Jiangping Zhao, Appl. Therm. Eng., **96**, 473-480 (2016), DOI: 10.1016/j.applthermaleng.2015.11.105
5. L. De Boeck, S. Verbeke, A. Audenaert, L. De Mesmaeker, Renew. Sust. Energ. Rev., **52**, 960-975 (2015), DOI: 10.1016/j.rser.2015.07.037
6. Surendra Singh Sisodia, Sandip Sarkar, Sandip K. Saha, International Journal of Thermal Sciences, **121**, 13-29 (2017), DOI: 10.1016/j.ijthermalsci.2017.06.027
7. T. Vit, M. Ren, Z. Travnicek, F. Marsik, C.C.M. Rindt, Exp. Therm. Fluid Sci., **31**, 751-760 (2007), DOI: 10.1016/j.expthermflusci.2006.08.002
8. A.-B. Wang, Z. Travnicek, K.-C. Chia, Phys. Fluids, **12**, 6 (2000), DOI: 10.1063/1.870391
9. S. Sarkar, A. Dalal, G. Biswas, Int. J. Heat Mass Transf., **54**, 3536-3551 (2011), DOI: 10.1016/j.ijheatmasstransfer.2011.03.032
10. M.A. Sheremet, Thermophys. Aeromechanics, **18**, 1 (2011), DOI: 10.1134/S0869864311010124

MATEC Web of Conferences **141**, 01005 (2017) DOI: 10.1051/matecconf/201714101005
Smart Grids 2017

Peculiarities of temperature fields formation in vapor channels of thermosyphons with heat carriers boiling at low temperatures

Atlant Nurpeiis[1,*], *Konstantin* Ponomarev[1], and *Tatyana* Nemova[1]

[1]National Research Tomsk Polytechnic University, 634050 Tomsk, Russia

Abstract. We conducted experiments on specially developed setup consisting of evaporation, transport and condensation parts. Heat was supplied to the evaporation part by the heating element which was supplied with voltage and alternating current from a single-phase transformer. Temperatures in the characteristic sections of each part were recorded by thermocouples. Junctions of thermocouples were mounted on the axis of symmetry in the liquid layer, at the lower boundary, in the middle part, and at the upper boundary of the vapor channel. To minimize the influence of the random factors (ambient air movement, operation of ventilation system, room temperature, etc.), we placed thermosyphon in a glass box. We used N-pentane as a heat carrier, and the filling ratio of the thermosyphon is equal to 4%.

1 Introduction

Closed two-phase thermosyphons as a device for transferring heat from the high-temperature zone of energy-saturated equipment have been known for a long time [1, 2]. But, despite the fact that in recent years there has been a noticeable increase in numbers of researches devoted to heat transfer processes in thermosyphons [3, 4], there are still few examples of their real usage in modern technologies Among the numerous problems of analyzing the relationships between the characteristics of thermosyphons and their design parameters, the most important tasks are to assess the "performance" of thermosyphon (heat transfer intensity) and analyze the influence of the main significant factors on this target function (using the terms of the experimental design theory). In many cases the main characteristic of thermosyphon operation is its effective thermal conductivity or thermal resistance. But these characteristics usually provide only integral assessment of the thermosyphon performance and are not a basis for analyzing the behavior of heat transfer in such complex heat exchangers.

The information about temperature fields of characteristic parts of thermosyphon is the most significant for physical analysis. However, this kind of measurements presents the objective difficulties. For this reason, the results of temperature measurements are given in many works only on separate parts of the outer surfaces (Ts) of such heat exchanger [7, 8]. As a result of a rather intensive heat transfer across the body of the thermosyphon along the

* Corresponding author: nurpeiis_atlant@mail.ru

 DOI: 10.1051/matecconf/201714101005

transverse and longitudinal coordinates, Ts measurements are not enough to analyze the processes in the evaporation and condensation parts, as well as in the vapor channel. A few results of temperature measurements inside the thermosyphon [9, 10] reflect their change only at certain points of the internal surface of this device. In order to conduct intrinsic analysis of heat transfer behavior, we need to obtain the information about temperature distributions or at least gradient T at separate parts corresponding to zones of vapor evaporation, transport and its condensation.

The purpose of this work is to determine experimentally temperature fields in the evaporation, transport and condensation parts in a closed two-phase thermosyphon using the example of a low-boiling heat carrier of N-pentane.

2 Results and discussion

We conducted experiments on specially-developed setup [11], where evaporation, transport and condensation parts were separated. Heat was supplied to the evaporation part by the heating element which was supplied with voltage and alternating current from a single-phase transformer. Temperatures at the characteristic sections were registered by thermocouples. Junctions of thermocouples were mounted on the axis of symmetry in the liquid layer, at the lower boundary, in the middle part, and at the upper boundary of the vapor channel. To minimize the influence of the random factors (ambient air movement, operation of ventilation system, room temperature, etc.), we placed thermosyphon in a glass box. The filling ratio of the thermosyphon with N-pentane was equal to 4%.

Figure 1 presents typical thermograms of cavity of thermosyphon in the range of heat load from 695 to 1568 W/m^2.

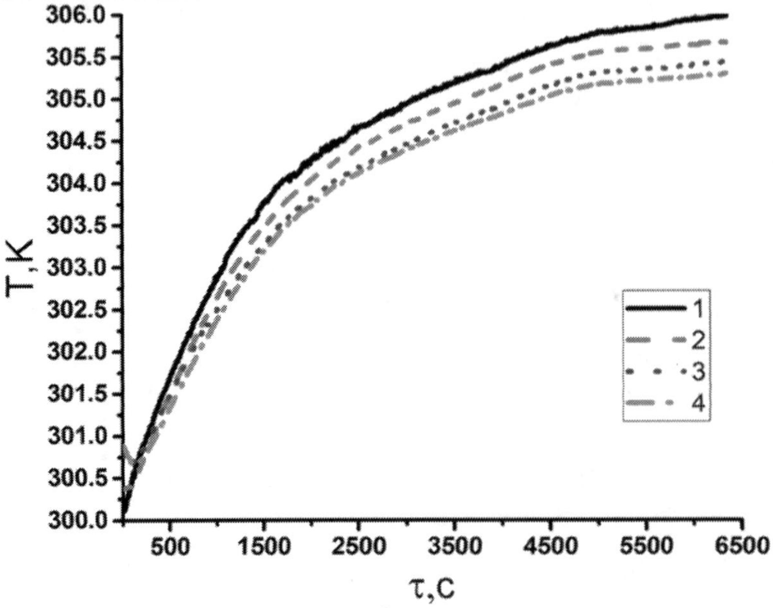

a)

DOI: 10.1051/mateconf/201714101005

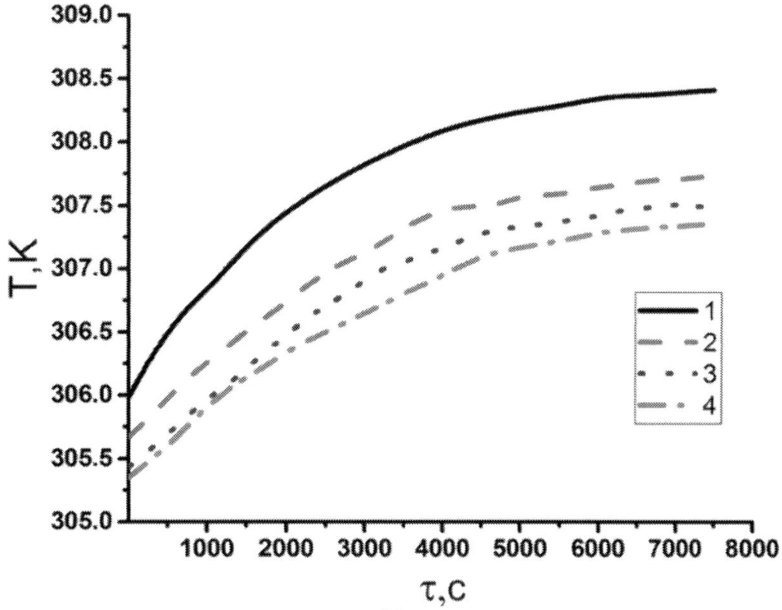

b)

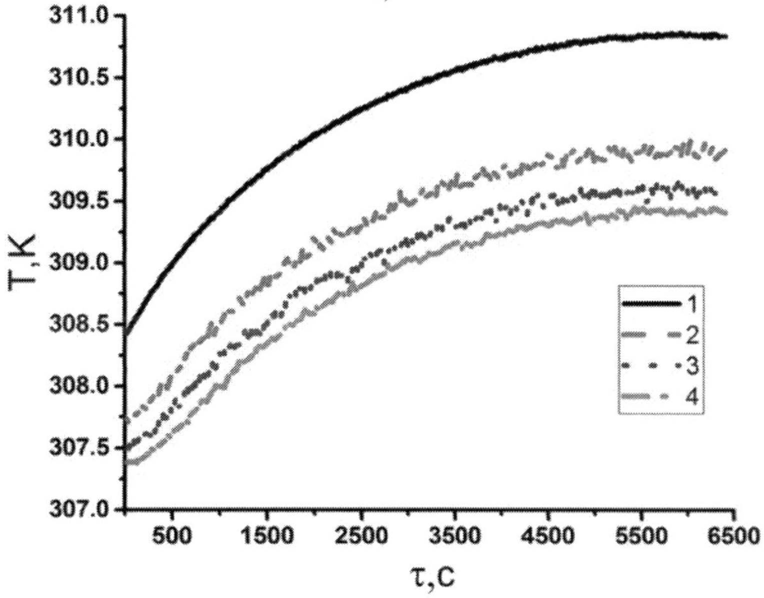

c)

Fig. 2. Time dependences of temperatures in the characteristic points, when the filling ratio is equal to 4% and the heat load is: a) q=695 W/m^2, b) q=1088 W/m^2, c) q=1568 W/m^2 (1 is a layer of heat carrier; 2 is a lower boundary of the vapor channel; 3 is a center of the vapor channel; 4 is an upper boundary of the vapor channel).

It should be noted that the steady-state temperature regime was achieved after almost two hours in the entire range of changes in heat fluxes to the evaporation region. This time decreased with increasing heat fluxes. This is most likely due to the accumulation of heat by the liquid heat carrier and material of the thermosyphon case. From Fig. 1, temperature fluctuations were detected when the heat flux is equal to q=1088 W/m^2 and higher. It is due to the fact that N-pentane has a low boiling point (36.1 ° C). When thermal energy is supplied, the evaporation process at "liquid-vapor" interfacial surface is intensified, and there is a fast

DOI: 10.1051/matecconf/201714101005

movement of vapor from the bottom cover to the upper one. These experimental results allow to make a conclusion that there is a high probability of a transition from laminar to turbulent flow during the vapor movement in the vapor channel. An increase in the heat flux after reaching the boiling point of the heat carrier is found not to lead to a significant change in the temperature distribution at the characteristic points. The pressure in the thermosyphon increases with the growth of the heat flux. As a result, saturation temperature of N-pentane also increases.

In conclusion, it should be noted that the observed temperature differences between the evaporation and condensation surfaces in the thermosyphon give grounds for the hypothesis that it is possible to simulate temperature fields in the thermosyphon with a sufficiently high reliability without describing the processes of vapor movement in the vapor channel with the use of sophisticated hydrodynamic models on the basis of the Navier-Stokes equations [12] or the Prandtl model [13].

The reported research was supported by Russian Federation President Grant for state support of the Russian Federation leading scientific schools SS-7538.2016.8 (No.14.Y31.16.7538-SS).

References

1. P. Byrne, J. Miriel, Y. Lenat, Appl. Therm. Eng., **88**, 5 (2011), DOI: 10.1016/j.apenergy.2010.12.009
2. M.A. Hakeem, M. Kamil, I. Arman, Appl. Therm. Eng., **28**, 13 (2008), DOI: 10.1016/j.applthermaleng.2007.10.002
3. D. Jafari, A. Franco, S. Filippeschi, P. Di Marco, Renewable and Sustainable Energy Reviews, **53**, 575-593 (2016), DOI: 10.1016/j.rser.2015.09.002
4. H. Jouhara, A.J. Robinson, Appl. Therm. Eng., **30**, 2-3 (2010), DOI: 10.1016/j.applthermaleng.2009.08.007
5. P. Naphon, P. Assadamongkol, T. Borirak, Int. Commun. Heat Mass Transf., **35**, 10 (2008), DOI: 10.1016/j.icheatmasstransfer.2008.07.010
6. S.H. Noie, Appl. Therm. Eng., **26**, 559-567 (2006), DOI: 10.1016/j.applthermaleng.2005.07.012
7. I. Khazaee, R. Hosseini, S.H. Noie, Appl. Therm. Eng., **30**, 406-412 (2010), DOI: 10.1016/j.applthermaleng.2009.09.012
8. E. Gedik, Energy and Buildings, **127**, 1096-1107 (2016), DOI: 10.1016/j.enbuild.2016.06.066
9. S.H. Noie, M.R. Sarmasti Emami, M. Khoshnoodi, Heat Transf. Eng., **28**, 4 (2007), DOI: 10.1080/01457630601122997
10. E. Ibrahim, M. Moawed, N.S. Berbish, Heat Mass Transf., **48**, 9 (2012), DOI: 10.1007/s00231-012-0995-9
11. D.V. Feoktistov, E.A. Vympin, A.E. Nurpeiis, MATEC Web of Conferences, **72**, 01081 (2016), DOI: 10.1051/matecconf/20167201081
12. G.V. Kuznetsov, A.E. Sitnikov, High Temp., **40**, 6 (2002), DOI: 10.1023/A:1021437502952
13. G.V. Kuznetsov, M.A. Al-Ani, M.A. Sheremet, J. Engin. Thermophys., **20**, 2 (2011), DOI: 10.1134/S1810232811020081

Experimental study of temperatures in characteristic sections of the working zone of a closed two-phase thermosyphon under the condition of a heat removal by external periphery

Atlant Nurpeiis[1,*], *Evgeniya* Orlova[1], and *Gennady* Mamontov[1]

[1]National Research Tomsk Polytechnic University, 634050 Tomsk, Russia

Abstract. We present the results of the experimental study of temperature fields in a closed two-phase thermosyphon. Operational modes of a thermosyphon with different heat supply conditions are studied experimentally using setup consisting of the copper case, systems of heat supply and removal in evaporation and condensation zones, and temperature recording facilities. The height of the heat exchanger is 161 mm, thickness of the side walls and bottom wall are 1.5 mm and 2 mm, respectively, inner diameter is 39 mm. Heat is supplied to the bottom wall by heating element. The heat carrier is distilled water. We obtained thermograms when heat fluxes to the bottom wall of the thermosyphon are 695 – 2136 W/m^2.

1 Introduction

Heat removal from the heat-generating parts and modules is the most important thing for successful operation of modern energy-saturated equipment [1–6]. Traditional cooling systems with ventilation and pump installations cannot be always applied because it is often necessary to create additional heat exchange surfaces to remove high heat fluxes. Overheating or failure of auxiliary equipment leads to emergency operation of energy-saturated equipment. For this reason, the use of heat-transfer devices independent of electrical power sources is topical. One of such autonomous heat-exchanger is a closed two-phase thermosyphon which can be used to thermostate and thermoregulate different technological processes [7, 8]. Evaporation and condensation in the heat exchangers under consideration are spatially separated, which makes it possible to transform the heat fluxes by changing the ratio of evaporation and condensation surfaces. However, at present thermosyphons are not widely spread in industry due to the fact that the physics of the joint heat transfer processes and phase transformations in the vapor channel, the heat carrier in the evaporation and condensation zones, the condensate film flowing along the side walls in the thermosyphon has not been sufficiently studied. The known results of a theoretical study of these processes, for example, [9, 10], were obtained using complex mathematical models. Solving problems of the type [9, 10] is laborious and time consuming. To simplify the same problems of type

* Corresponding author: nurpeiis_atlant@mail.ru

 DOI: 10.1051/matecconf/201714101006

[9, 10], the experimental data on the heat transfer in thermosyphons are not sufficient. Therefore, experimental studies of the temperature fields of thermosyphons are relevant.

The purpose of this work is to study experimentally temperature changes behavior in the characteristic sections of the working zone of the thermosyphon under the condition of a heat removal by external periphery.

2 Results and discussion

To conduct experimental studies of the operational modes of thermosyphon with different heat supply conditions, the setup [11] has been developed. It consists of the copper case, systems of heat supply and removal in the evaporation and condensation zones, and temperature recording facilities. The height of the heat exchanger is 161 mm, thickness of the side walls and bottom wall are 1.5 mm and 2 mm, respectively, inner diameter is 39 mm. Heat is supplied to the bottom wall by heating element. The heat carrier is distilled water. The choice of such a liquid is due to the fact that the distilled water is fire and explosion-proof, as well as easily accessible.

Figure 1 presents typical thermograms obtained when the part of thermosyphon cavity is filled with the heat carrier in the range of heat loads from 695 to 2136 W/m^2.

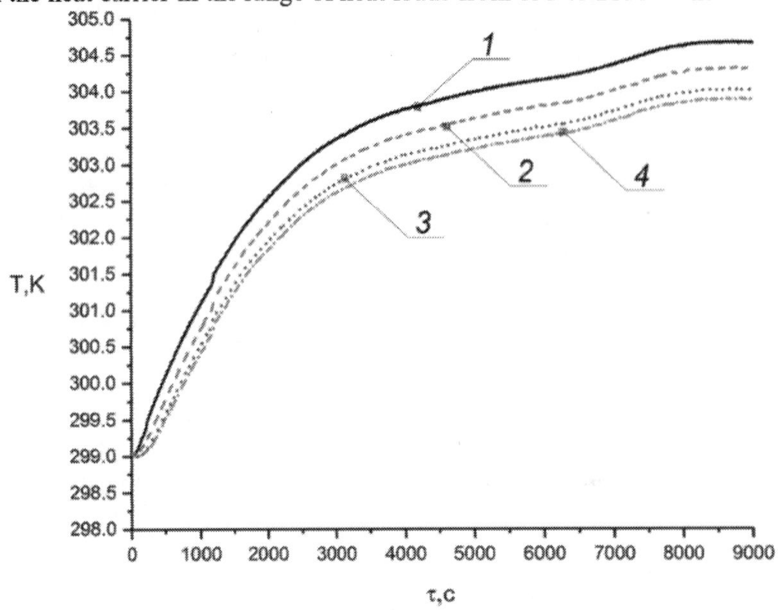

a)

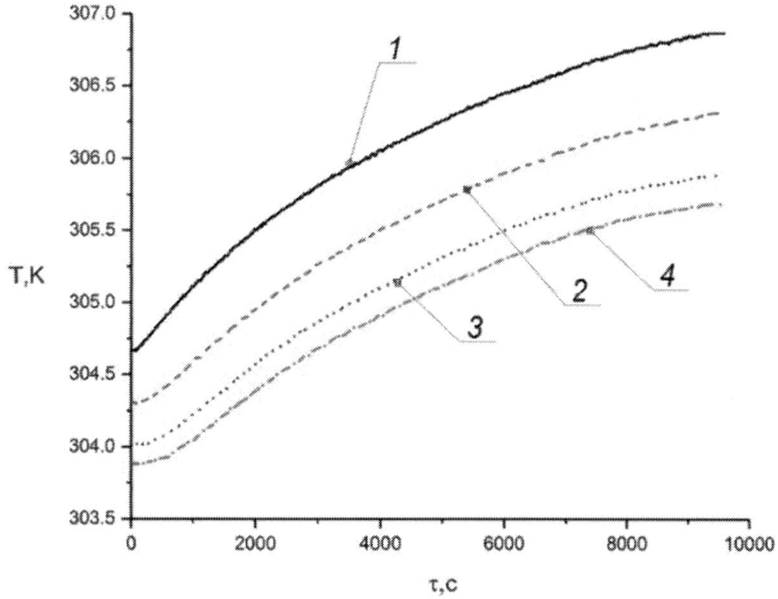
307.0
306.5
306.0
305.5
305.0
304.5
304.0
303.5
T,K
0
2000
4000
6000
8000
10000
τ,c
1
2
3
4

b)

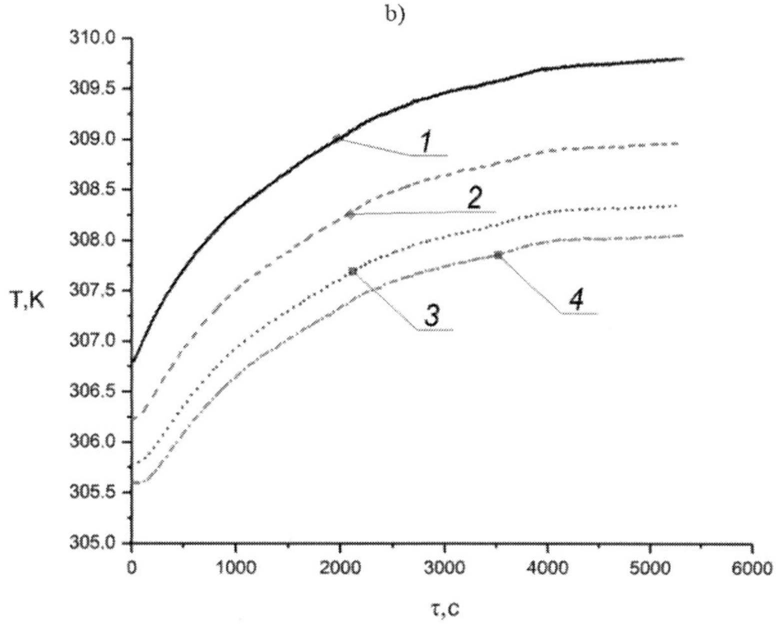
310.0
309.5
309.0
308.5
308.0
307.5
307.0
306.5
306.0
305.5
305.0
T,K
0
1000
2000
3000
4000
5000
6000
τ,c
1
2
3
4

c)

 DOI: 10.1051/matecconf/201714101006

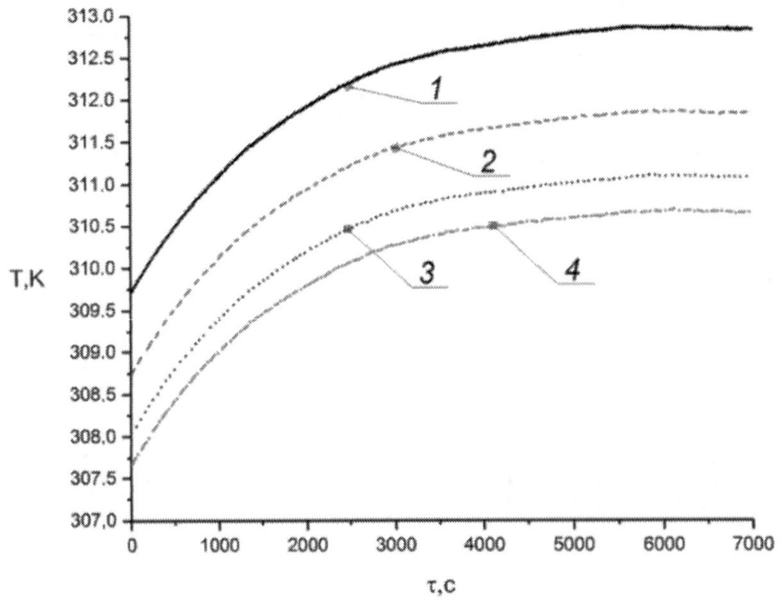

d)

Fig. 1. Time dependences of temperatures in the characteristic points when the filling ratio is equal to 4% and the heat load is: a) q=695 W/m^2, b) q=1088 W/m^2, c) q=1568 W/m^2, d) q=2136 W/m^2 (1 is a layer of heat carrier; 2 is a lower boundary of the vapor channel; 3 is a center of the vapor channel; 4 is an upper boundary of the vapor channel).

It can be concluded from Figure 1 that the process of reaching the stationary regime of the thermosyphon temperature fields is quite long in the whole range of the heat fluxes measurement on the heated boundary of the lower wall. Temperature in all points of measurements grow in time during 3000-5000 seconds, herewith the temperature rise depends on intensity of heating. When q=695 W/m^2, the temperatures rise by 4.5 K during 4000 seconds, and when q=2136 W/m^2, the rise is 3 K. The most important thing in such case is that the differences in the readings of the thermocouples located at different points of the vapor channel remain practically unchanged. That is, for example, the temperature difference between the evaporation and condensation zones (between the boundaries of these zones) is from 2 K to 2.3 K in the range $0<t<7000$ seconds when q=2136 W/m^2. It is worth noting that the non-stationarity of the temperature fields of the inner cavity of the thermosyphon registered in the experiment is not significant and can be neglected in the first approximation in the analysis of the thermosyphon operation. At the same time, it is necessary to clarify that the change of temperature of vapor and condensate in the vapor channel by 2-3 K during 2.5 hours of work is due to the gradual heating of the thermosyphon walls and cover. In addition, there is a rise in temperature (although insignificant) of water in the area of the bottom cover, respectively, the temperature of evaporation surface and vapor increase.

The reported research was supported by Russian Federation President Grant for state support of the Russian Federation leading scientific schools SS-7538.2016.8 (No.14.Y31.16.7538-SS).

References

1. V.I. Voronin, P.A. Byshnovich. *Electronics industry*. **9**, (1985). [in russian].
2. S.V. Garimella, V. Singhal, D. Liu, Proceedings of the IEEE, **94**, 8 (2006), DOI: 10.1109/JPROC.2006.879801

3. A.G. Korotkikh, K.V. Slyusarskiy, K.B. Larionov, V.I. Osipov, J. Phys. Conf. Ser., **754**, 5 (2016), DOI: 10.1088/1742-6596/754/5/052005
4. S. Yankovsky, A. Tolokolnikov, V. Zaytsev, MATEC Web of Conferences, **110**, 01080 (2017), DOI: 10.1051/matecconf/201711001080
5. K.V. Slyusarskiy, K.B. Larionov, V.I. Osipov, S.A. Yankovsky, V.E. Gubin, A.A. Gromov, Fuel, **191**, 383 (2017), DOI: 10.1016/j.fuel.2016.11.087
6. J.B. Marcinichen, J.R. Thome, B. Michel, International Journal of Refrigeration, **33**, 7 (2010), DOI: 10.1016/j.ijrefrig.2010.06.008
7. M.K. Tiwari, S. Zimmermann, C.S. Sharma, F. Alfieri, A. Renfer, T. Brunschwiler, I. Meijer, B. Michel, D. Poulikakos, 13th IEEE ITHERM Conference, (2012), DOI: 10.1109/ITHERM.2012.6231478
8. A.I. Isakeev, I.G. Kiselev, V.V. Filatov. *Effective ways to cool the power semiconductor devices* (Energoizdat, Leningrad, 1982) [in russian]
9. G.V. Kuznetsov, A.E. Sitnikov, High Temp., **40**, 6 (2002), DOI: 10.1023/A:1021437502952
10. G.V. Kuznetsov, M.A. Al-Ani, M.A. Sheremet, J. Engin. Thermophys., **20**, 2 (2011), DOI: 10.1134/S1810232811020081
11. D.V. Feoktistov, E.A. Vympin, A.E. Nurpeiis, MATEC Web of Conferences, **72**, 01081 (2016), DOI: 10.1051/matecconf/20167201081

Numerical investigation of heat transfer in thermosyphon under the emergency mode of operation of lithium-ion batteries of aircraft

Alexander Krasnoshlykov[1,*]

[1]National Research Tomsk Polytechnic University, 634050 Tomsk, Russia

Abstract. Numerical study of heat transfer in a closed two-phase thermosyphon using the software package ANSYS FLUENT is carried out. Characteristic temperature distributions, current lines and velocity vectors are obtained at critical thermal loads. The temperature difference in the investigated region decreases with increasing heat load was obtained.

1 Introduction

Rechargeable batteries are one of the main parts of the emergency power supply system for aircraft [1, 2]. Their resource characteristics determine the efficiency and safety of the entire power supply system. Nowadays, lithium-ion batteries are attracting the attention of researchers of power sources [3]. But the use of these batteries in aviation technology is associated with the danger of an emergency. This is confirmed by a number of fire-hazardous accidents on board the aircraft model Boeing 787 [4, 5]. The cause of the emergency situation is overheating of the battery.

During the operation of lithium-ion batteries, the thermal runaway effect [6] occurs, followed by a short circuit of the electrical circuit, when the temperature of the individual parts of the battery increases significantly from 10 °C to 150 °C in a short period of time (2 min) [3]. The main cause of thermal runaway of the battery is its overheating as a result of insufficient heat emission from the battery surface.

It should be noted that the system for removing heat from accumulator batteries based on two-phase closed thermosyphons has not been developed, since the theory of processes in the steam channel and the flowing fluid film has not been formulated. The models described in [7, 8] do not take into account the entire complex of heat transfer processes in the vapor and liquid phases of the working fluid.

Purpose of the work: mathematical modeling of the heat transfer process in two-phase closed thermosyphons at critical temperatures of rechargeable batteries of aircraft.

When setting the task, a number of assumptions are adopted. Steam was considered an ideal gas. It was assumed that the thickness of the liquid film does not vary in height.

* Corresponding author: krasnas@tpu.ru

2 Formulation of the problem

A schematic diagram of the thermosyphon is shown in Fig.1.

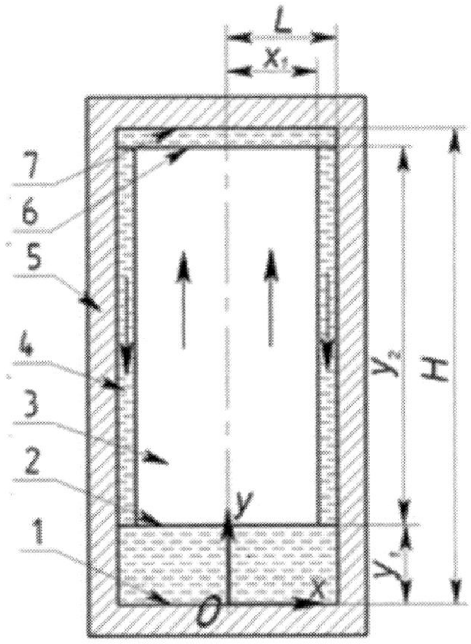

Fig.1. Schematic diagram of the thermosyphon. 1 – bottom cover; 2 – surface evaporation; 3 – vapor flow; 4 – liquid film; 5 – vertical wall; 6 – condensing surface; 7 – top cover.

Mathematical modeling of heat transfer in a two-phase thermosyphon of a rectangular cross-section was performed using the ANSYS FLUENT package [9].

$$\frac{\partial \rho}{\partial t}+\frac{\partial(\rho u)}{\partial x}+\frac{\partial(\rho \upsilon)}{\partial y}=0; \tag{1}$$

$$\frac{\partial \rho_1 u_1}{\partial t}+\frac{\partial(\rho_1 u_1 u_1)}{\partial x}+\frac{\partial(\rho_1 \upsilon_1 u_1)}{\partial y}=\rho_1 g_x-\frac{\partial P_1}{\partial x}+\frac{\partial}{\partial x}\left(\mu_1\frac{\partial u_1}{\partial x}\right)+\frac{\partial}{\partial y}\left(\mu_1\frac{\partial u_1}{\partial y}\right); \tag{2}$$

$$\frac{\partial \rho_1 \upsilon_1}{\partial t}+\frac{\partial(\rho_1 u_1 \upsilon_1)}{\partial x}+\frac{\partial(\rho_1 \upsilon_1 \upsilon_1)}{\partial y}=\frac{\partial P_1}{\partial y}+\frac{\partial}{\partial x}\left(\mu_1\frac{\partial \upsilon_1}{\partial x}\right)+\frac{\partial}{\partial y}\left(\mu_1\frac{\partial \upsilon_1}{\partial y}\right); \tag{3}$$

$$\frac{\partial \rho_2 u_2}{\partial t}+\frac{\partial(\rho_2 u_2 u_2)}{\partial x}+\frac{\partial(\rho_2 \upsilon_2 u_2)}{\partial y}=\rho_2 g_x-\frac{\partial P_2}{\partial x}+\frac{\partial}{\partial x}\left(\mu_2\frac{\partial u_2}{\partial x}\right)+\frac{\partial}{\partial y}\left(\mu_2\frac{\partial u_2}{\partial y}\right); \tag{4}$$

$$\frac{\partial \rho_2 \upsilon_2}{\partial t}+\frac{\partial(\rho_2 u_2 \upsilon_2)}{\partial x}+\frac{\partial(\rho_2 \upsilon_2 \upsilon_2)}{\partial y}=\frac{\partial P_2}{\partial y}+\frac{\partial}{\partial x}\left(\mu_2\frac{\partial \upsilon_2}{\partial x}\right)+\frac{\partial}{\partial y}\left(\mu_2\frac{\partial \upsilon_2}{\partial y}\right); \tag{5}$$

$$\rho_1 C_{p1}\left(\frac{\partial T_1}{\partial t}+u_1\frac{\partial T_1}{\partial x}+\upsilon_1\frac{\partial T_1}{\partial y}\right)=\lambda_1\left(\frac{\partial^2 T_1}{\partial x^2}+\frac{\partial^2 T_1}{\partial y^2}\right); \tag{6}$$

$$\rho_2 C_{p2}\left(\frac{\partial T_2}{\partial t}+u_2\frac{\partial T_2}{\partial x}+\upsilon_2\frac{\partial T_2}{\partial y}\right)=\lambda_2\left(\frac{\partial^2 T_2}{\partial x^2}+\frac{\partial^2 T_2}{\partial y^2}\right). \tag{7}$$

where u, υ – the velocity components in the projection on the axis x, y, respectively; P – pressure; T – temperature; ρ – density; x, y – Cartesian coordinates; t – time; Cp – heat capacity; g – acceleration of gravity; λ – coefficient of thermal conductivity; μ – dynamic viscosity; indices 1, 2 – properties of the liquid and vapor.

The initial conditions for the system of equations (1-7) are given in the form:

$$u(x,y)=0;\ T(x,y)=T_0;\ P(x,y)=P_0.$$

The boundary conditions for the equations (1-7) have the form:

$$x=0, 0<y<H,\quad \frac{\partial T_2}{\partial x}=0;\quad \frac{\partial u_2}{\partial x}=0,\quad \frac{\partial \upsilon_2}{\partial x}=0,$$

$$x=L,\ 0\le y\le H,\quad \lambda_1\frac{\partial T_1}{\partial x}=0;\quad u_1=\upsilon_1=0,$$

$$x=x_1, y_1\le y\le y_1+y_2,\quad \begin{cases} T_1=T_2 \\ \lambda_1\dfrac{\partial T_1}{\partial x}=\lambda_2\dfrac{\partial T_2}{\partial x} \end{cases};\quad \begin{cases} u_1=u_2=0 \\ \upsilon_2=0 \end{cases},\ \frac{\partial \upsilon_1}{\partial x}=0,$$

$$y=y_1, 0\le x\le x_1,\quad \begin{cases} T_1=T_2 \\ \lambda_1\dfrac{\partial T_1}{\partial y}=\lambda_2\dfrac{\partial T_2}{\partial y}-Q_e w_e-\upsilon_2 C_p\rho(T_2-T_0) \end{cases},\quad \upsilon_2=\frac{w_e}{\rho_2},\ \upsilon_1=\frac{w_e}{\rho_1},$$

$$y=y_1+y_2, 0\le x\le x_1,\quad \begin{cases} T_2=T_1 \\ \lambda_2\dfrac{\partial T_2}{\partial y}=\lambda_1\dfrac{\partial T_1}{\partial y}+Q_c w_c+\upsilon_2 C_p\rho(T_1-T_0) \end{cases},\quad \upsilon_1=\frac{w_c}{\rho_1},\ \upsilon_2=0,$$

$$y=0, 0\le x\le L,\quad \lambda_1\frac{\partial T_1}{\partial y}=q_h,\quad u_1=\upsilon_1=0,$$

where α – heat transfer coefficient, W/(m²·K).

Mathematical modeling of the task carried out in the package ANSYS FLUENT software. Mass rate of evaporation and condensation were calculated according to the formula [7]:

$$w_c=\beta\sqrt{\frac{M}{2\pi RT_H}}\left(P_2-P_H\right),$$

where $P_н$ – saturation pressure; $T_н$ – saturation temperature; R – universal gas constant; M – molar weight; β – accommodation coefficient.

To verify the solution method used, the results of mathematical modeling were compared with experimental studies.

An experimental study [10] was carried out for different values of the heat flux. To compare the results of numerical simulation with experimental data, two typical values of the heat flux to the outer walls of the thermosyphon body in the evaporation zone (100 W, 261 W) were chosen. Figure 2 shows that the results of mathematical modeling are in good agreement with the experimental data [10].

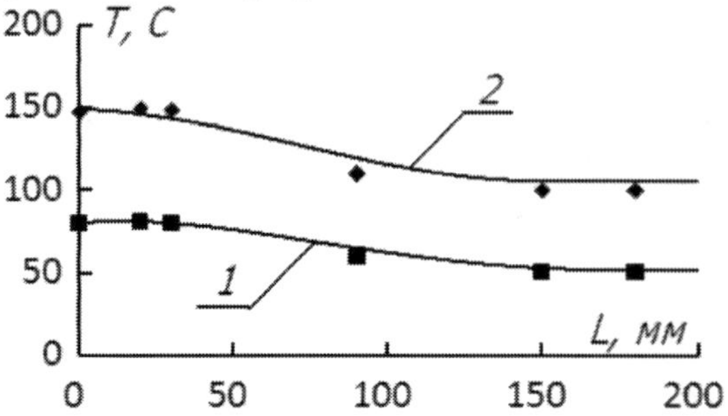

Fig.2. Temperature distributions for the height of the thermosyphon: numerical simulation (solid curves), experimental data [10]. Curve 1 - 100 W, curve 2 - 261 W.

3 Results and Discussion

Numerical modeling of heat and mass transfer processes in a two-phase closed thermosyphon of a rectangular shape is performed in the emergency mode of operation (Fig.1) with dimensions: height H = 200 mm, transverse dimension L = 100 mm. The choice of geometric dimensions is due to the characteristic dimensions of a typical lithium-ion battery of the Boeing 787 aircraft. The following variants of the heat flux densities on the bottom cover of the thermosyphon in the section y = 0: $0.82 \cdot 10^3$ W/m^2; $1.76 \cdot 10^3$ W/m^2 in accordance with the critical temperatures obtained in [3].

Figure 3a shows the temperature distributions, current lines, and velocity vectors that characterize the main regularities of the processes under consideration in the region under study at a thermal load q_h=$0.82 \cdot 10^3$ W/m^2. It is seen that in the region of vertical walls symmetrical circulation flows are formed. The lines of constant temperatures here are oriented in the direction *y*. It should be noted that in the region of the vortex currents, low velocities of steam have been obtained, increasing toward the center of the steam channel. Analysis of the isotherms shows that the temperature difference in the region under study in the direction of y is about 4 °K.

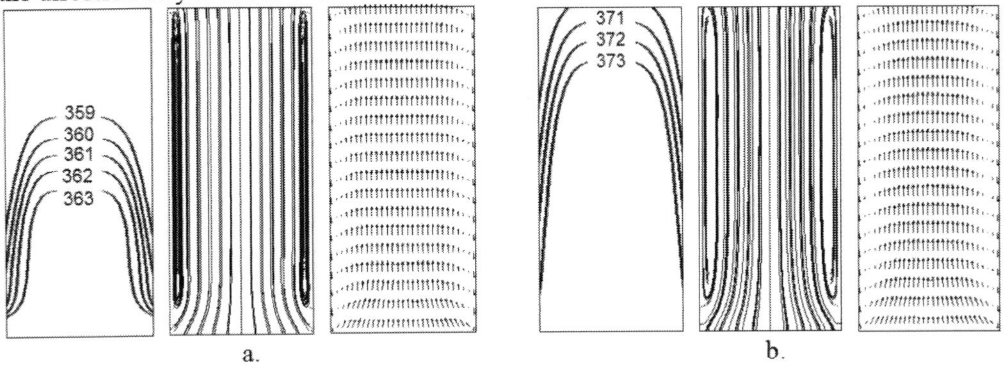

Fig.3. Lines of constant temperature, streamlines and velocity vectors in the study area. a - q_h=$0.82 \cdot 10^3$ W/m^2, b - q_h=$1.76 \cdot 10^3$ W/m^2.

It should be noted that with the increase of q_h to $1.76 \cdot 10^3$ W/m^2 (Fig. 3b), the configuration of the lines corresponding to the main characteristics of the steam flow does not change significantly. The temperature difference along the height of the steam channel decreases by approximately 2 °K due to the intensification of heat transfer in the direction of *y* due to the more intensive evaporation of the coolant on the lower lid.

4 Conclusion

As a result of numerical investigation, characteristic lines of constant temperatures, current and velocity vectors were obtained.

In the practice of using closed two-phase thermosyphons, the temperature drop from the evaporation zone to the condensation zone is one of the main parameters characterizing the efficiency of its operation. It was found that the increase in the thermal load on the lower cover from $0.82 \cdot 10^3$ W/m^2 to $1.76 \cdot 10^3$ W/m^2 leads to a decrease in the temperature difference in the investigated region by 2 °K. This is probably due to the intensification of heat transfer and the increase in the vapor velocity in the device under study.

The reported research was supported by Russian Federation President Grant for state support of the Russian Federation leading scientific schools SS-7538.2016.8 (No.14.Y31.16.7538-SS).

DOI: 10.1051/matecconf/201714101007

References

1. D.E. Bruskin, *Electric power supply of aircrafts* (Higher School, Moscow, 1988) [in Russian]
2. D.A. Khrustalyov, *Rechargeable batteries* (Izumrud, Moscow, 2003) [in Russian]
3. S.G. Rebrov, S.V. Yanchur, V.S. Mansurov, S.A. Moskovkin, Proceedings of the MAI, **72** (2014) [in Russian]
4. R. Slayton, G. Spinardi, Technovation, **47**, 47-58 (2016), DOI: 10.1016/j.technovation.2015.08.004
5. T. Song, Y. Li, J. Song, Z. Zhang, Procedia Engineering, **80**, 628-637 (2014), DOI: 10.1016/j.proeng.2014.09.118
6. A. Sambaa, N. Omar, Electrochimica Acta, **117**, 246-254 (2014), DOI: 10.1016/j.electacta.2013.11.113
7. G.V. Kuznetsov, A.E. Sitnikov, High Temp., **40**, 6 (2002), DOI: 10.1023/A:1021437502952
8. Kuznetsov G.V., Al-Ani, M.A., Sheremet M.A., J. Eng. Thermophys., **20**, 2 (2011), DOI: 10.1134/S1810232811020081
9. Ansys Help. FLUENT Theory Guide.
10. H. Jouhara, A.J. Robinson, Appl. Therm. Eng., **30**, 2 (2010), DOI: 10.1016/j.applthermaleng.2009.08.007

 DOI: 10.1051/matecconf/201714101008

Special features of evaporation of water-salt solutions on a horizontal heating surface

Nina Mironova[1,*] , *Vladimir* Morozov[2], and *Sergey* Elistratov[1]

[1]Novosibirsk State Technical University, 630073 Novosibirsk, Russia
[2]Kutateladze Institute of Thermophysics SB RAS, 630090 Novosibirsk, Russia

Abstract. Experimental investigate the process of evaporation of water-salt solutions on the horizontal heating surface at a temperature of 90 °C. It was revealed that there are differences in the process of evaporation of hydrous solutions of salts forming and not forming water-salt hydrates. The absence of hydrates of salts in hydrous solutions NaCl, $CaCl_2$ and CsCl made it possible to fix the rate of formation of a solid phase (nucleation) in the volume of the supersaturated solution.

1 Problem Statement

The efficiency of absorption thermotransformers (ATT) depends directly on the ability of the working fluid to absorb steam. As working fluid for ATT, water-salt solutions are most widely used. The higher their absorption capacity, the better the performance of absorption chillers and heat pumps.

Water-salt solutions have the greatest capacity for absorption near the equilibrium crystallization line. In Fig. 1 shows the solubility curves of various salts as a function of temperature. In practice, it is customary to produce work processes in ATT with less concentrated solutions to eliminate the phenomenon of solution crystallization, especially in regenerative heat exchangers. In general, this reduces the efficiency of the desorption and absorption processes, increases the total specific quantity of metal of the design.

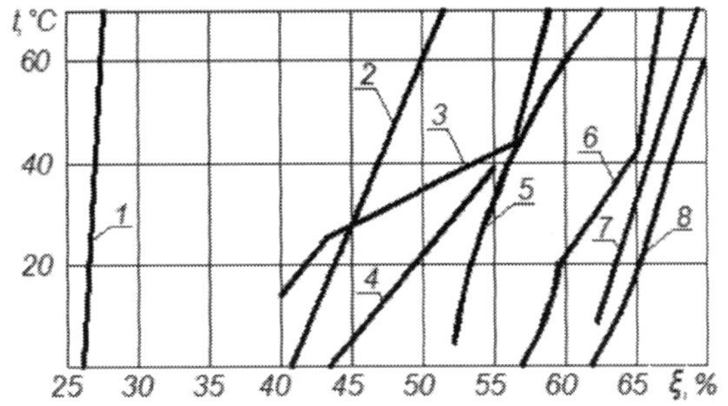

Fig. 1. Dependence of the mass fraction of salts solubility in water on temperature [1]: 1 – NaCl; 2 – LiCl; 3 – $CaCl_2$; 4 – LiCl-$CaCl_2$; 5 – LiCl-$LiNO_3$; 6 – LiBr; 7 – LiCl-CsCl; 8 – CsCl.

* Corresponding author: mironovaNina@inbox.ru

DOI: 10.1051/matecconf/201714101008

However, for batch machines (fig. 2) with combined "absorber-desorber" and "condenser-evaporator" devices, the formation of salt crystals in solution as a solid dispersed phase (volumetric nucleation process), can on the contrary result in improved effectiveness.

It can be seen (fig. 1) that there except lithium bromide are a number of salts and their mixtures that have a degree of solubility in water higher than lithium bromide. In addition, in an aqueous solution, widely used in practice, the salt LiBr exists as hydrates $LiBr \cdot H_2O$, $LiBr \cdot 2H_2O$, $LiBr \cdot 3H_2O$, $LiBr \cdot 5H_2O$. Salts of CsCl, NaCl and $CaCl_2$ do not form hydrates, which allows In contrast to the hydrate-forming salts (LiBr, LiCl, $LiNO_3$) the solution to be evaporated (desorbed) to visualization of the process of formation of the smallest salt particles in the solution volume. Salts of CsCl, NaCl and $CaCl_2$ do not form hydrates, which allows unlike hydrate forming salts (LiBr, LiCl, $LiNO_3$), to evaporate (desorb) solutions to visualization of the process of formation of the smallest particles of salt in the solution volume.

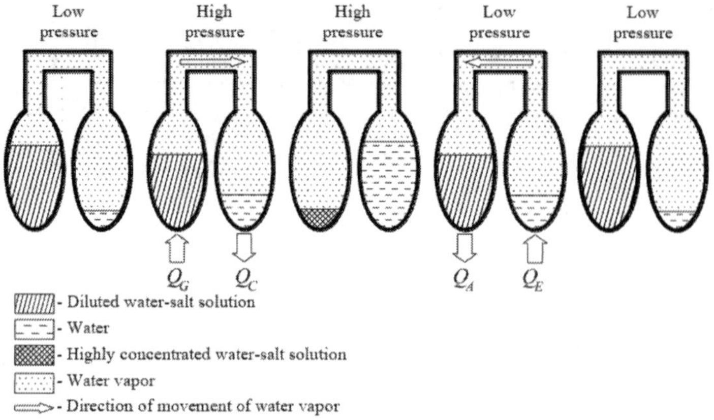

Fig. 2. Operation scheme of ATT of periodic action: Q_G – heat input to the generator (desorber); Q_C – removal of heat of condensation; Q_A – removal of heat of absorption; Q_E – heat input to the evaporator.

The laws of spontaneous formation (nucleation) and growth in supersaturated solutions of solid-phase nuclei can be easily studied by examining the desorption of thin films of aqueous-salt solutions on a horizontal heating surface. Depending on the temperature of the heating surface, three basic desorption modes can be realized (Figure 3): evaporation in a layer with a heating surface temperature (T_{SUR}) is less than the boiling point (T_B) of the solution; bubble boiling, when the surface temperature is equal to or higher than the boiling point, but below the Leidenfrost temperature (T_L); evaporation in a spheroidal state with separation of the solution from the heating surface by a continuous vapor film at a surface temperature above the Leidenfrost temperature [2].

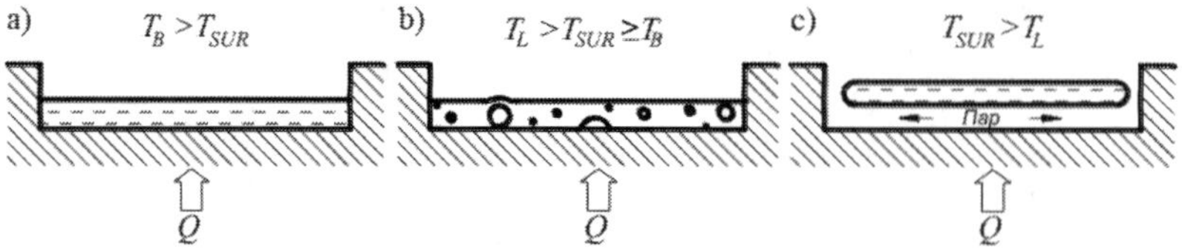

Fig. 3. Basic modes of desorption of water-salt solution: a) evaporation in a layer, b) bubble boiling, c) evaporation in a spheroidal state.

 DOI: 10.1051/matecconf/201714101008

2 Results

The nucleation process in thin films of aqueous-salt solutions was investigated at temperatures of the heating surface lower than the boiling point of the solution (fig. 3a).

In Fig. 4 shows photographs of the process of desorption of a thin film of an aqueous solution of CsCl on the horizontal heating surface of the experimental setup [3, 4], with a working area made of titanium. The working area was 70 mm in diameter and the working surface temperature was maintained at 90 °C.

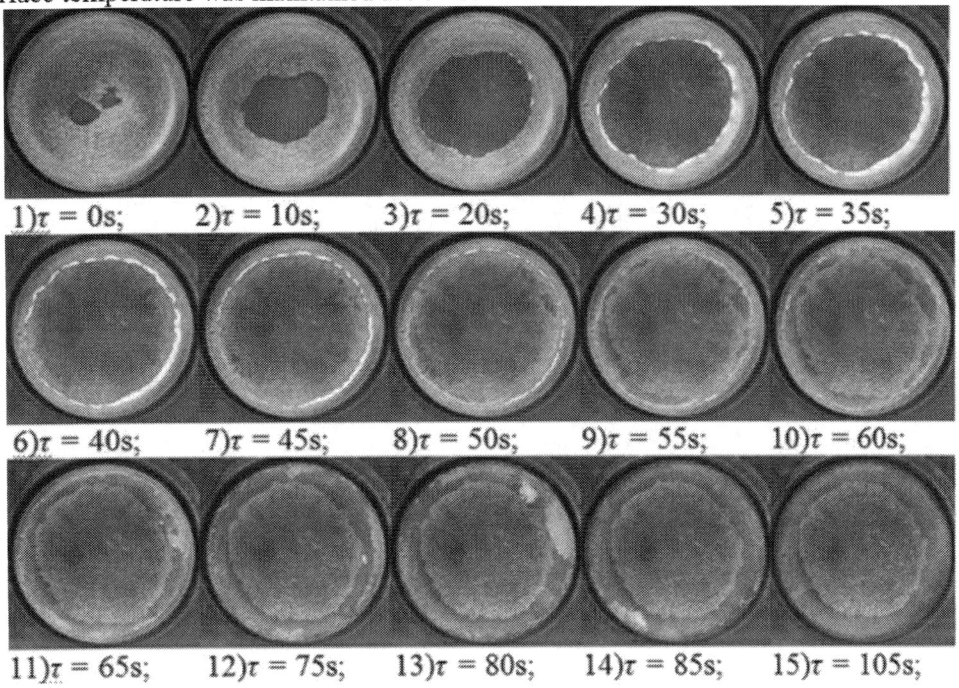

Fig. 4. Development of the process of desorption of water-salt solution CsCl at T_{SUR} = 90 °C.

When the maximum thickness is reached, determined by the surface tension forces, the thin continuous film of the solution breaks and begins to gradually contract from the center to the periphery of the working area (fig. 4, frames 1-4), forming an zone free of solution. At the same time on the "dry" surface, free from the liquid solution, an uneven trace of the smallest particles of salt is formed. The apparent surface concentration of these particles increases in the direction from the center to the periphery. As the concentration of the solution increases, in places of contact with the heating surface due to a decrease in the thickness of the solution film, the maximum evaporation rate is observed, which leads to the formation of the first nucleation zones (fig. 4, frames 3-4 (light zones at the internal boundary of the solution)). As the concentration increases, a spontaneous process of redistribution of steam across the entire surface of the solution is observed (fig. 4, frames 11-14 (light zones on the surface of the solution)), which is expressed as the appearance of visible areas of salt deposition and their dissolution. Steam moves from the peripheral part of the working area to the inner boundary of the solution film. Relative stabilization of this process for supersaturated solution is achieved through 105s. (fig. 4, frame 15). When the thin vapor-air jet is directed to the surface of the supersaturated solution, a mobile zone of the solution forms in which salt crystals are missing for a while (fig. 5, frames 2-6).

 DOI: 10.1051/matecconf/201714101008

However, the visible boundary of the crystallization front disappears after 3.5 seconds (fig. 5, frame 7) and the process of dynamic equilibrium again stabilizes in time.

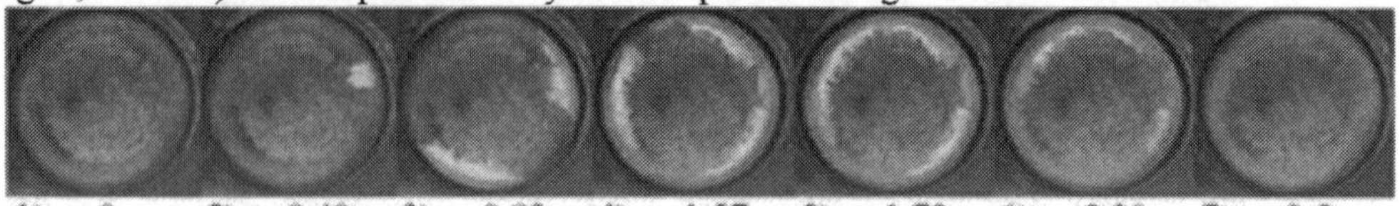

1)τ = 0s; 2)τ = 0.40s; 3)τ = 0.93s; 4)τ = 1.57s; 5)τ = 1.73s; 6)τ = 2.30s; 7)τ = 3.5s;

Fig. 5. Dynamics of the motion of the crystallization front in a thin film of a solution.

A similar picture can be observed with the desorption of thin films of aqueous solutions of salts NaCl and $CaCl_2$. However, the size of single and aggregated crystals formed for these salts is much larger than the size of CsCl crystals (fig. 6).

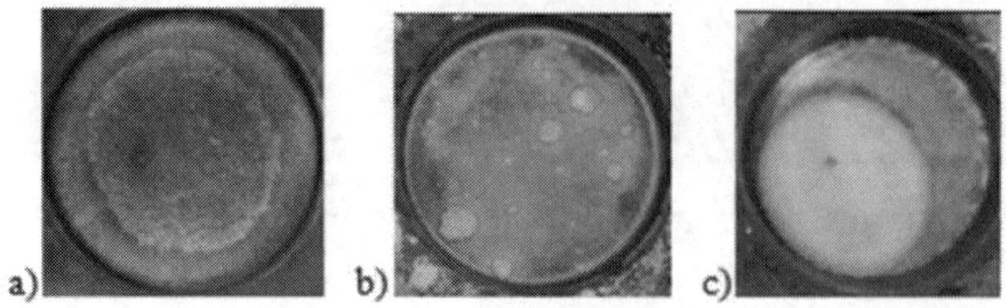

Fig. 6. Crystals of salts on the heating surface T_{SUR} = 90 ° C: a) CsCl; b) $CaCl_2$; c) NaCl.

It is important that, in contrast to supersaturated solutions of salts with a relatively low solubility (NaCl, $CaCl_2$, etc.), supersaturated CsCl solutions retained the flowability property in the presence of a large number of small unaggregated particles of the solid phase.

Conclusion

The results of the experiments make it possible to identify a new vector of studies on the intensification of heat and mass transfer processes in devices of absorption thermotransformers: the possibility of operating under the supersaturation of aqueous solutions of highly soluble salts, for example CsCl, which has a higher solubility in water than the widely known LiBr salt.

The work was financially supported by a grant from the Ministry of Science and Education of the Russian Federation, agreement No. 14.Z50.31.0041 of February 13, 2017.

References

1. A.V. Baranenko, S.V. Karavan, O.A. Pinchuk, D.V. Karavan, *Water solutions of absorption thermo-transformers* (Pero Publ., Moscow, 2014) [in Russian]
2. S.S. Kutayeladze, *Fundamentals of the theory of heat transfer* (Atomizdat Publ., Moscow, 1979) [in Russian]
3. V.E. Nakoryakov, S.L. Elistratov, J. Eng. Thermophys., **22**, 2 (2009), DOI: 10.1134/S181023281302001X
4. V.E. Nakoryakov, S.L. Elistratov, J. Eng. Thermophys., **18**, 3 (2009), DOI: 10.1134/S1810232809030011

MATEC Web of Conferences **141**, 01009 (2017) DOI: 10.1051/matecconf/201714101009
Smart Grids 2017

Ignition conditions and characteristics for distillant fuel under to local conductive heating

Arkadiy Zakharevich[1,*], *Pavel* Panchenko[1] , and *Yekaterina* Panchenko[1]

[1]National Research Tomsk Polytechnic University, 634050 Tomsk, Russia

Abstract. Describes the methodology and results of experimental studies of ignition of diesel fuel and a cloth impregnated with this fuel by a hot steel particle are outlined. The layer thickness of liquid fuel was changed from 1 mm to 11 mm. It was established dependence of numerical values of the ignition time delay from the thickness layer of diesel fuel. The revealed ignition mode and the limiting layer thickness for diesel fuel when the ignition is missing. As experimental study of the ignition conditions and characteristics for fabric impregnated with diesel fuel under local conductive heating is reported.

1 Introduction

Statistics of fires for 2011-2016 [1] shows a tendency to reduction of the number of fires in Russia. According to EMERCOM of Russia, only in 2016 there were 139.7 thousand fires, they caused material damage in the amount of 14.3 billion rubles, 8760 people died and 9909 people suffered injuries of varying severity. This problem is especially serious for the chemical, petrochemical, oil refining and related branches of industry, which are characterized by significant concentrations of combustible substances and the existence of various ignition sources of their own nature [2-5]. In order to implement measures for prevention of explosions and fires, it is necessary to know the conditions and characteristics of the formation of combustible systems capable of exothermic decomposition. The difficulty of analyzing the process of ignition of liquids is due to the joint flow of interconnected processes of different physical nature: evaporation, heat conduction, crystallization, diffusion, convection under the mutual influence of thermal, kinetic and hydrodynamic conditions [6,7]. Therefore, experimental studies of the processes of ignition of a layer of distillate fuels, as well as of a tissue partially or completely impregnated with this fuel by a single hot particle, are relevant in the modern world.

The task of the paper is to study experimentally the effect of the thickness of a layer of diesel fuel and fabric impregnated with this fuel on the characteristics and conditions of inertia of the ignition process by a local energy source.

* Corresponding author: bet@tpu.ru

 DOI: 10.1051/matecconf/201714101009

2 Experimental technique

Experimental studies were carried out at the stand, which basic scheme is given in the work [8]. The source of heating were monolithic steel particles in the form of a disk of a typical size (diameter $d_p = 6 \cdot 10^{-3}$ m and height $h_p = 5 \cdot 10^{-3}$ m). The objects of the study were presented by winter diesel fuel DT-Z-K3-38 and fabric impregnated with this fuel. The experiments were performed at room temperature of 298 K and relative humidity of 60-70%. A video recording of the process was carried out with the help of high-speed Phantom v411 camera fitted with "TemaAutomotive" software. There were registered moments of contact "particle - slick of diesel fuel", "particle - damp fabric", "a particle - dry fabric" and the appearance of a flame. After processing of the video frames, the ignition delay times (τ_{ign}) were determined. A series (6-8) of experiments was performed for each case with a fixed initial temperature of the steel particle (T_p). After that, average values of τ_{ign} and the root-mean-square deviations were determined for each constant value of T_p. As a result of the analysis of the ignition delay times for the experiments, it was established that variation coefficients has varied from $\pm$4.5% (at high T_p) to $\pm$9% (at relatively low T_p). The systematic error in measuring of the initial temperature of the particle, estimated by the method [9,10], has not exceeded 2%. Obtained mean-square deviations and coefficients of variation can be considered acceptable for the conducted experiments in regard with the complexity of the mechanism for realizing the process, which is under study.

3 Results and discussion

Figure 1 shows the ignition delay time of diesel fuel the dependences on the layer thickness (h) obtained in the experiments.

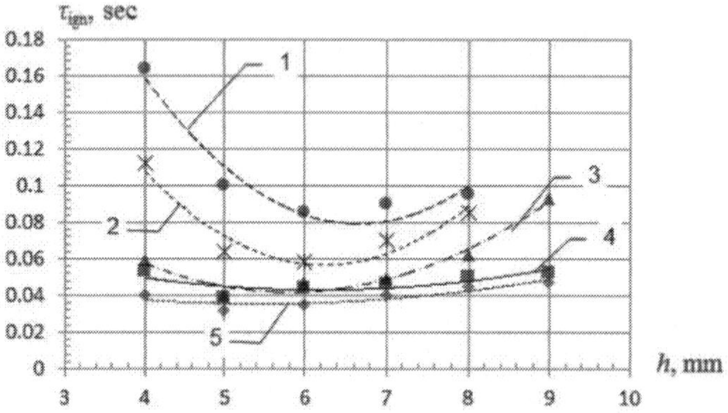

Fig. 1. The experimental dependences of the ignition delay time on the thickness of the diesel fuel layer (h): 1 – a steel particle with a temperature of 1273 K, 2 – a steel particle with a temperature of 1293 K, 3 – a steel particle with a temperature of 1333 K, 4 - a steel particle with a temperature of 1353 K, 5 – a steel particle with a temperature of 1373 K.

The curves correspond to different values of steel particle temperature, which was used as a source of heating (ignition). The minimum and maximum values of the layer thickness correspond to the limiting conditions for initiating the combustion reaction. If the layer thickness is less than 4 mm, when a hot particle falls into a fuel, a dry spot forms. As a result, there is no active evaporation in the immediate vicinity of the heated surfaces of the particle. Consequently, the fuel concentration is insufficient to initiate the fuel ignition process. At sufficiently extreme values of the fuel layer thickness, the particle is completely immersed and there is no ignition. At sufficiently extreme values of the thickness of the fuel layer, the particle is completely immersed. The resulting pairs of diesel fuel are heated

 DOI: 10.1051/matecconf/201714101009

when the flow around the steel particle. At h> 9 mm, the energy from the steel particle is insufficient to initiate combustion due to subsequent cooling (when the vapors move upward). Fuel vapours rise up, when the height of vertical channels in the layer is from 4 to 9 mm, and leave out of direct contact with the source of heat energy. As a result fuel vapours shifted toward to particle axis of symmetry and heated by heat exchange with the upper horizontal surface of the particle. In this case the upper surface was source of active heating of diesel vapours. Experimental video analysis shows that ignition reaction proceeds on symmetry axis of system diesel layer – particle – channel filled with fuel vapours.

The resulting pairs of diesel fuel are heated when the flow around the steel particle. At h> 9 mm, the energy from the steel particle is insufficient to initiate combustion due to subsequent cooling (when the vapors move upward). Fuel vapours rise up, when the height of vertical channels in the layer is from 4 to 9 mm, and leave out of direct contact with the source of heat energy. As a result fuel vapours shifted toward to particle axis of symmetry and heated by heat exchange with the upper horizontal surface of the particle. In this case the upper surface was source of active heating of diesel vapours. Experimental video analysis shows that ignition reaction proceeds on symmetry axis of system diesel layer – particle – channel filled with fuel vapours.

From the approximation dependences (Fig. 2), it can be seen that the ignition of the fabric impregnated with fuel occurs at temperatures lower than the films of diesel fuel. This is due to the fact that the fabric has a lower thermal conductivity, which reduces the intensity of heat removal from the steel particle. Consequently, most of the heat is expended directly on the heating and ignition of fuel vapors.

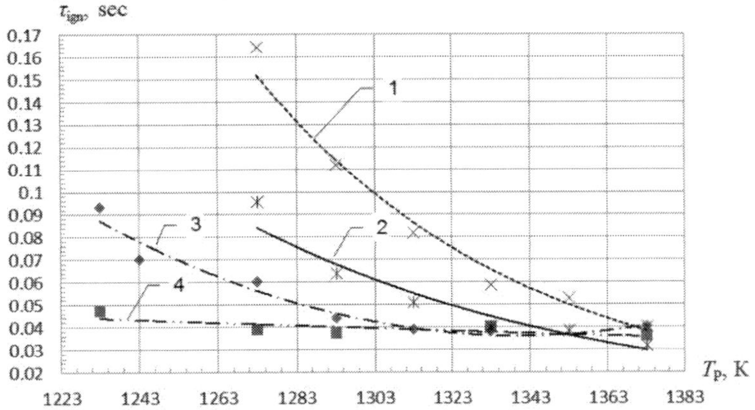

Fig. 2. The experimental dependences of the ignition delay time on the initial temperature of the hot steel particle: 1 – the thickness of the layer of diesel fuel is 4 mm, 2 – the thickness of the layer of diesel fuel is 6 mm, 3 – the fabric is impregnated with fuel with a relative humidity of 85%, 4 – the fabric is impregnated with fuel with a relative humidity of 95%.

On the basis of experimental studies, it is possible to formulate a physical model of the process of ignition of diesel fuel. It is established that ignition of fuel occurs in a small neighborhood of the interface between two media: a steel particle and fuel. Such a mechanism of the ignition process is typical for any layer of the fuel film. Diesel fuel, when interacting with a steel particle, receives heat from it, which is consumed to heat the fuel, namely the phase transition and the heating of the vapors. The resulting vapor, when moving into the oxidant (air) zone, heats up when the particle is flowing around and ignites with a sufficient amount of oxidizing agent. Based on the video recording of the ignition process, the flame appears in the vicinity of the steel particle upper face, having the maximum temperature.

At a thickness of a layer of fuel more than 9 mm, the steel particle falls in a layer of fuel, exceeding its sizes. There is a complete immersion of the particle. The fuel actively evaporates, and actively removes heat from the particle. To ignite the fuel layer, direct contact of the heated fuel vapors with the heated surface of the steel particle and the amount of oxygen necessary for ignition is necessary. However, with a complete immersion, there are no such conditions, hence there is no possibility of ignition.

If the thickness of the fuel layer is less than 4 mm, then when a hot steel particle falls into fuel, a dry spot forms. As a result, there is no active evaporation (in the immediate vicinity of heated surfaces) and as a result there is no intensive heating of the vapors. As a result, there is no ignition.

4 Conclusion

On the basis of experimental studies threshold ignition mode layer of diesel fuel. The obtained experimental dependences of ignition delay time cloth impregnated with diesel fuel from the initial temperature of the steel particles. The significant difference of the inertia of the investigated process ignition "particle fuel", "particle – wet tissue" from the main local source of heating.

The work was supported by the Russian President's grant (Scientific School project - 7538.2016.8).

References

1. M.A. Chebukhanov, A.A. Kozlov, Yu.A. Matyushin, A.G. Firsov, V.I. Sibirko, T.A Chechetin, *Fire and fire safety in 2016: Statistical compilation* (VNIIPO, Moscow, 2017) [in Russian]
2. M.V. Beschastnov, *Industrial explosions. Estimation and prevention* (Chimiya, Moscow, 1991)
3. P. Dagaut, M. Cathonnet, Prog. Energy Combust. Sci., **32**, 48-92 (2006), DOI: 10.1016/j.pecs.2005.10.003
4. S. Kumagai, *Combustion* (Khimiya, Moskow, 1979) [in Russian]
5. G.V. Kuznetsov, P.A. Strizhak, J. Eng.Thermophysics **18**, 162-167 (2009)
6. V.N. Vilynov, *Theory of ignition of condensed substances* (Soc. Siberian Branch, Novosibirsk, 1984) [in Russian]
7. D.A. Frank-Kamenetsky, *Diffusion and Heat Transfer in Chemical Kinetics* (Nauka, Moscow, 1987) [in Russian]
8. A.V. Zakharevich, M.S. Zygin, P.V. Panchenko, Y.A. Vlasova, MATEC Web of Conferences **72**, 01128 (2016), DOI: 10.1051/matecconf/20167201128
9. A.N. Zaidel, *Errors of measurements of physical quantities* (Nauka, Leningrad, 2009) [in Russian]
10. V.E. Gmurman, *The theory of probability and mathematical statistics* (Vyssh. Shk., Moscow, 2003) [in Russian]

Natural convection of water in a partially porous enclosure with a heat-generating element

Marina Astanina[1,*], and *Mikhail* Sheremet[1]

[1]Tomsk State University, 634050 Tomsk, Russia

Abstract. Paper presents the numeral results of transient natural convection in a partially porous enclosure with a heat-conducting and heat-generating element. The horizontal walls of the cavity are adiabatic, while the vertical ones are isothermal and have low temperature. A mathematical model formulated in dimensionless variables such as vorticity, stream function and temperature have been solved by means of the finite difference method. The governing parameters are the porous layer thickness and dimensionless time. The effects of these parameters are shown using streamlines, isotherms and average temperature of the heat-conducting element. The results demonstrate an opportunity to intensify the passive cooling system using the porous layer.

1 Introduction

Study of convective heat transfer in different cavities plays an important role for science and engineering. Research of heat transfer in technical systems helps to enhance the energy parameters.

At present, interest in the problems of natural convection is increasing in connection with the appearance of real task in production. Thus, Chen et al. [1] have conducted a numeral investigation for enhanced heat transfer from multiple discrete heated sources in a horizontal channel by metal-foam porous layer. They have showed that an increase in the solid-fluid interfacial heat exchange results in a decrease in the temperatures difference between the solid and fluid phases for fixed Reynolds number, where the porous media tend to reach local thermal equilibrium (LTE) with the fluid and a larger cooling augmentation of heaters is obtained. Astanina et al. [2] have numerically studied natural convection with temperature-dependent viscosity in a porous cavity. They have considered in detail the effect of viscosity variation parameter on heat transfer.

The objective of this study is a numeral analysis of the fluid flow and heat transfer patterns in a square partially porous enclosure with a heated element under the effect of internal heat generation.

* Corresponding author: astanina.marina@bk.ru

2 Mathematical model and numeral method

Physical model is schematically shown in Figure 1. Consider the natural convection in a square partially porous enclosure with a heat-generating and heat-conducting element. The left and right walls of the cavity are maintained at a constant cooling temperature T_c, the top and bottom walls are adiabatic. A heat-generating and heat-conducting element is placed in the centre of the bottom wall and has a constant density of volumetric heat flux q. The interior of the cavity is divided into two parts, namely, a water saturated porous layer of height γ and a clear water under the porous layer. It is assumed, that the thermophysical properties of the medium do not depend on temperature. The fluid is Newtonian, heat-conducting and the flow is laminar; the Boussinesq approximation is valid.

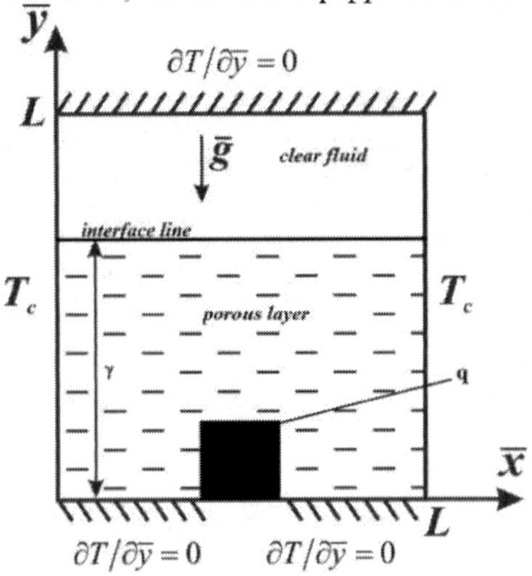

Fig. 1. Physical model.

Under such assumpions the convervation equations for mass, momentum and energy can be written for the two-demensional natural convection in terms « stream function – vorticity – temperature » in the following form [2–4]:

- for the clear fluid layer

$$\frac{\partial^2 \Psi}{\partial X^2} + \frac{\partial^2 \Psi}{\partial Y^2} = -\Omega \tag{1}$$

$$\frac{\partial \Omega}{\partial \tau} + U\frac{\partial \Omega}{\partial \tilde{O}} + V\frac{\partial \Omega}{\partial Y} = \sqrt{\frac{Pr}{Ra}}\left(\frac{\partial^2 \Omega}{\partial \tilde{O}^2} + \frac{\partial^2 \Omega}{\partial Y^2}\right) + \frac{\partial \Theta}{\partial X} \tag{2}$$

$$\frac{\partial \Theta}{\partial \tau} + U\frac{\partial \Theta}{\partial X} + V\frac{\partial \Theta}{\partial Y} = \frac{1}{\sqrt{Ra \cdot Pr}}\left(\frac{\partial^2 \Theta}{\partial X^2} + \frac{\partial^2 \Theta}{\partial Y^2}\right) \tag{3}$$

- for the porous layer on the basis of the Brinkman-extended Darcy model [2]

$$\frac{\partial^2 \Psi}{\partial X^2} + \frac{\partial^2 \Psi}{\partial Y^2} = -\Omega \tag{4}$$

$$\frac{\partial \Omega}{\partial \tau} + U\frac{\partial \Omega}{\partial X} + V\frac{\partial \Omega}{\partial Y} = \sqrt{\frac{Pr}{Ra}}\left(\varepsilon\frac{\partial^2 \Omega}{\partial X^2} + \varepsilon\frac{\partial^2 \Omega}{\partial Y^2} - \varepsilon^2\frac{\Omega}{Da}\right) + \varepsilon^2\frac{\partial \Theta}{\partial X} \tag{5}$$

DOI: 10.1051/matecconf/201714101010

$$\frac{\partial \Theta}{\partial \tau}+U\frac{\partial \Theta}{\partial X}+V\frac{\partial \Theta}{\partial Y}=\frac{a_{pm}/a_f}{\sqrt{Ra\cdot Pr}}\left(\frac{\partial^2 \Theta}{\partial X^2}+\frac{\partial^2 \Theta}{\partial Y^2}\right) \tag{6}$$

- for the heat-generating element

$$\frac{\partial \Theta}{\partial \tau}+U\frac{\partial \Theta}{\partial X}+V\frac{\partial \Theta}{\partial Y}=\frac{a_{hs}/a_f}{\sqrt{Ra\cdot Pr}}\left(\frac{\partial^2 \Theta}{\partial X^2}+\frac{\partial^2 \Theta}{\partial Y^2}+Os\right) \tag{7}$$

In these equations Pr is the Prandtl number, Ra is the Rayleigh number, Da is the Darcy number, Os is the Ostrogradsky number. They are defined as follows:

$$Pr=\frac{\nu}{\alpha},\quad Ra=\frac{g\beta\Delta TL^3}{\alpha\nu},\quad Da=\frac{K}{L^2},\quad Os=\frac{qL^2}{\lambda_{hs}\Delta T}. \tag{8}$$

The initial and boundary conditions for the equations (1) – (7) are as follows :

$$\begin{aligned}
&\tau=0:\ 0\le X\le 1,\ 0\le Y\le 1:\ \Psi=\Omega=\Theta=0\\
&\tau>0:\ X=0,\ 0\le Y\le 1:\ \Psi=0,\ \Omega=-\frac{\partial^2\Psi}{\partial X^2},\ \Theta=0\\
&\tau>0:\ X=1,\ 0\le Y\le 1:\ \Psi=0,\ \Omega=-\frac{\partial^2\Psi}{\partial X^2},\ \Theta=0\\
&\tau>0:\ Y=0,\ 0\le X\le 1:\ \Psi=0,\ \Omega=-\frac{\partial^2\Psi}{\partial Y^2},\ \frac{\partial\Theta}{\partial Y}=0\\
&\tau>0:\ \text{at heat source surface:}\ \Psi=0,\ \Omega=-\frac{\partial^2\Psi}{\partial n^2},\ \begin{cases}\Theta_{hs}=\Theta_f\\ \frac{\lambda_{hs}}{\lambda_{pm}}\frac{\partial\Theta_{hs}}{\partial n}=\frac{\partial\Theta_{pm}}{\partial n}\end{cases}\\
&\tau>0:\ \text{at internal fluid-porous interface:}\ \begin{cases}\Psi_{pm}=\Psi_f\\ \frac{\partial\Psi_{pm}}{\partial y}=\frac{\partial\Psi_f}{\partial y}\end{cases}\begin{cases}\Omega_{pm}=\Omega_f\\ \frac{\partial\Omega_{pm}}{\partial y}=\frac{\partial\Omega_f}{\partial y}\end{cases}\begin{cases}\Theta_{pm}=\Theta_f\\ \frac{\lambda_{pm}}{\lambda_f}\frac{\partial\Theta_{pm}}{\partial y}=\frac{\partial\Theta_f}{\partial y}\end{cases}
\end{aligned} \tag{9}$$

The partial differential equations (1) – (7) with corresponding initial and boundary conditions (9) have been solved by the finite difference method using the uniform grid [2, 4]. For the approximation of the convective and diffusive terms we used the difference scheme of the second order accuracy. The parabolic equations have been solved on the basis of Samarskii locally one-dimensional scheme [2, 4, 5]. The discretized equations have been solved by Thomas algorithm. The equations for the stream function and have been discretized using the five-point difference scheme on the basis of central differences for the second derivatives. The obtained difference equations have been solved by the successive over relaxation method. Optimum value of the relaxation parameter has been chosen on the basis of computing experiments.

 DOI: 10.1051/matecconf/201714101010

3 Results and discussion

The numeral analysis has been conducted for the following values of the governing parameters: $Ra = 10^5$, $Pr = 7.0$, $Da = 10^{-3}$, $Os = 10.0$, $\gamma = 0.25 - 1.0$, $\tau = 0.0 - 100.0$. The simulation results are illustrated in Figs. 2 and 3.

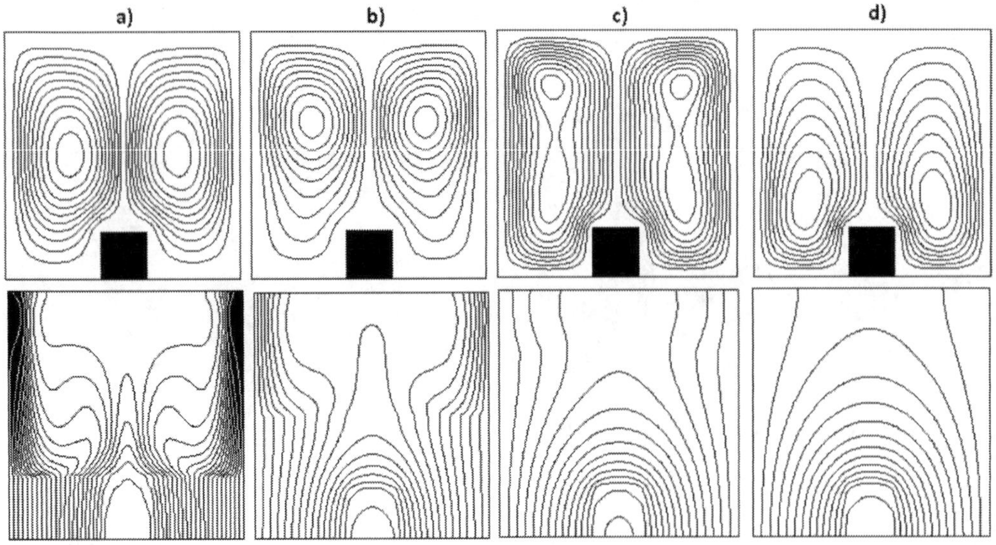

Fig.2. Streamlines and isotherms for $Os = 10$, $Da = 10^{-3}$: $\gamma = 0.25$ – a, $\gamma = 0.5$ – b, $\gamma = 0.75$ – c, $\gamma = 1.0$ – d.

Fig. 2 shows the streamlines and isotherms for different values of the porous layer height γ. It can be seen, that two convection cells are formed in the cavity, while their cores are located above the interface line, which is clearly seen in Figs. 2a, 2b, 2c. The observed circulations reflect the formation of the ascending flows in the central part of the cavity and the descending ones along the cooling vertical walls. It should be noted, that the heat-generating element reflects a uniform heating of the cavity in the zone of the porous medium due to low circulation with additional resistance from the solid porous matrix and non-uniform heating in the zone of clear water. In the case of a completely porous enclosure (see Fig. 2d) the cavity is heated uniformly everywhere.

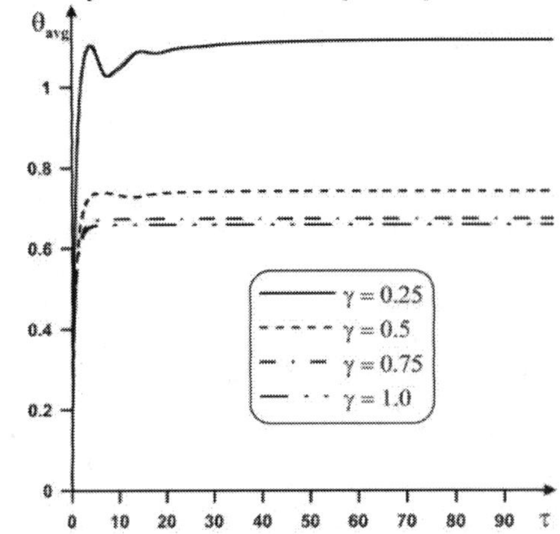

Fig.3. Dependences of the average temperature inside the heater on the porous layer thickness.

The effect of the porous layer thickness on the average temperature within the heat source is presented in Figure 3. It is possible to conclude that the porous layer allows to intensify the passive cooling system inside the closed domain with a heat-generating element. Moreover, one can find an optimum value of the porous layer thickness to have an effective cooling and optimal fluid flow patterns inside the cavity.

4 Conclusion

Numerical simulation of natural convection of water in a square cavity partially filled with a porous medium under the effect of heat-generating and heat-conducting element has been performed. Obtained results have shown that porous insertion can be an effective parameter for an intensification of heat removal from the heat-generating element. Such technique can be used for cooling of electronic elements in the case of impermeable electronic cabinets with one or several heaters.

This work was supported by the Russian Scientific Fond (Project No. 17-79-2014).

References

1. C.C. Chen, P.C. Huang, H.Y. Hwang, Int. J. Heat Mass Transf., **58**, (2013), DOI: 10.1016 / j.ijheatmasstransfer.2012.11.041
2. M.S. Astanina, M.A. Sheremet, J.C. Umavathi, Transp. Porous Media, **110**, 1 (2015), DOI: 10.1007/s11242-015-0558-x
3. D.A. Nield, A. Bejan, *Convection in Porous Media* (Springer, New York, 2006)
4. A. Shenoy, M.A. Sheremet, I. Pop, *Convective flow and heat transfer from wavy surfaces: viscous fluids, porous media and nanofluids* (CRC Press, Boca Raton, 2016)
5. A.A. Samarsky, *Theory of difference schemes* (Nauka, Moscow, 1977)

Experimental modelling of evaporation and boiling processes in a two-phase thermosyphon

Konstantin Ponomarev[1,*], *Siddharth Raj* Gupta[2] , *Evgeniya* Orlova[1] , and *Dmitry* Feoktistov[1]

[1]National Research Tomsk Polytechnic University, 634050 Tomsk, Russia
[2]Indian Institute of Technology Kanpur, Kanpur, India

Abstract. A model of a closed two-phase thermosyphon was developed to study thermo-physical processes in a thermosyphon thermal condition systems of energy-saturated equipment. This model differs from the known by reasoned choice of thermocouples installation providing sealing the device during operation. In addition, the program VI (Virtual Instrument) was created in graphical environment Labview for registration of the temperatures and pressure in the thermosyphon.

1 Introduction

A decrease in the heat losses, an increase in the efficiency of the heat exchange equipment and supply of safety operation of the technological installations are relevant problems of power engineering [1, 2]. One of the solutions is to apply –the passive cooling systems of thermally loaded elements based on two-phase thermosyphons [3, 4]. These devices have variety of advantages (independence, ease, availability, high heat-transfer coefficient) compared with traditional cooling systems [5]. Promising ways of increasing efficiency of the thermosyphons are intensification of the boiling processes [6], turbulization of flows [7], coating of the definite structure on inside wall of thermosyphon [8], for example, by laser emission [9] or plasma [10]. Thermosyphons are not widely used in industry due to the absence of scientific –bases of their design and construction.

The aim of this work is to develop the experimental model of a closed two-phase thermosyphon for studying thermo-physical processes in a thermosyphon thermal condition systems of energy-saturated equipment.

2 Experimental technique

In experimental model of the thermosyphon, 15 thermocouples TT type K by OMEGA with diameter of the junction of 130 μm were used. For eleven of them, holes (d = 1.5 mm) were drilled at the bottom and top chamber towards flanged fitting of the copper cylinder of the by the engraver of Sparky MK 135E fixed on the micrometric moving element. Thermocouples were modified for fixed them into the thermosyphon. In order to avoid the

* Corresponding author: kop.tpu@gmail.com

condensate movement from the vapor channel along the thermocouple housing, the connection of the junction and the housing was treated with heat-resisting silicone sealant. Eleven thermocouples were installed in steel casings with diameter of 1.2 mm and canned on both sides. The ends of the needles and the wires protruding from them were wrapped by insulated tape to strengthen the developed constructions of needle thermocouples, which were further installed in the thermosyphon (Fig. 1).

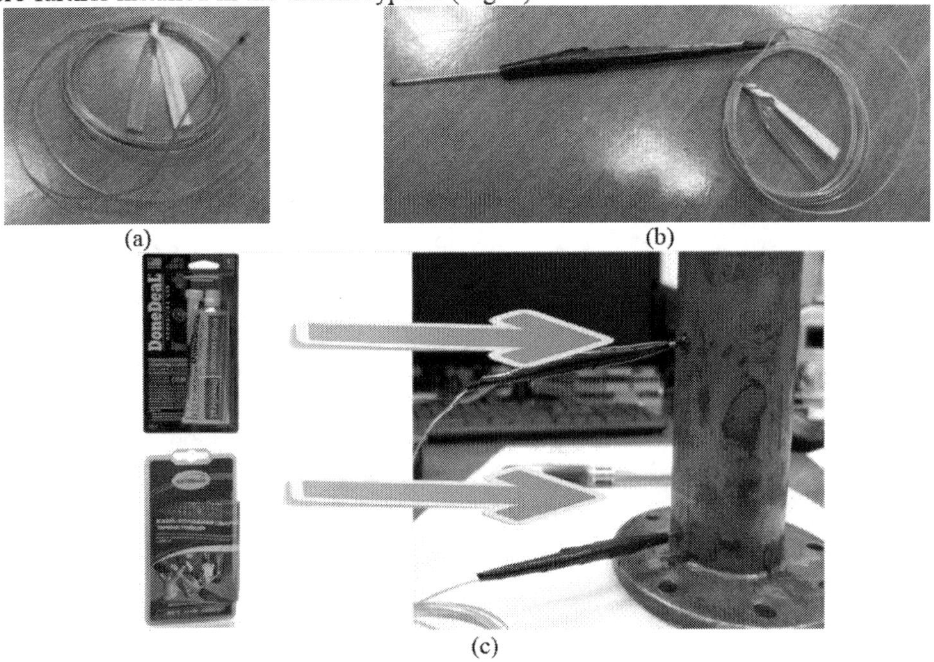

Fig. 1. Thermocouple before (a) and after (b) modification, thermocouples fixation with sealant and cold welding (c).

The cold welding by Astrohim is found to be preferable for fixing thermocouples into thermosyphon than the sealant by Done Deal, because there was not change of geometric position of the thermocouple due to gravitational forces and applied external forces in time.

Often the vacuum degree of the thermosyphon body does not play a big role, since the thermosyphon successfully works if there is an uncertain volume of air in its cavity. The presence of air results in a partial decrease of the condensation surface and an increase of the initial pressure. The higher the working pressure in the thermosyphon, the air will fill less volume. When the heat transfer characteristics of thermosyphons are under research, a vacuum (6.6×10^{-3} Pa) is especially necessary in studying the condensation process, since even a small amount of air can decrease the heat-transfer coefficient from steam to the thermosyphon wall. Since the experimental model of the thermosyphon was developed to study the processes of boiling and evaporation, vacuum was not created in the cavity of the thermosyphon.

The conditions for the end-to-end supply of heat to the thermosyphon were adopted. These conditions are close to real, when the thermosyphons are installed directly on the heat-loaded elements. The source of heat was a flat silicone heater (51 mm x 51 mm) with an operating temperature in the range from -60 to +250 °C (maximum short-term temperature is +300 °C) (Fig. 2 (a)).

Using the thermal imaging camera by TESTO (with an error of no more than 2% in the temperature range from 0 to +350 °C), the temperature field of the heating element was measured (Fig. 2 (b)). It was established that the heating element is heated uniform, the

DOI: 10.1051/matecconf/201714101011

exceptions are the edges of the heater due to the technical design of the heating element (Fig. 2 (c)).

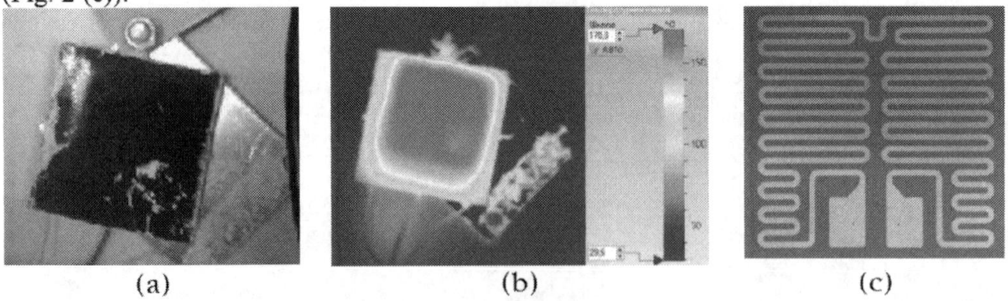

Fig. 2. Silicon heater: photo (a), thermogram (b), etched heating element (c).

The silicone heater is located between two aluminum plates, fastened with four bolts. The bottom and lateral sides of the heating cell are thermally insulated with a foam rubber by Rockwool. Operating temperature is in the range from -180 to +550 °C. Thermal conductivity of the foam rubber at room temperature is 0.04 W/(m·K). The external frame is made of foamed polystyrene by POLPAN. Operating temperature is in the range from -50 to +75 °C, thermal conductivity is 0.03 W/(m·K) (Fig.3 (a)). Two copper substrates and a thermosyphon were installed on the heating cell. In each of the sections 1-1, 2-2, 3-3 (Fig.3 (b)), two thermocouples were located, the error of which was 0.1 degrees.

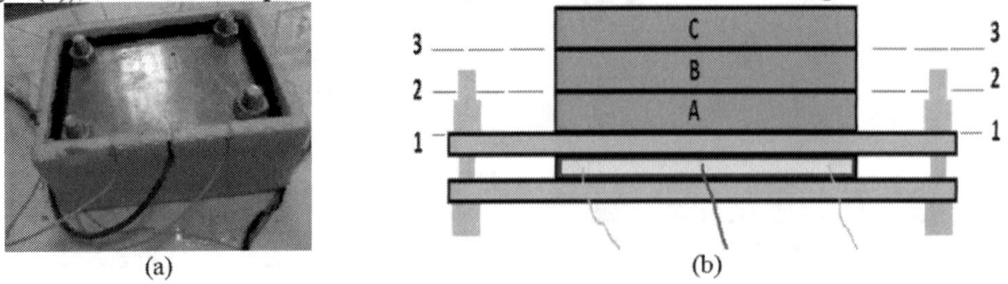

Fig. 3. Photo of the heating cell (a) and the heating cell diagram (b). A, B - copper substrates, C - bottom of the thermosyphon.

Due to poor contact, there were large gaps in the sections, in which the thermocouples were in contact with air. When a load of 80 V was installed on a laboratory autotransformer, after 5000 seconds, when a quasi-steady operation occurred, the difference between the temperatures recording by thermocouples located in one section (1-1, 2-2, 3-3) or at different edges of the substrate could exceed 10 degrees. In order to reduce the interaction with air and add stability to the design, the grooves for the thermocouples were cut in the substrate with a depth of 0.6 mm, and a silicon organic thermal paste (KPT 8) was used.

The heat-flux density was calculated according to the Fourier's law:

$$q = \frac{\Delta T}{(\delta / \lambda)}, W / m^2 \tag{1}$$

where ΔT is the temperature difference between the edges of the substrate, $°C$; δ is the thickness of the substrate, m; λ is the thermal conductivity of copper, $W/(m \cdot K)$.

The temperature difference and the heat-flux density (Eq. 1) over the average values of the parameters (for 10 minutes) in the quasi-steady operation are presented in Table 1.

Table 1. The results of a preliminary series of the experiments.

Parameters	Without termoconductive paste	With termoconductive paste
The difference between the two thermocouples (section 1-1), Δt, °C	0.1	0.1
The difference between the two thermocouples (section 2-2), Δt, °C	0.2	0
The difference between the two thermocouples (section 3-3), Δt, °C	0.4	0.2
The temperature difference between the edges of the first substrate "A", Δt, °C	5.1	1.6
The temperature difference between the edges of the second substrate "A", Δt, °C	5.0	1.3
Heat flux density through the first substrate "A", q, MW/m^2	0.47	0.15
Heat flux density through the second substrate "B", q, MW/m^2	0.43	0.11
The difference between the heat-flux density (heat losses), Δq, MW/m^2	0.04	0.04

According to the analysis of Table 1, the difference between the two thermocouples installed without the use of termoconductive paste located in the one section in the groove was less than half a degree, but exceeded the error of the thermocouples. The temperature difference between the edges of the substrates decreased twofold and was approximately 5 °C.

After applying termoconductive paste with the thermal conductivity of more than of air ($\lambda_m = 1\, W/(m \cdot K)$ > $\lambda_a = 0.027\, W/(m \cdot K)$), the difference between the two thermocouples in two of the three sections did not exceed their error. The temperature difference between the edges in the substrates decreased more than three times. Despite this fact, the heat flux density through the substrates was 150 and 110 kW/m^2. And according to the Joule-Lenz's law it is 9.6 kW/m^2.

It is established that when using termoconductive paste, the measurement error is less than without it. However, it was not possible to approach the ideal contact between the two substrates, that is why the heat-flux density input to the bottom of the thermosyphon was determined according to the Joule-Lenz law, taking into account the 15% loss to the environment:

$$q = 0.85 \cdot \frac{U \cdot I}{S}, W/m^2 \tag{2}$$

where U is voltage, V; I is current, A; S is internal area of the cross-section of the thermosyphon, m^2.

To record the temperatures and pressure, the VI program (Virtual Instrument) was developed in the LabView graphical environment.

By setting the input voltage and internal diameter of the thermosyphon in the program, the power supplied to the heating element and the heat-flux density were determined. Calibration of the thermocouples was carried out in the developed program which allows recording temperatures and pressure in the thermosyphon.

The developed experimental model of the closed two-phase thermosyphon with heat removal from the local heat source through the lower cover of the heat exchanger is shown in Fig. 4.

Fig. 4. Photo of the experimental setup.

3 Conclusion

An experimental model of a closed two-phase thermosyphon with heat removal from a local heat source through the lower cover of the heat exchange device was developed to study the efficiency of such cooling systems. The cold welding is found to be preferable for fixing thermocouples in the thermosyphon than a sealant. It is established that when using termoconductive paste, the error in measuring the temperature is less than without it. The heat-flux density input to the bottom of the thermosyphon is more accurately determined according to the Joule-Lenz law, taking into account losses of 15% to the environment than to the Fourier law.

The reported research was supported by Russian Federation President Grant for state support of the Russian Federation leading scientific schools SS-7538.2016.8 (No.14.Y31.16.7538-SS).

References

1. D. Glushkov, G. Kuznetsov, P. Strizhak, Math. Prob. Eng., **2014**, 920480 (2014), DOI: 10.1155/2014/920480
2. S. Juengjaroennirachon, N. Pratinthong, P. Namprakai, T. Suparos, J. Mech. Sc. Tech., **31**, 1 (2017), DOI: 10.1007/s12206-016-1241-z
3. O. Kabov, E. Chinnov, Journal of Enhanced Heat Transfer, **9**, 2 (2002), DOI: 10.1615/JEnhHeatTransf.v9.i2.10
4. D. Feoktistov, E. Vympin, A. Nurpeiis, MATEC Web of Conferences, **72**, 01081 (2016), DOI: 10.1051/matecconf/20167201081
5. K. Ponomarev, E. Orlova, A. Nurpeiis, MATEC Web of Conferences, **110**, 01062 (2017), DOI: 10.1051/matecconf/201711001062
6. V. Nakoryakov, S. Misyura, J. Eng. Thermophys., **25**, 1 (2016), DOI: 10.1134/S1810232816010033
7. V. Lebedev, V. Lemanov, S. Misyura, V. Terekhov, Fluid dynamics, **28**, 5 (1993), DOI: 10.1007/BF01050044
8. S. Misyura, Applied Thermal Engineering, **113**, 472-480 (2017), DOI: 10.1016/j.applthermaleng.2016.11.072

9. K. Batishcheva, D. Feoktistov, Y. Shanenkova, E. Kirichenko, MATEC Web of Conferences, **110**, 01013 (2017), DOI: 10.1051/matecconf/201711001013
10. A. Sivkov, A. Pak, D. Nikitin, I. Rakhmatullin, I. Shanenkov, Nanotechnol. Russ., **8**, 7-8 (2013), DOI: 10.1134/S1995078013040174

MATEC Web of Conferences **141**, 01012 (2017) DOI: 10.1051/matecconf/201714101012
Smart Grids 2017

Heat-transfer, inside of the ground heat-transfer units, from liquid, additionally cooling the oil-immersed transformer

Ilnur Madyshev[1*], *Oksana* Dmitrieva[1], and *Andrey* Dmitriev[2]

[1]Kazan National Research Technological University, 420016 Kazan, Russia
[2]Kazan State Power Engineering University, 420066 Kazan, Russia

Abstract. A device for cooling the oil-immersed transformer by means of thermoelectric transducers has been developed. Their operating principle is based on the fact that at night at minimum ambient temperature, the cold is accumulated inside of a vessel due to formation of water ice layer on the developed surfaces of thermoelectric transducers` ribs. Melting of this layer is used in the hottest time of a day for additional cooling of transformer oil in the cooling system of the transformer. Pre-cooling of the circulating water takes place in vertical ground heat-transfer units (VGHT units), mounted into ground drilled boreholes. Certain estimate calculations of the heat-transfer process have been carried out.

1 Introduction

Power oil-immersed transformers are essential elements of electrical engineering units and quality of distributed electrical power depends upon reliability and operating safety of these units [1]. One of the main factors, determining the reliability of transformers, is an efficient cooling of them. However, existing cooling systems do not always ensure a sufficiently effective heat removal from the winding insulation material at short-term significant overloads [2]. In this regard, relevant objective is the development of additional cooling system of transformers, capable of decreasing the peak load and its uniform distribution throughout operational period due to cold accumulation within hours of minimum heat load.

2 Description of the cooling system

An example of a constructive design of this system is application of additional vessel, equipped with a cascade of semiconductor thermoelectric modules (TEM cells), cold surfaces of which, generate the cold and are used for cooling the liquid in vessel, while hot surfaces are equipped with a system of heat discharge into the environment [3]. A coil pipe *7* is mounted into the lower part of oil-immersed transformer`s tank 1 for additional cooling of power transformer by liquid (water) from vessel *2*. TEM cells *8* are fixed to the outer

* Corresponding author: ilnyr_91@mail.ru

surface of one of the vessel walls *2*. Hence, ribbing *9* has been executed from the side of TEM cells *8* in order to form the greatest thickness of ice layer on the inner surface of a vessel *2* (Fig. 1). A square or right angle shaped vertical channel *5* is installed for effective heat removal from heated sides of TEM cells *8*. Pre-cooling of the circulating water takes place in VGHT units *3* and *4*, mounted into ground drilled boreholes of 10-100 m deep [4].

Additional cooling system of power oil-immersed transformers has two operation modes. The first operation mode - accumulation of cold – is used within operating hours of transformer, corresponding to small heat loads. By means of pump 6 with closed shut-off valves *11*, water from vessel *2* is supplied for a preliminary cooling into the first section of VGHT units *3*. After that, the water enters a vertical channel *5*, where heat removal from TEM cells` heated sur-faces *8* takes place. Liquid film, coming down along a vertical surface of a channel *5* most effectively removes the heat excess, providing lower temperature of the TEM cells` cold sides *8*. Due to this, water crystallization and ice layer buildup take place on the inner ribbed surface of a vessel *2*. After that, water, heated in a vertical channel *5*, is supplied for additional cooling into the second section of VGHT units *4*, and then enters the upper part of a vessel *2*. Thus, the cold is accumulated inside of a vessel *2* at a constant circulation of cooling water due to the work of TEM cells *8*.

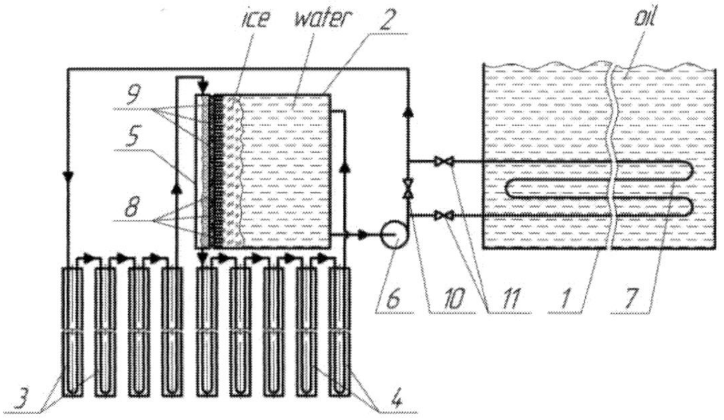

Fig. 1. Additional cooling system of power transformer: *1* – transformer tank; *2* – vessel; *3*, *4* – VGHT units; *5* – square or right-angle shaped vertical channel; *6* – pump; *7* – coil pipe; *8* – TEM cells; *9* – ribs for ice layer formation; *10*, *11* – shut-off valves.

The second operation mode corresponds to high heat loads of the transformer operation, as a rule, at high ambient temperatures. In this case, by means of pump *6* with closed shut-off valves *10*, cooled water from vessel *2* is supplied into the coil pipe *7* and cools the transformer oil in the tank 1. After that, this water enters the same devices, as at the first operation mode. Thus, additional liquid cooling of tank's bottom part of transformer increases the efficiency of cooling the transformer oil within operating hours, corresponding to the greatest heat loads, preventing the failure of power transformers.

3 Calculation procedure and results

An arithmetic model, proposed in this study, is used for determination of design and operating mode parameters of VGHT units, designed for the water pre-cooling before its supply into vessel [5]. Heat transfer in the apparatus is described by the system of equations:

$$Q = G_m c_p \left(t_{L1} - t_{L2}\right) = 2k_l \pi L n \left(t_L - t_G\right), \tag{1}$$

 DOI: 10.1051/matecconf/201714101012

where Q – heat flow, W; G_m – mass rate of heat-transfer agent (water), кg/s; c_p – mass specific heat capacity of water, J/(кg·K); t_{L1} – water temperature at inlet of VGHT units, K; t_{L2} – water temperature at outlet from VGHT units, K; k_l – linear factor of heat transfer, W/(m·K); L – length of VGHT units (borehole depth), m; n – number of VGHT units; t_L – average water temperature in VGHT units, K; t_G – ground temperature, K.

Linear factor of heat transfer, considering heat-transfer resistance of the filler layer of borehole, can be determined by the following formula:

$$k_l = \frac{1}{\dfrac{1}{\alpha d_1} + \dfrac{1}{2\lambda_s}\ln\dfrac{d_2}{d_1} + \dfrac{1}{2\lambda_z}\ln\dfrac{d_2 + 2\delta_z}{d_2}}, \qquad (2)$$

where α – heat-transfer coefficient, W/(m²·K); λ_s – heat conductivity coefficient of material of VGHT unit's pipe, W/(m·K); λ_z – heat conductivity coefficient of borehole filler layer, W/(m·K); d_1 – inner diameter of VGHT unit's pipe, m; d_2 – outer diameter of VGHT unit's pipe, m; δ_z – borehole filler layer thickness, m.

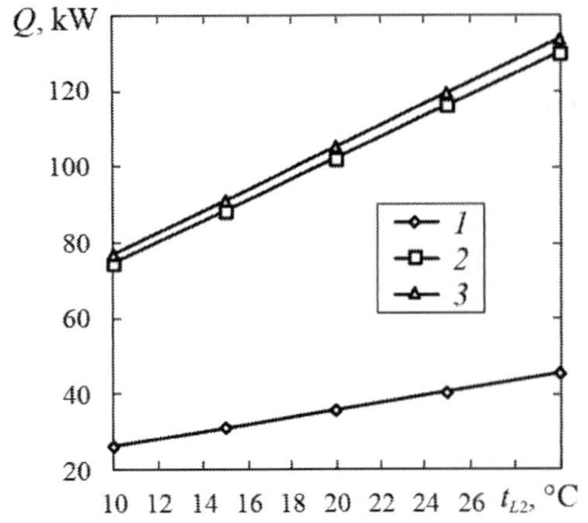

Fig. 2. Heat flow change, removed into VGHT units, depending on the water temperature at outlet from VGHT units for pipes, made of various materials: *1* – polyethylene (λ_s = 0.38 W/(m·K)); *2* – stainless steel AISI 321 (λ_s = 17.5 W/(m·K)), *3* – cast iron BS200 (λ_s = 54.0 W/(m·K)).

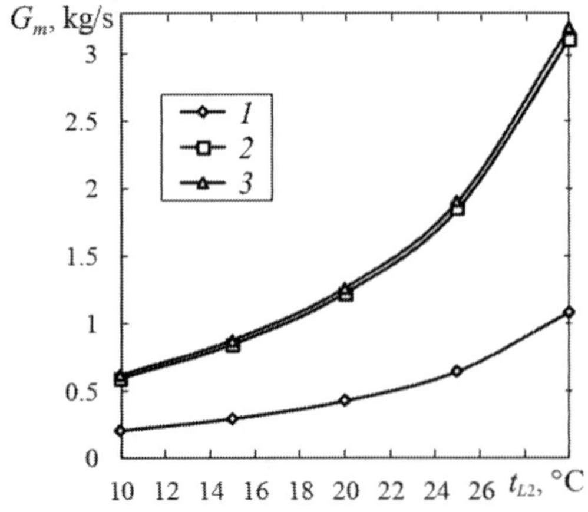

Fig. 3. Mass flow rate change of cooled water, depending on its temperature at outlet from VGHT units for pipes, made of various materials: *1* – polyethylene (λ_s = 0.38 W/(m·K)); *2* – stainless steel AISI 321 (λ_s = 17.5 W/(m·K)), *3* – cast iron BS200 (λ_s = 54.0 W/(m·K)).

The heat transfer coefficient of water, flowing around the straight pipes, can be determined by equations in work [6]. U-shaped steel pipes with a diameter of 32×3.0 mm have been taken as VGHT units. The heat conductivity of pipes has been taken equal to 17.5 W/(m·K). Pipes are located in the borehole with a diameter of 120 mm. Material heat conductivity of filler layer of borehole is 2.3 W/(m·K). Design calculation of VGHT units has been carried out considering the possibility of additional heat removal from the oil-immersed transformer, equal to 100 kW. Meanwhile, ground design temperature was 8°C, while temperatures of heat-transfer agent at inlet of and at outlet from VGHT units were 40°C and 20°C respectively. Under given conditions, the estimated total length of VGHT units was 213.2 m. In this regard, when carrying out research calculations, length of VGHT units (depth of a borehole) was 40 m, quantity – 3. The estimated calculations have been carried out for pipes, made of polyethylene, stainless steel and cast iron. The temperature difference of heat-transfer agent at inlet of and at outlet from VGHT unit has been varied from 10°C to 30°C. The water temperature at inlet of VGHT unit has been taken constant (t_{L1} = 40°C). Ground design temperature was 10°C.

The results of numerical studies show that heat flow magnitude, additionally removed from power transformer, depends essentially on temperature characteristics of heat-transfer agent and material of U-shaped pipes of VGHT units at the same dimensions.

In Fig. 2 we may see that substitution of polyethylene pipes by steel pipes increases the heat flow by 2.84–2.87 times under the studied operating con-ditions of VGHT units. Increase of heat amount, transferred into VGHT units, leads to an increase of mass flow rate of the cooled agent (Fig. 3). It should be noted that further increase of heat conductivity coefficient of pipes' walls of VGHT units, for example by 3.086 times, leads to the heat flow increase only by 2.86–2.89%. In this case, the main heat transfer resistance is concentrated in the filler layer of borehole and heat transfer resistance contribution of this layer to the total heat transfer resistance reaches up to 97.1%.

4 Conclusion

Performed numerical studies show that in order to increase the heat flow, additionally removed from power transformer, we should choose VGHT unit's steel pipes and thickness of filler layer of borehole should be minimal. Thus, it can be said that application of the proposed liquid cooling system of oil-immersed transformers will let additionally remove up to 50% of heat at the moments of peak heat loads.

The work was performed with funding from the grant of the President of the Russian Federation No. MK-5215.2016.8 (contract No. 14.Z56.16.5215-MK dated March 14, 2016).

References

1. L. Kish, *Heating and cooling of transformers* (Energy, Moscow, 1980) [in Russian]
2. A.M. Golunov, N.S. Seshenko, *Cooling devices of oil-immersed transformers* (Energy, Moscow, 1976) [in Russian]
3. A.V. Dmitriev, O.S. Dmitrieva, I.N. Madyshev, MATEC Web of Conferences, **95**, 15008 (2017), DOI: 10.1051/matecconf/20179515008
4. O.S. Dmitrieva, G.R. Patrakova, A.V. Dmitriev, 2017 International Conference on Industrial Engineering, Applications and Manufacturing (ICIEAM), 1-4, (2017), DOI: 10.1109/ICIEAM.2017.8076305
5. S.O. Filatov, V.I. Volodin, Proceedings of BSTU. Chem. and tech. of inorganic substances, **3**, 179 (2011) [in Russian]
6. K. Savina, A. Abramovskih, V. Haymin, S. Lavrinenko, MATEC Web of Conferences, **72**, 01042 (2016), DOI: 10.1051/matecconf/20167201042

MATEC Web of Conferences **141**, 01013 (2017) DOI: 10.1051/matecconf/201714101013
Smart Grids 2017

Modeling the evolution of the aerosol cloud of toxicants in the atmosphere

Ivan Bondarchuk[1], *Sergei* Bondarchuk[2] , and *Boris* Borisov[3,*]

[1]Tomsk State University, 634050 Tomsk, Russia
[2]Tomsk State Pedagogical University, 634061 Tomsk, Russia
[3]Tomsk Polytechnic University, 634050 Tomsk, Russia

Abstract. Using the methods of mathematical modeling, the formation and evolution of aerosol clouds of toxicants in the atmosphere from the chemical industry enterprises, thermal power engineering and rocket carriers of space vehicles is analyzed. The processes of dynamic interaction of drops between themselves and a two-phase flow, processes of agglomeration, crushing and evaporation of aerosol particles are taken into account. The results of numerical calculations are presented.

1 Introduction

Increasing the impact of human life on the environment requires constant improvement of the means of monitoring this impact and analyzing its consequences. One of the most important objects of the described impact influence is the atmosphere. It is accepts the toxicants flow ejected from the pipes of thermal power plants, chemical plants, etc. of. A significant contribution to the violation of the environment is made by space vehicles and aircraft, especially when forcibly dumping the remains of fuel components before a forced, unforeseen or emergency landing of the apparatus.

The present work is devoted to mathematical modeling of the formation, evolution and distribution of toxic aerosols in the atmosphere with the localization of the site of their deposition, the analysis of the possible degree of contamination for a particular area, taking into account various physical and meteorological factors in order to predict possible environmental risks. When formulating a mathematical model, the features of the processes of dynamic interaction of moving deformable particles (droplets) of a dispersed phase in the general case with a two-phase dispersion medium are considered [1-5]; the influence of meteorological features of the region is analyzed; the set of interrelated processes - deformation and fragmentation of drops with the analysis of processes of their evaporation and heat and mass exchange with a turbulent atmosphere [6-7] is taken into account. Reviews on this problem are presented in [8-9].

The paper presents the results of mathematical modeling of the evolution and dynamics of a liquid-drop cloud in the case of an emergency dump of liquid aviation or rocket fuel.

* Corresponding author: bvborisov@tpu.ru

2 Physical and mathematical models of processes

As shown by the analysis of the experimental data [10] in the range of moderate (Re≈10÷100) and small (Re≤1) Reynolds numbers in the region of large Bond numbers Bo typical for the aerosol droplets under consideration, the deformation of the droplet from the Weber number We [11] is determined by the relation:

$$\varepsilon = \frac{D_m}{D_0} = 1 + 0.027\,\mathrm{We}\,, \tag{1}$$

where D_m is the diameter of the average cross section of the drop; D_0 - diameter of an equivalent spherical drop (whose volume is equal to the volume of the deformed drop).

The similarity criteria for the conditions of specific experiments using the following formulas were calculated:

$$\mathrm{Re} = \frac{\rho\left|U - U_p\right| D}{\mu}, \quad \mathrm{We} = \frac{\rho\left|U - U_p\right|^2 D}{\sigma}, \quad \mathrm{Bo} = \frac{\rho_p D^2}{\sigma}\left|\omega\right|, \tag{2}$$

where σ is the coefficient of surface tension; ρ_p is the density of the drop.

The process of coagulation and fragmentation of aerosol particles on accordance with the generalized approximation formula for the collision parameter of droplets of a two-phase flow of similar sizes [11, 12] was modelled:

$$\Phi_{ji} = 1.4 + 1.979\beta + 0.507(2\beta^2 - 1)\,, \tag{3}$$

where $\beta = \left(\frac{\mathrm{Re}_{ji}}{383.6}\right)^{0.572}\left(\frac{\mathrm{Lp}_i}{370.37}\right)^{-0.153}\left(\frac{\gamma_{ji}}{2.37}\right)^{-0.597}$; Φ_{ji} is the mathematical expectation of the ratio of the change in the mass of the target drop D_i, caused by the interaction of *j-i*, to the total mass of the projectile droplets D_j ($D_j < D_i$) that collide with it. Re_{ji} = 6 to 385; Lp_i = 0.2 to 600; Re_{ji} =1 ÷ 12; Re_{ji} - Reynolds number, calculated by the diameter of the drop-projectile D_j; Lp is the Laplace number; γ_{ji} is the ratio of the diameters of the colliding drops.

The Reynolds number from the diameter of the D_j projectile was calculated. Equations of motion of a droplet of radius rp for the its velocity componens in a Cartesian coordinate system relative to the surface (the z axis is directed along the gravitational acceleration vector of gravity of gravity) has the form [11]:

$$\frac{du_p}{dt} = \varphi_1(u - u_p), \quad \frac{dv_p}{dt} = \varphi_1(v - v_p), \quad \frac{dw_p}{dt} = \varphi_1(w - w_p) + g\,, \tag{4}$$

where $\varphi_1 = \frac{3\rho}{8\rho_p r_p} C_D\left|U - U_p\right|$; $\left|U - U_p\right| = \sqrt{(u-u_p)^2 + (v-v_p)^2 + (w-w_p)^2}$; *u*, *v*, *w* are the components of the wind speed vector ***U***; u_p, v_p, w_p are the components of the velocity vector of the droplet U_p; C_D is the coefficient of aerodynamic resistance.

At Re = (1÷700) (intermediate flow regime), the Klyachko formula to determine the coefficient of aerodynamic drag was used [13]:

$$C_D = \frac{24}{\text{Re}} + \frac{4}{\sqrt[3]{\text{Re}}}. \qquad (5)$$

For Re = (700 ÷ 3·105) (self-similar flow) [14], C_D = 0.44 was assumed. The process of aerodynamic crushing of droplets was determined in accordance with the calculated values of the Weber and Bond criteria. In accordance with the analysis of the experimental data, it was assumed that when the critical Weber number We=We*=17 or the critical value of the Bond number is Bo=Bo*=22.5, the drop is split into two spherical drops of equal mass. The physical properties of air due to vertical coordinate of account for a specific date using the parameters of the standard atmosphere and the data of weather stations was taken on. The coefficient of dynamic viscosity of air using the Sutherland formula [15] was calculated:

$$\mu = \frac{0.68\cdot 10^{-2}}{T+122}\left(\frac{T}{273}\right)^{3/2},$$

where T is the absolute temperature.

The drop is a thermally thin body, and its temperature averaged over the volume T_p in accordance with the following equation was determined:

$$\frac{dT_p}{dt} = \varphi_2 \text{Nu}(T - T_p) - \frac{q_r}{mc_p} G, \qquad (6)$$

where $\varphi_2 = \frac{3}{2}\frac{\lambda}{r_p^2 \rho_p c_p}$; c_p is the specific heat of the liquid; λ is the coefficient of thermal conductivity of air; m is the mass of the drop; G is the mass of the evaporated liquid from the drop surface per unit time (evaporation rate); q_r is the specific heat of evaporation of the liquid.

To calculate the Nusselt number, the following dependence was used [16]: $\text{Nu}=2+0.6\text{Re}^{1/2}\text{Pr}^{1/3}$, where Pr is the Prandtl number. The change in the droplet radius due to evaporation on the equation was determined:

$$\frac{dr_p}{dt} = -\frac{G}{4\pi\rho_p r_p^2} \qquad (7)$$

The evaporation rate by the formula was determined:

$$G = 4\pi r_p^2 k \frac{X}{1-X}, \qquad (8)$$

where k is the mass-transfer coefficient; X is the mole fraction of the vapor near the drop surface.

Equation (7) in accordance with the relationship between X and the partial vapor pressure p_0 was written:

$$\frac{dr_p}{dt} = -\frac{k}{\rho_p}\frac{p_0}{p - p_0}, \qquad (9)$$

where p is the atmospheric pressure. The mass-transfer coefficient k from the formula [17] was calculated:

$$k = \frac{c_f D_f M_p}{2r_p}\left[2+0.6\left(\frac{2r_p\left|U-U_p\right|\rho_f}{\mu_f}\right)^{1/2}\left(\frac{\mu_f}{\rho_f D_f}\right)^{1/3}\right], \tag{10}$$

where D_f is the binary diffusion coefficient at the film temperature $T_f = (T_f + T)/2$; c_f and μ_f are the density and coefficient of dynamic viscosity of air at T_f.

The diffusion coefficient in accordance with the Fuller-Schletter-Giddings [18] proposals for binary gas systems at low pressure was calculated:

$$D = \frac{T^{1.75}[(M_A + M_B)/(M_A M_B)]^{0.5}}{p\left[\left(\sum V_A\right)^{1/3} + \left(\sum V_B\right)^{1/3}\right]^2}, \tag{11}$$

where $[D_f]$ = cm2/sec; $[p]$ = atm; $[T]$ = K; M_A and M_B are the molecular masses of components A and B (drops of liquid and air); $\sum V_A$, $\sum V_B$ are the diffusion volumes of molecules.

The trajectory of droplet motion was determined in accordance with kinematic equations:

$$u_p = \frac{dx}{dt}, v_p = \frac{dy}{dt}, w_p = \frac{dz}{dt}. \tag{12}$$

3 Simulation results

For numerical integration of the system (4) - (12), the implicit difference scheme for the linearized right-hand sides was used [11]. In the calculations, the dynamics of each droplet fraction with a fixed size in the range 0-3 mm was separately analysed. At each time step, the kinematics of the motion is separately considered, taking into account the possibility of aerodynamic fragmentation of the droplet, the new droplet coordinates and the components of its velocity vector are determined, the new droplet temperature is determined, and finally, the droplet size are modelled taking into account evaporation. Integration of the system of equations for each fraction until the droplet evaporates completely or when it reaches, the surface of the Earth is carried out. The reliability of the results of numerical modelling of the formation and the dynamics of the change in aerosol clouds of toxicants in the atmosphere comparison with typical experimental data of the results of the launch of the "Proton" carrier rocket on October 26, 2007 [10] was verified by. In this case, the toxicant was an asymmetric dimethyl hydrazine (component of liquid fuel), and the height of the formation of the toxic aerosol cloud was about 30 103 m. The projections of the trajectories of aerosol particles of the main calculated fractions shows in Fig. 1. The experimentally confirmed discrepancy between the trajectories of different particle sizes the influence of the wind speed varying in altitude was explained by.

The change in the rate of evaporation of particles in height corresponds due to the change of the atmospheric temperature on altitud (Fig. 2).

Figure 2 shows the rate of evaporation G in height, defined as the mass of evaporated toxicant per meter of fall height and the corresponding atmospheric temperature t. Completely evaporation of drops of a maximum initial size of 3 mm took place at an altitude of 2200 m. Good qualitative and satisfactory quantitative coincidence of the calculations and analysis contained in [10] allows us to recommend the calculation methodology developed by the authors for its use in assessing the impact of thermal energy facilities, chemical plants, etc. on the environment.

 DOI: 10.1051/matecconf/201714101013

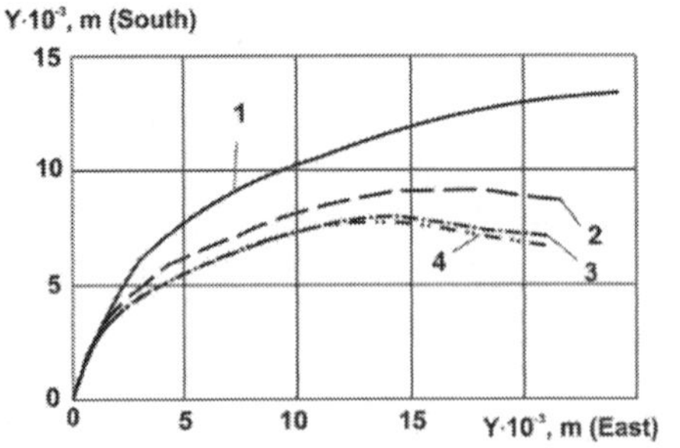

Fig. 1. Projections of trajectories of particles of different initial radius on a horizontal plane. 2 mm - curve number 1; 3 mm - curve number 2; 1.5 - curve number 3; 1 mm - curve number 4.

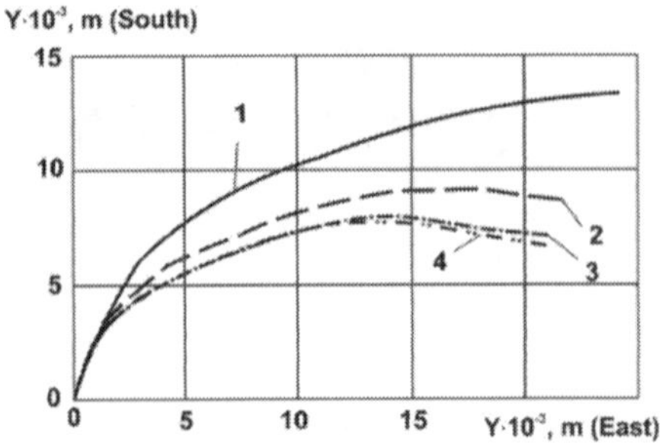

Fig. 2. Mass evaporation rate G and temperature of the atmosphere t in height.

Ministry of Education of Russia sponsored the study within the base part No. 6.8356.2017.

References

1. G. K. Batchelor, *An introduction to fluid dynamics* (Cambridge Univ. Press, London and New York, 1967)
2. G. Hetsroni, *Handbook of multiphase systems* (Hemisphere Publishing Corporation, Washington, New York, London, 1982)
3. J. F. Brady, G. Bossis, Annu. Rev. Fluid. Mech., **20**, 1 (1988), DOI: 10.1146/annurev.fl.20.010188.000551
4. C.T. Crowe, R. Johnson, A. Prosperetti, M. Sommerfeld, Y. Tsuji, Society of Mechanical Engineers, **236**, 1 (1996)
5. J.D. Schwarzkopf, C.T. Crowe, P. Dutta, AIChE Journal, **55**, 6 (2009), DOI: 10.1002/aic.11773
6. V.I. Terekhov, V.V. Terekhov, N.E. Shishkin, K.Ch. Bi, J. Eng. Phys. Thermophys., **83**, 5 (2010), DOI: 10.1007/s10891-010-0410-7
7. M.A. Pakhomov, V.I. Terekhov, International Journal of Heat and Mass Transfer, **66**, 210-222 (2013), DOI: 10.1016/j.ijheatmasstransfer.2013.07.013
8. E.L. Aleksandrov, Meteorol. Hydrol., **4**, 36-45 (1993)
9. E.L. Aleksandrov, *Behavior of liquid rocket fuels in atmosphere* (ANKIL, Moscow, 2000)
10. V.A. Arxipov, E.A. Kozlov, S.S. Titov, A.S. Tkachenko, A.S. Usanina, I.K. Zharova, Arab. J. Geosci., **9**, 2 (2016), DOI: 10.1007/s12517-015-2161-4
11. I.M. Vasenin, V.A. Arkhipov, V.G. Butov, A.A. Glazunov, V.F. Trofimov, *Gas dynamics of two-phase flows in orifices* (TSU, Tomsk, 1986)

12. V. A. Arkhipov, V.G. Butov, I.M. Vasenin, F.G. Gaponich, A.M. Podvysotsky, V.F. Trofimov, A.A. Shrayber, App. Mech. Appl. Phys., **5**, 90-94 (1985)
13. L.S. Klyachko, Otoplen. Ventil., **4**, 27-29 (1934)
14. R. Clift, J.R. Grase, M.E. Weber, Bubbles, drops and particles (Acad. Press., New York, 1978)
15. G.N. Abramovich, Applied gas dynamics (Nauka, Moscow, 1991)
16. S.K. Aggarwal, AI. Tong, W.A. Sirignano, AIAA Journal, **22**, 1448-1457 (1985), DOI: 10.2514/3.8802
17. R.B. Bird, W.E. Stewart, E.N. Lightfoot, *Transport Phenomena* (John Wiley and Sons, New York, 1974)
18. R.C. Reid, J.M. Prausnitz, T.K. Sherwood, *The properties of gases and liquids* (Chemistry, Leningrad, 1982)

MATEC Web of Conferences **141**, 01014 (2017) DOI: 10.1051/matecconf/201714101014
Smart Grids 2017

Experimental investigation of the processes of dehumidification of coniferous biomass

Elena Bulba[1*], and *Natalya* Ivanova[1]

[1]National Research Tomsk Polytechnic University, 634050 Tomsk, Russia

Abstract. This work includes the results of experimental studies of the moisture removal processes in the temperature range from 333 K to 413 K from coniferous woods which are typical for many regions. There are obtained the dependences of the mass rate of moisture removal on time and temperature. The effect of the evaporation of bound moisture was identified for the wood species studied. There are calculated the accommodation coefficient and the partial pressure at the evaporation surface for each type of biomass.

1 Introduction

The preparation of wood biomass for burning is an energy-consuming process due to the fact that it is always more or less saturated with moisture [1-3] because of its high porosity. Though, the calculation of the characteristics of the moisture output from porous materials is very complicated. The description of the processes of the evaporation front motion from the surface of the heated material to its deep layers is currently being carried out in many cases (for example, [4-5]) using the mathematical expression of the Hertz-Knudsen-Langmuir law [6-7]. That is why it is necessary to know the accommodation coefficient, which is not a constant value and in each specific case it depends on the heating and outflow conditions of the liquid vapor (for example, [4-5]) from the surface of the phase transformations. The definition of this characteristic in all cases of its using is a complex task, which is solved in the course of experimental studies. Therefore, the study of regularities in the processes of moisture removal is one of the main tasks in the development of technologies for burning biofuel in furnaces of steam and hot water boilers [2].

2 Experiment, results and discussion

There were studied the samples of coniferous species of wood, which are most common in the territory of Siberia: pine, spruce, cedar, fir. The methodology of the experiment was previously described in detail in [8-9].

* Corresponding author: bulba@tpu.ru

DOI: 10.1051/matecconf/201714101014

After the experimental determination of the evaporated during the time interval Δt moisture mass in a drying chamber was found the mass flow rate of dehumidification of fluid from the sample studied wood

$$W = \frac{\Delta m}{S \cdot \Delta t}, \qquad (1)$$

where W - the mass flow rate of dehumidification, kg/(m^2 s); Δm – the change in the mass of sample, kg; S - the surface area of the evaporation ($S = 0.0227$ m^2); Δt is the drying time.

According to the obtained values of W there are constructed the dependences of the mass rate of dehumidification on time and temperature, which are presented in Figures 1-5. Vertical bars are confidence intervals determining W.

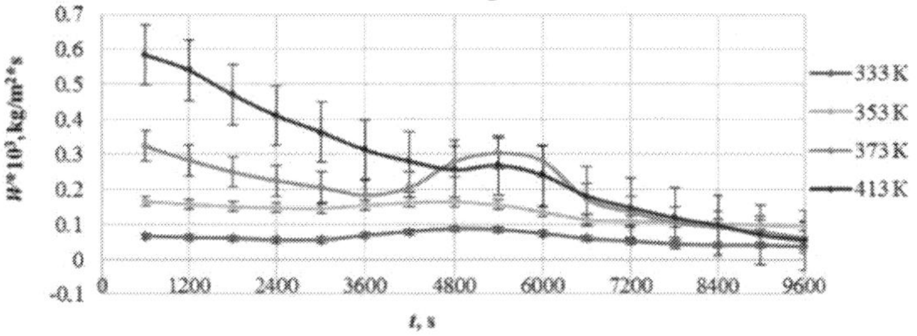

Fig. 1. Dependences of the mass evaporation rate on time for cedar.

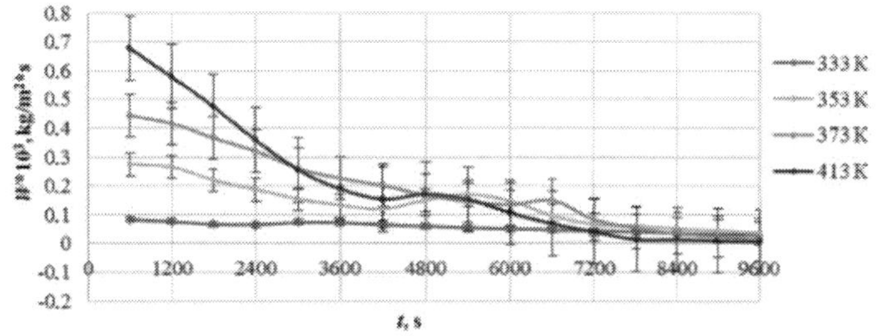

Fig. 2. Dependences of the mass evaporation rate on time for fir.

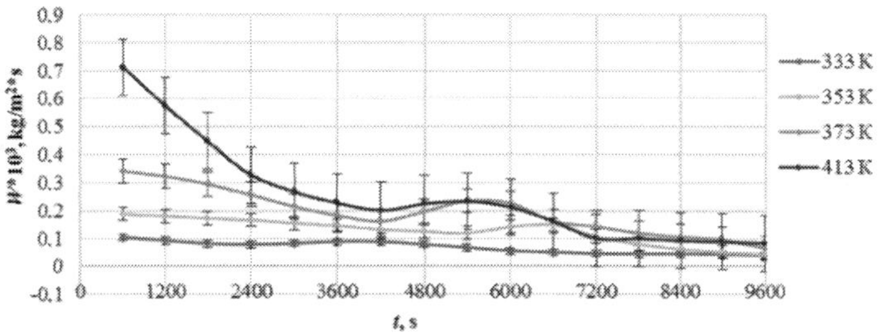

Fig. 3. Dependences of the mass evaporation rate on time for eat.

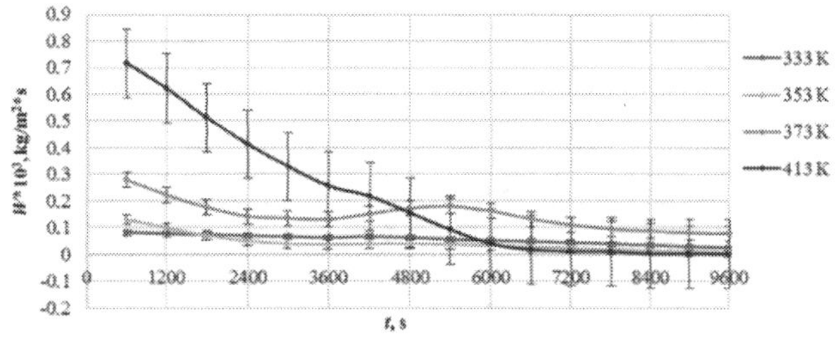

Fig. 4. Dependences of the mass evaporation rate on time for pine.

As can be seen from the obtained dependences in the temperature range 333 - 353 K, the process of removal of bound moisture does not begin for 9600 seconds. At temperatures of 373 - 413 K the evaporation rate decreases during the time interval of about 60 minutes is the first drying period. It corresponds to the removal of free moisture, distributed in the pores of the material. Then begins the process of removal of bound moisture. This period lasts for about 2 hours. The increase in evaporation rate before the onset of the second period is explained by the fact that free and bound moisture begin to evaporate together and this increases the overall evaporation rate.

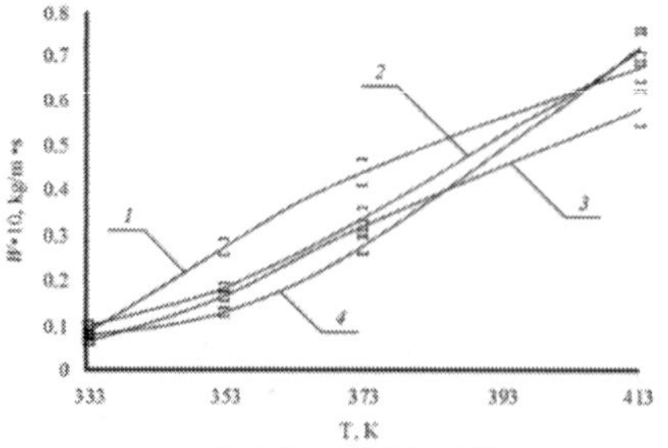

Fig. 5. Dependence of the mass evaporation rate on temperature

The analysis of the received dependences showed that the highest rate of dehumidification corresponds to the fir. This is due to the fact that the density of this wood is lower than the density of the other samples (spruce 15% pine and 25%). Accordingly, pores occupy the largest part in the structure of this breed of coniferous trees, in comparison with the other parts. So water vapor with less resistance comes to the heating surface. It is installed that the pine has the lowest rate of dehumidification as it corresponds to a higher density in comparison with the other materials. It can be concluded that the drying of fir and spruce occur with less time and energy costs.

The accommodation coefficient and partial pressure were determined using the experimental results and kinetic theory equation of Hertz-Knudsen [7].

$$W = \frac{A\cdot(P-P^{*})}{\sqrt{\frac{2\pi RT}{M}}}, \qquad (2)$$

where P - saturated vapor pressure, PA; P^{*}– partial pressure, PA; R– universal gas constant, R = 8314 J / Kmol·K; M – molecular mass of moist air, T - temperature, K; A - coefficient of accommodation.

In equation (2) for two unknowns – the accommodation coefficient and the partial pressure.

The second equation, which allows to close the system and calculate the partial pressure and the accommodation coefficient is the equation of state.

$$P_a \cdot V = (m_1 + m_2)\cdot\frac{R}{M}\cdot T, \qquad (3)$$

where P_a – pressure of humid air, which is assumed equal to atmospheric, P_a = 101325 PA; V - volume of drying chamber, V = 0,36 m^3; m_1 - mass of dry air, kg. m_2 - mass of water vapor, kg.

The molecular mass of moist air is present in the form of the expression.

$$M = \frac{m_1+m_2}{\frac{m_1}{M_1}+\frac{m_2}{M_2}}, \qquad (4)$$

where M_1 is the molecular mass of dry air, M_1 = 29 g/mol; M_2 - molecular weight of water vapor, M_2 = 18 g/mol.

Solving the system of equations (3) and (4) we obtain the expression for the mass of dry air.

$$m_1 = [\frac{P_a \cdot V}{R \cdot T} - \frac{m_2}{M_2}] \cdot M_1. \quad (5)$$

The total mass of moisture evaporated from the sample, determined by knowing the mass of moist air at the initial moment of time in the drying chamber.

$$m_2^0 = d \cdot m_1^0, \quad (6)$$

where d - moisture content of moist air; m_1^0 - mass of dry air in the chamber at the initial moment of time.

From the equation of state of an ideal gas calculate the moisture content of moist air.

$$d = 0{,}622 \cdot \frac{(\varphi \cdot P)}{P_a - (\varphi \cdot P)}, \quad (7)$$

where φ - relative humidity at the time of the experiment (40%).

The mass of dry air in the chamber determined from the following expression:

$$m_1^0 = V \cdot \rho, \quad (8)$$

where ρ - density of dry air, kg/m^3.

The density of the dry air in the room determined by the formula:

$$\rho = \frac{P_a - \varphi \cdot P}{\frac{R}{M} T}. \quad (9)$$

Then we find the total mass of water vapor in the chamber.

$$m_2 = \Delta m + m_2^0. \quad (10)$$

From equations (3) and (5) we Express the molecular weight of moist air.

$$M = \frac{m_2 + [\frac{P_a \cdot V}{R \cdot T} - \frac{m_2}{M_2}] \cdot M_1}{\frac{m_2}{M_2} + \frac{P_a \cdot V}{R \cdot T} - \frac{m_2}{M_2}} = \frac{m_2 + [\frac{P_a \cdot V}{R \cdot T} - \frac{m_2}{M_2}] \cdot M_1}{\frac{P_a \cdot V}{R \cdot T}} \quad (11)$$

Find the volume and mass fraction of water vapor:

- mass fraction of water vapor

$$g = \frac{m_2}{m_2 + [\frac{P_a \cdot V}{R \cdot T} - \frac{m_2}{M_2}] \cdot M_1}; \quad (12)$$

- volume fraction of water vapor

$$r = \frac{g \cdot M}{M_2}. \quad (13)$$

Knowing the volume fraction of evaporated moisture and vapor pressure, we find the partial pressure at the surface of evaporation at the dripping for each type of biomass

$$P^* = r \cdot P. \quad (14)$$

Then we determine the values of the accommodation coefficient in the appropriate temperature range for all studied materials, using the Hertz-Knudsen equation (2).

3 Conclusion

According to the results of the experimental studies of moisture removal processes typical of many regions of coniferous species of wood in the temperature range from 333 K to 413 K, there were established the dependences of the mass rate of moisture removal on time and temperature for four species of wood. There was proposed a method for calculating the accommodation coefficient and partial pressures at the evaporation surface for each type of wood.

The established regularities can be used with further refinement of the mechanism of moisture removal from the porous structure of moisture-containing materials and development of the general theory of drying. Also, the results of the research can be used in testing mathematical models for drying woody biomass.

Another possible application of the experimental data obtained can be their use in analyzing the processes of gas cleaning from wet wood waste (for example, [10-11]).

References

1. T. Nussbaumer, Proc. of the 10th Europ. Bioenergy Conf., 1142-1145 (1998)
2. Y. Solantausta, 8th EC Conf., Vienna, Austria, **2**, 1579-1591 (1994)
3. H. Dienhart, Proc. of the 10th Europ. Bioenergy Conf., 1460-1463 (1998)
4. S.V. Syrodoy, G.V. Kuznetsov, V.V. Salomatov, Thermal Engineering, **62**, A002 (2015), DOI: 10.1134/S0040601515100092
5. G.V. Kuznetsov, V.V. Salomatov, S.V. Syrodoy, Combust. Explos., **51**, 4 (2015), DOI: 10.1134/S0010508215040024
6. V.L. Strakhov, A.N. Garashchenko, G.V. Kuznetsov, V.P. Rudzinskii, Combust. Explos., **37**, 2 (2001), DOI: 10.1023/A:1017557726294
7. G.V. Kuznetsov, P.A. Kuybin, P.A. Strizhak, High Temp., **53**, 2 (2015), DOI: 10.1134/S0018151X15020133
8. E.E. Bulba, A.A. Malinovsky, MATEC Web of Conferences, **23**, 01053 (2015), DOI: 10.1051/matecconf/20152301053
9. E.E. Bulba, N.A. Ivanova, MATEC Web of Conferences, **110**, 01020 (2017), DOI: 10.1051/matecconf/201711001020
10. M.V.Vasilevsky, E.G.Zykov, A.S.Razva, Theoretical Foundations of Chemical Engineering, **45**, 3 (2011), DOI: 10.1134/S0040579511030146
11. A.S. Razva, M.V. Vasilevsky, V. Rykov, MATEC of Conferences, **23**, 01035 (2015), DOI: 10.1051/matecconf/20152301035

MATEC Web of Conferences **141**, 01015 (2017) DOI: 10.1051/matecconf/201714101015
Smart Grids 2017

Theoretical assessment of evaporation rate of isolated water drop under the conditions of cooling tower of thermal power plant

Sergey Shevelev[1,*]

[1]National Research Tomsk Polytechnic University, 634050 Tomsk, Russia

Abstract. The purpose of the work is numerical modelling of heat and mass transfer at evaporation of water drops under the conditions which are typical for a modern chimney-type cooling tower of a thermal power plant. The dual task of heat and mass transfer with movable boundary at convective cooling and evaporation for a 'drop–humid air' system in a spherical coordinate system has been solved. It has been shown that there is a rapid decline of water evaporation rate at the initial stage of the process according to temperature decrease of its surface. It has been stated that the effect of evaporation rate decrease appears greatly in the area of small radiuses.

1 Introduction

Today cooling towers are the cheapest way of low potential heat transfer from steam turbine plant condensers in schemes of thermal and atomic power plants (thermal power plant and atomic power plant). Increase of pressure in a condenser of steam turbine plant at 0.001 MPa from nominal value at other permanent parameters decreases electric power at 0.6-1%. Thus, on the whole, cost effectiveness of a power plant greatly depends on operational efficiency of a cooling tower. Besides, their exploitation allows saving up to 95% of fresh water [1].

The function of a cooling tower is to cool flow of water entering it. In modern cooling towers used at thermal power plants and atomic power plants this can be achieved via heat transfer from heated water drops to cold atmospheric air at their mutual motion. Water cooling mainly occurs because of convection and surface evaporation. The role of irradiation in the conditions specific for modern energetic cooling towers is small and, as a rule, is not taken into consideration at calculations. The ratio between a quantity of heat transferred by convection and evaporation is enormously determined by external factors. In the hot summer evaporation of heat evaporated by water is up to 90% and more and in cold winter months heat transfer by convection can reach 50% [1]. Formed water vapor mix with atmospheric air and heat it up thus worsening the conditions of cooling water. Evaporated part of cooled water is taken away from a cooling tower with atmospheric air stream. This water relates to losses as it is irrevocably lost in the atmosphere and needs replenishment

* Corresponding author: wevelev@tpu.ru

from the external source. Thus, it can be said that there is enormous negative influence of the process of surface evaporation of drops on cooling tower efficiency which appears in heat transfer worsening between heated water and cooling air and also in transfer of formed water vapor with atmospheric air. A choice of such modes of cooling towers work and also parameters of drops which provide minimal quantity of evaporated water and its maximum cooling represents a considerable interest.

Choice of effective modes of water drops cooling is possible at mathematical model creation which adequately defines the processes of heat and mass transfer at water evaporation under the conditions of drops moving in cooling towers. Nowadays there are no theories fully describing the accompanying processes.

The main factor determining water evaporation rate and quantity of removed moisture, accordingly, is different pressures of saturated vapor and vapor close to evaporation boundary [2]. Size of drops is also important for determining surface area. Influence of different sizes of drops on their evaporation rate in conditions typical for a standard powerful cooling tower used at modern thermal power plants and atomic power plants has a great importance for studying it nowadays.

The present work presents the results of numerical modelling of heat and mass transfer at water drop evaporation under the conditions which are typical for modern cooling towers of thermal power plants.

2 Formulation of the problem

The problem under consideration was heat and mass transfer at cooling of an isolated spherical water drop in air stream under the conditions typical for very effective modern cooling towers. It was supposed that a drop moved in the area filled with humid air. The initial temperature of a drop T_0 was a shade higher than the temperature of humid air T_{air} (at 15..30^{0}C). Cooling of a drop occurs at convection and also because of vapor formation (evaporation) on the boundary of phase change 'water drop – humid air'. Furthermore, the function of irradiation is small and it can be neglected [1]. Formed water vapor is blown into humid air and mix with it. Dimensions of a drop decrease at vapor formation.

Deformation of a drop is possible in real conditions of drop motion under the influence of forth of gravity. It has been experimentally stated [3, 4] that a drop is continuously deformed in the air, gradually obtaining different forms (sphere; ellipsoid, driven in the direction of travel; ellipsoid, driven in cross direction, 'pancake' etc.). Therefore, as a result of the problem statement acceptance of a spherical form of a drop is as much relevant as any other.

Thermophysical parameters of water and air were accepted regardless of temperature. Gaseous atmosphere is binary (a mixture of air and water vapor).

Scheme of region of problem solution is shown in figure1.

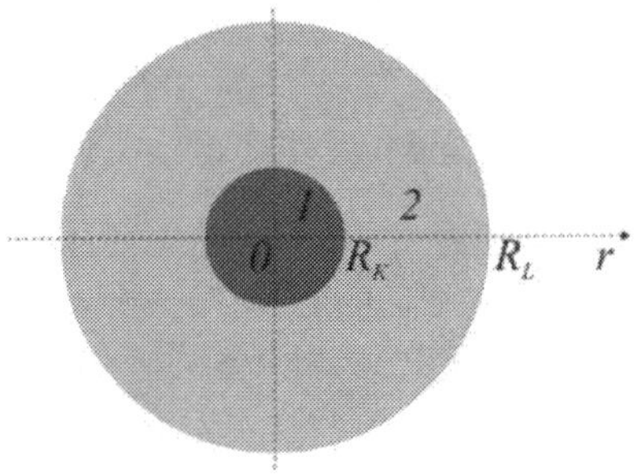

Fig. 1. Scheme of region of problem solution. 1 – water drop, 2 – humid air.

The process of heat and mass transfer the conditions of which are under consideration can be described with the help of the following system of differential equations in differential coefficient, written in spherical coordinate system:

$$\rho_1 c_1 \frac{\partial T_1}{\partial t} = \lambda_1 \left(\frac{\partial^2 T_1}{\partial r^2} + \frac{2}{r}\frac{\partial T_1}{\partial r} \right), 0 < r < R_K$$

$$\rho_2 c_2 \frac{\partial T_2}{\partial t} = \lambda_2 \left(\frac{\partial^2 T_2}{\partial r^2} + \frac{2}{r}\frac{\partial T_2}{\partial r} \right), R_K < r < R_L$$

$$\frac{\partial C_{w.v.}}{\partial t} = D \left(\frac{\partial^2 C_{w.v.}}{\partial r^2} + \frac{2}{r}\frac{\partial C_{w.v.}}{\partial r} \right), R_K < r < R_L$$

Here: T, K – temperature; ρ, kg / m^3 – density, c, $J / (kg{\cdot}K)$ – heat capacity; λ, $W / (m{\cdot}K)$ – heat conductivity, $C_{w.v.}$- water vapor concentration, D, m^2 / s – difusion coefficient. Index 1 is correspondent to a water drop, index 2 – to humid air.

Equations (1) – (3) were solved under the following initial and boundary conditions:

$$t = 0: T_2 = T_{air}, R_K \le r \le R_L$$

$$t = 0: C_{w.v.} = C_0, R_K \le r \le R_L$$

$$r = 0: \frac{\partial T_1}{\partial r} = 0$$

$$\begin{cases} T_1(t, R_K) = T_2(t, R_K) \\ -\lambda_1 \left.\frac{\partial T_1}{\partial r}\right|_{r=R_K} = -\lambda_2 \left.\frac{\partial T_2}{\partial r}\right|_{r=R_K} + QW, \end{cases}$$

$$r = R_L: \frac{\partial T_2}{\partial r} = 0$$

$$r = R_K: -D\frac{\partial C_{w.v.}}{\partial r} = \frac{W}{\rho_2}$$

$$r = R_L: \frac{\partial C_{w.v.}}{\partial r} = 0 .$$

Here Q, J / kg - heat of phase change; W, $kg / (m^2. s)$ - evaporation mass rate, found by applying the formula [2]:

$$W = \frac{A(P_S - P)}{\sqrt{2\pi R T_{bound} / M}},$$

where A - accommodation coefficient, P_S, Pa - saturated water vapor pressure, P, Pa - vapor pressure close to evaporation boundary, M, $kg / kmol$ - molecular mass of water, T_{bound}, K - temperature of a drop close to evaporation boundary. Accommodation coefficient is found according to the following data [5].

System of equations (1) – (3) was solved upon the application of the finite-difference method [6] by means of implicit difference scheme use. Nonlinear equations were solved by means of iteration method [7], different analogues of differential equations and boundary conditions by applying elimination method by utilizing analogous algorithm [7]. Motion computation of a boundary was made by applying evaporation front tracking in space network node.

It should be noted that the solution of the defined problem is complicated as there is a movable boundary and breakup of heat conduction coefficient on it. That is why we used the algorithm [7] with a second-order approximation for boundary conditions of the forth type on the boundary. Besides, time stem of difference mesh was chosen so that to implement a balance of heat when leaving the stationary mode.

3 Results and discussion

The problem was solved with the following basic data:
$R_K = 1,0-5,0\ mm$; $R_L = 100\ mm$; $\lambda_1 = 0,599\ W/(m\,.\,K)$; $c_1 = 4183\ J/(kg.\ K)$; $\rho_1 = 998,2\ kg/m^3$; $\lambda_2 = 0,0283\ W/(m.K)$; $c_2 = 1005\ J/(kg.K)$; $\rho_2 = 1,205\,kg/m^3$; $T_1 = 323\ K$; $T_2 = 288\ K$; $D = 0,0026\ m^2/s$; $C_0 = 0,022$.

Typical radial temperature distribution on a drop is shown in figure 2.

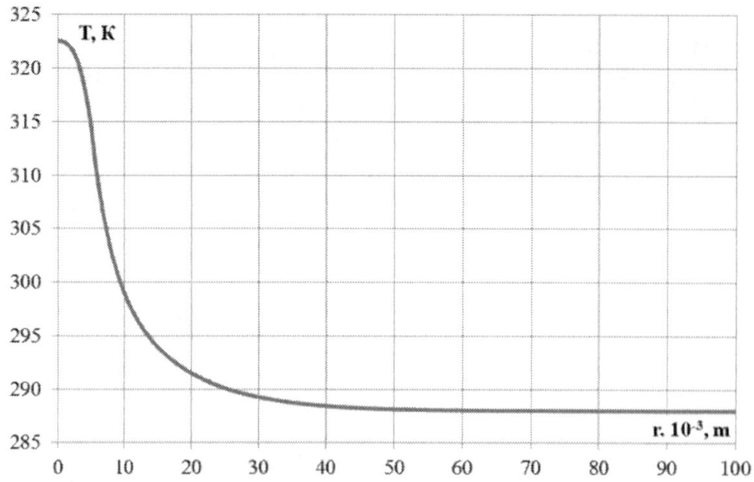

Fig. 2. Temperature distribution in a 'drop-humid air' system at $t = 10\ s,\ R_K = 5\ mm$.

Analysis of temperature change of a drop in the process of cooling it allows concluding about great temperature gradient along a radius of a drop (figure 2). It follows that surface temperature of a drop governs the processes of convective heat exchange and evaporation.

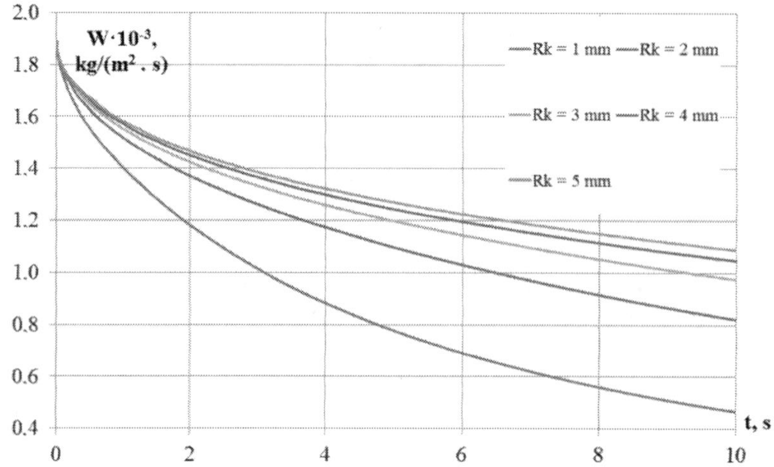

Fig. 3. Time change of drop evaporation rate.

Dependences of mass rate of evaporation on time for drops of different sizes shown in figure 3 allow us to make the following conclusion. At the initial stage (apart from drop radius) water evaporation rate greatly decreases due to decrease of its surface temperature. Moreover, the less drop radius is the faster its cooling is and, therefore, the more intensively decreases its evaporation rate.

References

1. V.S. Ponamarenko, *Gradirni promyshlennyh i jenergeticheskih predprijatij* (Energoizdat, Moscow, 1998) [in Russian]
2. Ju.V. Polezhaev, F.B. Jur'evich, *Teplovaja zashhita* (Jenergija, Moscow, 1976) [in Russian]
3. R.S. Volkov, G.V. Kuznetsov, P.A. Stizhak, Tech. Phys., **59**, 7 (2014), DOI: 10.1134/S1063784214070263
4. G.V. Kuznetsov, P.A. Stizhak, The motion of a manifold of finely dispersed liquid droplets in the counter flow of high-temperature gases Tech. Phys. Lett., **40**, 6 (2014), DOI: 10.1134/S1063785014060200
5. G.V. Kuznetsov, P.A. Kuybin, P.A. Stizhak, High Temp., **53**, 2 (2015), DOI: 10.1134/S0018151X15020133
6. Samarskij A.A. *Teorija raznostnyh shem* (Nauka, Moscow, 1983) [in Russian]
7. P.J. Roache *Fundamentals of Computational Fluid Dynamics* (Hermosa Publishers, Albuquerque, 2003)
8. G.V. Kuznetsov, M.A. Sheremet, Computational Thermal Sciences, **1**, 3 (2009), DOI: 10.1615/ComputThermalScien.v1.i3.60

MATEC Web of Conferences **141**, 01016 (2017) DOI: 10.1051/matecconf/201714101016
Smart Grids 2017

Investigation of the influence of the symmetry of membership functions in the fuzzy controller on the quality of MPPT regulation in the photovoltaic system

Ruslan Gimazov[1], and *Stanislav* Shidlovskiy[1,*]

[1]National Research Tomsk Polytechnic University, 634050 Tomsk, Russia

Abstract. The paper presents the results of the investigation of the influence of asymmetric membership functions in a fuzzy controller on the quality of MPPT regulation in the photovoltaic system. Comparison of the quality of fuzzy MPPT regulation with symmetrical and asymmetric membership functions is carried out. The results of the research showed the advantage of using asymmetric membership functions in the MPPT algorithm with fuzzy logic.

1 Maximum Power Point Tracking (MPPT)

MPPT - maximum power point tracking - one of the most effective ways to increase the efficiency of photovoltaic systems. For the implementation of MPPT digital devices are used. The devices analyze the current vs. voltage curve (CvV) by certain algorithms and determine the optimal current-voltage pair providing the maximum output power.

Figure 1 shows an example of the current-voltage characteristic of a photoelectric module with a maximum power point marked on it.

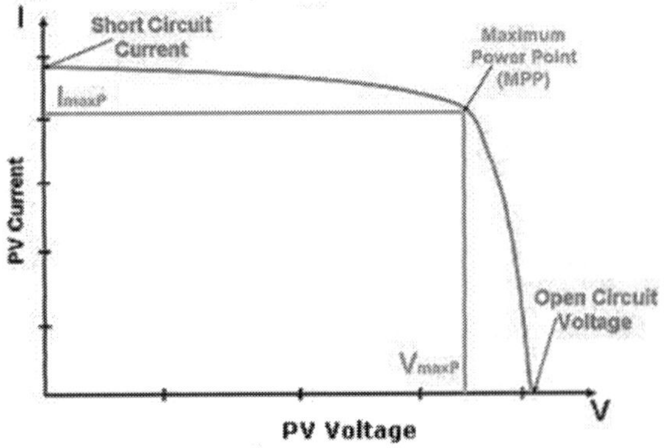

Fig. 1. MPPT on the current vs. voltage curve of the photovoltaic module.

* Corresponding author: shidlovskiy@tpu.ru

The most common MPPT algorithm is the "perturbation-observation" algorithm. In this method, the MPPT device modifies the input resistance by a small value. As a result, the voltage of the solar installation varies. Calculating the electrical power and comparing its current value with the previous value, you can get the direction of searching for the point of maximum power.

The principle of the "perturbation-observation" algorithm on the power vs. voltage (PvV) curve of a photocell is shown in Fig. 2.

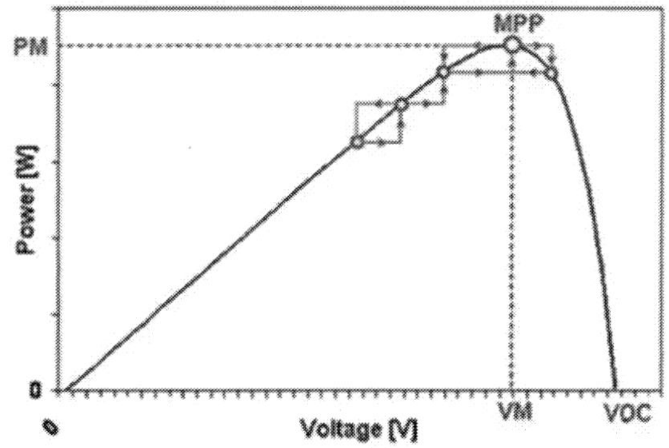

Fig. 2. The principle of the maximum power point tracking according to the perturbation and observation algorithm.

Disadvantages of the "perturbation-observation" algorithm are power fluctuations and a fixed time to performance. These problems can be solved with the help of adaptive algorithms, for example algorithms with fuzzy logic [1].

2 Fuzzy logic

The math theory of fuzzy logic allows you to generate fuzzy conclusions, through which more flexible control systems are implemented. It has been experimentally shown that the use of fuzzy logic in control (fuzzy control) gives better results, compared with those obtained with traditional control algorithms [2]. Figure 3 shows the principle of implementing a fuzzy controller.

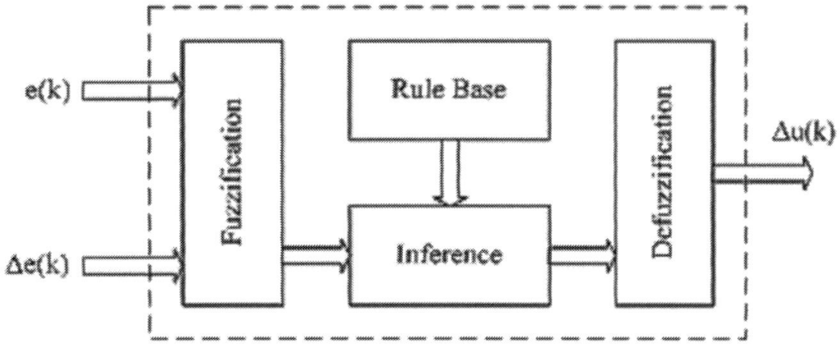

Fig. 3. Schematic diagram of the fuzzy controller.

A set of fuzzy input and output elements in the fuzzy controller database is regulated by the membership function area. In this paper, the influence of symmetric and asymmetric membership functions on the quality of MPPT regulation is investigated.

DOI: 10.1051/matecconf/201714101016

3 Investigation and conclusion

To implement the study, a simulation model of photovoltaic system was developed [3].

The symmetry of membership functions in a fuzzy controller is determined by the shape of the fuzzy subset area. The shape of such area is determined by the base of fuzzy rules. The base of fuzzy rules for the system under investigation is presented in Table 1.

Table 1. Rules of the fuzzy controller.

E↓ \ dE→	**low**	**middle**	**high**
low	low	low	low
middleL	midL	midL	low
middleH	midH	midL	low
high	high	midH	midL

As input linguistic variables of the fuzzy controller, we use the ratio of solar power change to solar voltage change (E), as well as the behavior of E (dE):

$$E(k) = (P(k)–P(k-1))/(V(k –V(k-1)) \quad (1)$$

$$dE(k) = E(k)–E(k-1) \quad (2)$$

In the presented table, the values in the leftmost column and in the upper line define the functional affiliation of the input variables (E, dE). The remaining positions of the table determine the function of the output quantity, depending on the combination of the input values. If all the areas (subsets) of the membership functions of some of the variables are symmetric, then such a variable will be called symmetrically inscribed, and the membership function is symmetric. If there exists at least one area of the membership function that is not symmetric, then such a variable will be called asymmetrically inscribed, and the membership function is asymmetric. Figure 4 shows an example of symmetric membership functions. Figure 5 shows an example of asymmetric membership functions.

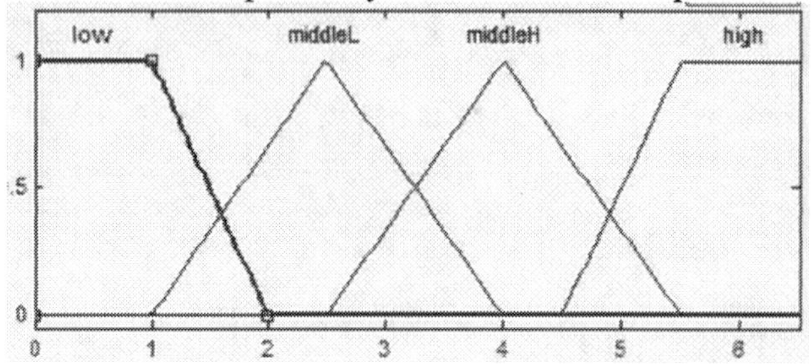

Fig. 4. Symmetric membership functions.

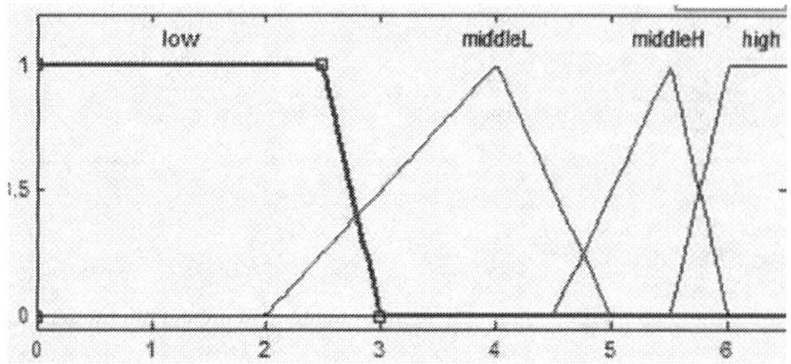

Fig. 5. Asymmetric membership functions.

In the course of the study, comparative graphs of the process of searching for the maximum power point were obtained. Figure 6 shows the voltage variation graph using the "perturbation-observation" algorithm. Figure 7 shows comparative graphs of the change in the output power of solar cells during MPPT regulation.

The graphs indicate: 1 - MPPT algorithm without fuzzy logic, 2 - symmetrical fuzzy MPPT algorithm, 3 - asymmetric fuzzy MPPT algorithm.

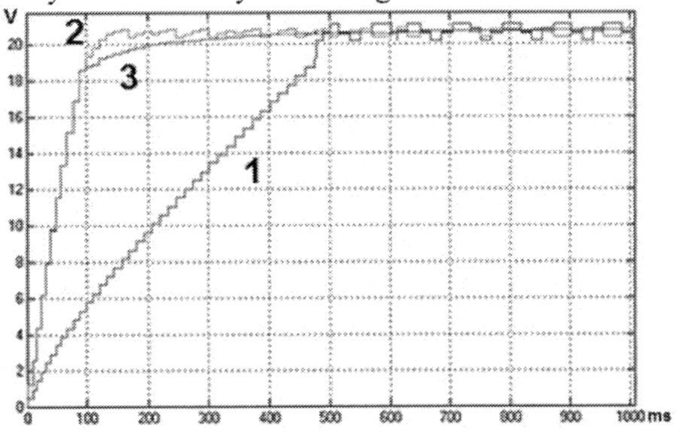

Fig. 6. Voltage variation.

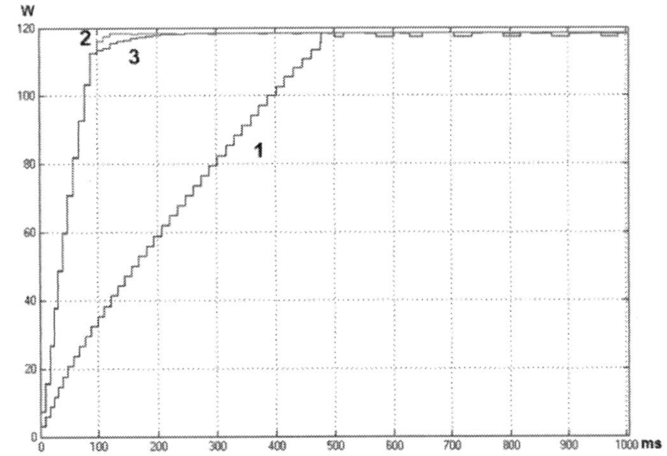

Fig. 7. Power variation.

A comparative analysis of the graphs shows that the use of fuzzy logic significantly improves the characteristics of the MPPT algorithm: decreases the search time for the extreme point, decreases the amplitude of voltage and power fluctuations. The use of asymmetrical membership functions to neutralize fluctuations in voltage and power. The use of symmetric functions minimizes the time of finding the extreme point.

References

1. S. Shidlovskiy, *Automatic control. Reconfigurable structures* (Tomsk State University, Tomsk, 2006)
2. K. M. Passino, S. Yurkovich, *Fuzzy Control* (Addison Wesley, Menlo Park, 1998)
3. S. Shidlovskiy, R. Gimazov, Cognitive Robotics, **1**, 51 (2016)

MATEC Web of Conferences **141**, 01017 (2017) DOI: 10.1051/matecconf/201714101017
Smart Grids 2017

Suppression of forest fuel thermal decomposition under the influence of liquid aerosol and water droplets with additives

Nikita Shlegel[1], *Alena* Zhdanova [1,*], *Svetlana* Kralinova[1], and *Ivan* Voytkov[1]

[1]National Research Tomsk Polytechnic University, 634050 Tomsk, Russia

Abstract. This article presents findings of experimental research into suppressing the thermal decomposition of forest fuel (birch leaves, spruce needles, kindling wood and mixed forest fuel) under conditions of exposure water aerosol and single water droplets, as well as water with specialized additives. We used typical model of fire source with fixed height 0.04 m and varied diameter 0.02-0.1m. Water spraying process, as well as interaction between water and forest fuel was registered using high-speed video camera (recording frequency $6 \cdot 10^5$ fps) and panoramic optical methods ("Particle Image Velocimetry", "Shadow Photography"). We established principles of water additions' influence on termination of combustion and thermal decomposition of fire source model.

1 Introduction

Forest fire fighting is an urgent problem for the entire world community. Especially it is necessary to distinguish brush fires, because the dry forest litter is often ignited. Huge areas of the forest burn down during brush fires [1, 2]. The steady spread of ignition along the lower tiers of forest vegetation and litter is often the cause crown fire. Deceleration of the fire front and elimination of forest fires involves dropping the extinguishing liquid directly into the combustion zone [3–5]. The aviation park for firefighting in boreal zone often includes tankers, seaplanes and helicopters, equipped with special tanks for collecting, transporting and draining the water. Increasing the effectiveness of using such method is one of the urgent tasks. Additive component are often used in order to enhance fire-extinguishing properties of water. Nowadays there is a large database of specialized additives to water. Such components can increase the viscosity of quenching liquids (fire-retardant) or reduce her interracial tension (wetting agents). Also, it is important to analyze the influence of the forest fuel (FF) structure on the process under consideration. These reasons lead to experimental studies of the process of quenching model of fire source in laboratory conditions in order to justify the choice of a chemical reagent for fire extinguishing. Such study also will help show how wetting and fire retardant chemical compounds effect on suppression of FF thermal decomposition.

* Corresponding author: zhdanovaao@tpu.ru

The aim of this work is an experimental study of the integral characteristics of flame burning suppression and model of fire source thermal decomposition with water, which contains specialized additives.

2 Experimental setup and procedure

Fig.1 shows the scheme of experimental setup used in carrying out researches. As the main recording equipment we chose a high-speed digital CMOS video camera Phantom V411 *1* with a recording frequency is of $6 \cdot 10^6$ fps and maximum resolution of 1280×1280 pixels), a high-speed (the minimum polling rate is 0.1 s) analog input module "National Instruments NI 9216" *5* for temperature sensors connection, XA (K) thermoelectric converters *7*; personal computer *8*, ViBRA HT 84RCE laboratory microbalance (with an accuracy of 10^{-5} g).

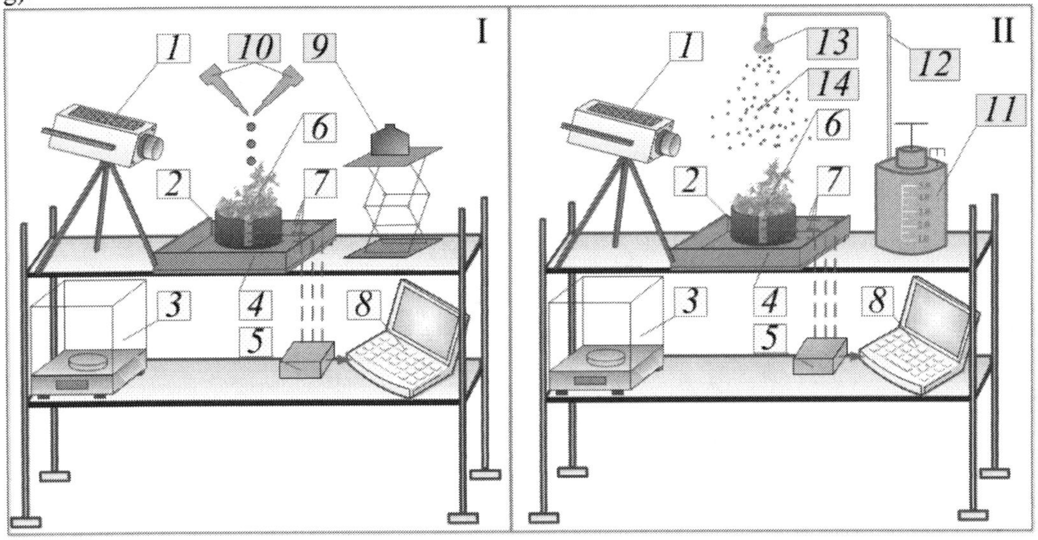

Fig. 1. A scheme of experimental setup: *1* – high-speed video camera; *2* – cylinder with a FF; *3* – analytical microbalance; *4* – metal tray; *5* – high-speed analog input module; *6* – FF; *7* – thermocouples; *8* – workstation (PC); *9* – quenching tank; *10* – single-channel dispencer; *11* – bulb with quenching liquid ; *12* – quenching liquid supply channel; *13* – spraying nozzle; *14* – liquid aerosol.

We used typical forest fuel, such as birch leaves, spruce needles, kindling wood and mixed FF(with mass ratio: 25% birch leaves, 15% spruce needles, 60% aspen branch) for our researches. The relative FF moisture in our experiments was 5–8 % for birch leaves, 7-10% for spruce needles, 10-14% for aspen branch. We used hollow cylinders of corrugated aluminum to create model of fire sources. Cylinders had height h_f and diameter d_f.

We picked distilled water (GOST 6709-72), water-based formulations with the "Progress" wetting agent addition and special-purpose foaming agents (type "AR") as quenching liquids. The relative mass concentration of the compositions used in the aqueous solution was 1%. Quenching liquids were prepared directly in the vessel *11*, which was full of distilled water. The solution with the "Progress" wetter was mixed until the foam formed on the surface of the liquid. The wetting agent solubility was controlled visually. Similar actions were carried out for the preparation of a quenching liquid based on water and a foaming agent "AR".

Ignition of the model of fire source was carried out uniformly over the entire area of the open FF surface. We used three piezoelectric gas burners simultaneously for this purpose.

The procedure for conducting the experiments included three stages. During the first stage of experiments the features and times of complete model of fire sources burn-up (t_b) for each FF were recorded. At the second stage of our experiments we registered the times of suppression FF thermal decomposition ($t_{e(d)}$) during successive feeding of single droplets of quenching liquid. The third stage was analogous to [6] and included determining the extinguishing times ($t_{e(ae)}$) of model of fire sources in the conditions of the liquid aerosol supply.

The first stage of experiments consisted of several operations. The model of fire source was ignited. We started the countdown on the electronic stopwatch. Three type-XA (K) needle thermocouples were placed on cylinder's symmetry axis (temperature measurement range 223–1473 K, accuracy ± 3 K, heat retention no more than 0.1 s). The value t_b was registered. The choice of this value was determined by the corresponding limiting temperatures of the forest fuel thermal decomposition. The criterion for stable FF burning was the excess of the thermocouple readings above 370 K (FF thermal decomposition temperature).

At the second stage we determined the times of suppression of the FF thermal decomposition $t_{e(d)}$ during successive delivery quenching liquid droplets. Quenching was carried out by successive feeding of single droplets with radii R_d≈1.5 mm (the drops' volume was V_d≈15 μl). Sequential droplet generation was realized using two electronic single-channel dispensers *10* Finnpipette Novus (the minimum and maximum withdrawable volumes are 10 μl and 100 μl, the variation step is 0.1 μl). The frequency of generation is 1 drop per 2 seconds. The droplet delivery was realized until the model fire was completely suppressed. The conditions for suppressing the combustion were fixed by the results of the thermocouple readings. Afterward the obtained results were refined when analyzing the video of the quenching process from a high-speed video camera *1*.

The third stage of the experiments (quenching of the model of fire sources) included the liquid aerosol supply to the FF surface. The macroscopic principles of the interaction between water vapor and thermal decomposition products were recorded using optical methods of diagnostics ("Particle Image Velocimetry", "Shadow Photography", similar to [6, 7].) After 10-15 seconds (sufficient time to light the entire FF hinge), at the moment when the readings of the two upper thermocouples were higher than the thermal decomposition temperature (T_{if}≈370 K), the shut-off valve was opened and water from the tank *11* was supplied through the channel *12* to the inlet of the spraying nozzle *13*. The spraying continued until the full suppression of model of fire source combustion. The stopping point of combustion was fixed by the thermocouple readings *7*. The values $t_{e(ae)}$ were recorded.

We conducted from 15 to 20 experiments for each model of fire source with corresponding FF. After that we chose the experimental values of t_b or $t_{e(ae)}$, $t_{e(d)}$, differed by less than 5% for identical experimental conditions. Amount of results excluded from consideration in this case did not exceed 10-15% of the total. The error in determining the values of t_b and t_e was due to the thermal delay time and was 1 s.

3 Results and discussion

Conducted researches enable us to make the conclusion about efficiency of extinguishing by a liquid aerosol and single water droplets. The recorded termination times of the thermal decomposition reaction are much less than the burn-out times of models of fire source. In addition, the obtained experimental results (Fig. 2) for determining the time of the FF pyrolysis reaction suppression (for example for birch leaves) make it possible to conclude that foaming agents and wetting agents are need to added to the water. Fire-retardant (for example, a foaming agent of the "AR" type), chemical compositions allow increasing the

 DOI: 10.1051/matecconf/201714101017

quenching liquid viscosity, and, as a consequence, increasing the its fire-extinguishing properties. Adding the same surface-active additives (wetting agents, in particular, "Progress"), on the contrary, can significantly reduce the water surface tension, which contributes to a significant increase in its spreadability over the surface and promotes crushing into small droplets.

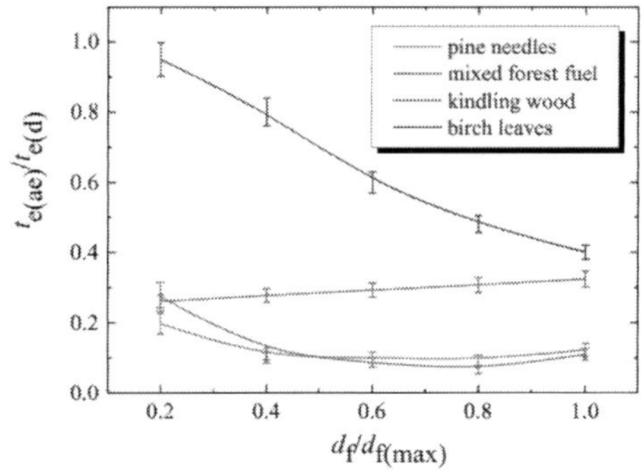

a

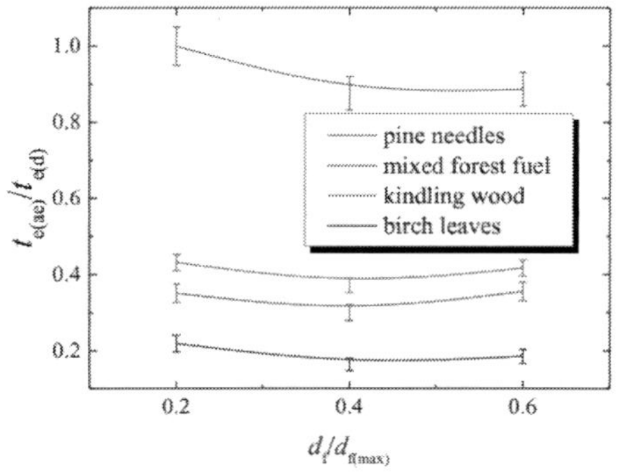

b

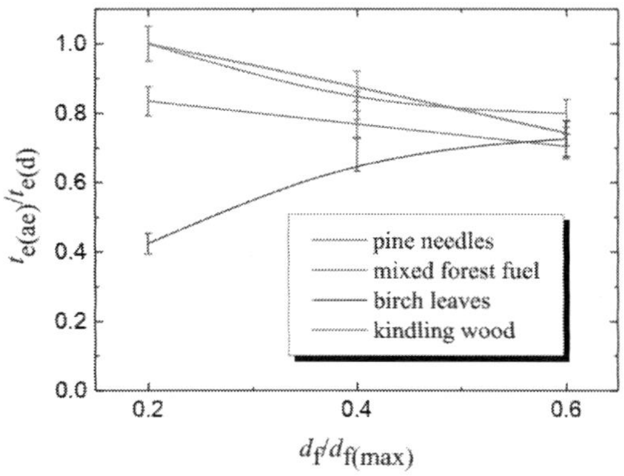

c

Fig. 2. Dependence $t_{e(ae)}/t_{e(d)}$ on $d_f/d_{f(max)}$ for typical FF under the influence of liquid aerosol and single droplets: *a* –water, *b* – wetting agent "Progress", *c* – foaming agent type "AR".

Moreover, chemical compounds added to water affects the extinguishing liquid evaporation completeness as it passes through flaming and flameless combustion zones

(Fig. 2) [8]. The most typical mechanism for suppressing the reaction of FF destructiom is realized by increasing the quenching liquid surface tension. Droplets of a sprayed liquid, draining on the FF surface, form a liquid film and block the heated pyrolysis gas products output. When water was mixed with the appropriate chemical compounds, we observed foam formation, which led to the creation of a stable buffer zone between the heated FF pyrolysis products and the oxidizer.

4 Conclusion

As a result of the experimental studies, we established times of suppression of forest fuel thermal decomposition under the influence of water and its compounds. We established, how model of fire sources' transverse dimension effects on the process of extinguishing the ground cover. The performed experiments make it possible to draw a conclusion on the expediency of adding "AR" type foaming agents to water. Close values of the suppression the FF thermal decomposition reaction time during extinguishing with steam and droplets of quenching compositions based on a foaming agent are established.

Investigations of high-temperature evaporation of water droplets were carried out with the support of a grant from the President of the Russian Federation (MD–1221.2017.8). Experiments to determine the specific differences between the processes of thermal decomposition and combustion of various types of forest fuel were carried out within the framework of grant from the President of the Russian Federation (MK-1684.2017.8).

References

1. C. C. Remy, M. Lavoie, M. P. Girardin, H. Asselin, A. A. Ali, J. Biogeogr., **1** (2017)
2. A. O. Zhdanova, G. V. Kuznetsov, J. C. Legros, P. A. Strizhak, Thermal Science, **21**, 6A, 2559 (2017), DOI: 10.2298/TSCI151006121Z
3. R. S. Volkov, A. O. Zhdanova, G. V. Kuznetsov, P. A. Strizhak, Tech. Phys., **60**, 1443 (2015), DOI: 10.1134/S1063784215100291
4. D. Morvan, Fire Saf. J., **71**, 34 (2015), DOI: 10.1016/j.firesaf.2014.11.012
5. O. P. Korobeinichev, A. G. Shmakov, V. M. Shvartsberg, A. A. Chernov, S. A. Yakimov, K. P. Koutsenogii, V. I. Makarov, Fire Saf. J., **51**, 102 (2012), DOI: 10.1016/j.firesaf.2012.04.003
6. V. M. Gumerov, A. O. Zhdanova, M. M. Osmolovskaya, P. A. Strizhak, MATEC Web of Conferences **37**, 01022 (2015), DOI: 10.1051/matecconf/20153701022
7. R. S. Volkov, O. V. Vysokomornaya, A. O. Zhdanova, P. A. Strizhak, EPJ Web of Conferences, **82**, 01059, (2015), DOI: 10.1051/epjconf/20158201059
8. R. S. Volkov, A. O. Zhdanova, G. V. Kuznetsov, P. A. Strizhak, Thermal Engineering (English translation of Teploenergetika), **62**, 586 (2015), Article number A008, DOI: 10.1134/S0040601515080091

Research of rheological properties improvement methods of coal-water fuel based on low-grade coal

Andrey Zenkov[1,]*, *Kirill* Larionov[1], *Stanislav* Yankovsky[1], and *Sergey* Lavrinenko[1]

[1]National Research Tomsk polytechnic university, 634050 Tomsk, Russia

Abstract. Experimental studies of coal-water fuel (CWF) rheological properties based on 3B brown coal have been conducted using different processing methods, such as rotary flows modulation device (RFMD), sodium hydroxide and lignosulfonate. Physicochemical properties of initial solid fuel have been determined using JEOL JCM 6000 microscope. Optimal method of coal-water treatment has been determined based on obtained data considering its influence on viscosity and sedimentation stability of coal-water slurry (CWS).

1 Introduction

The problem solution of catching the sulphur dioxide and nitrogen oxides out of the flue gases from the thermal power plants is connected with the processing of large volumes of gas emissions, exceeding in some cases million cubic meters per hour [1-4]. As a rule, high-quality power plant coals produced in the Russian Federation are exported abroad. Share of coal supplied to domestic TPPs is about 30%, 90% of which are low-quality coals [5].

Thus, domestic power plants are forced to use low-grade coal, which leads to environmental issues that require development and implementation of new efficient technologies that ensure full use of produced fuel and significant economic effect [6].

One of the most promising directions of low-grade coal application in power engineering is coal-water fuel technology (CWF). This technology represents composite dispersed system that consists of solid phase in the form of fine coal, liquid medium (water) and plasticizing agent [7].

2 Experimental setup

3B brown coal from "Balakhtinskoe" deposit was considered as initial sample of solid fuel. After initial grinding, the coal was sieved through a sieve with a mesh size of 80 μm.

Coal structure may vary greatly even within different layers of one deposit [8]. Due to importance of knowledge of coal properties for further evaluation of CWF characteristics [9], micrograph of initial coal (Figure 1) has been obtained with JEOL JCM 6000 microscope. It shows that coal particles have rough porous surface. It is known [10] that coals with such

* Corresponding author: andreyzenkov@mail.ru

DOI: 10.1051/matecconf/201714101018

structure have higher grindability. In addition, large number of pores leads to an increase in sorption volume, which leads to an increase in moisture capacity and hydrophilicity of the coal, thereby viscosity of CWS increases [11].

Fig. 1. Microphotograph of initial coal particle.

Application of rotary flows modulation device (RFMD) is proposed in this paper as additional stage of slurry processing prior to combustion for CWF rheological characteristics improvement. There are also proposed ways of chemical influence on coal-water slurry (CWS) - application of sodium hydroxide (NaOH), lignosulfonate (LST) and a mixture of these two additives.

RFMD is a liquid mixer with rotary siren, in which input raw material is subjected to mechanical, acoustic and hydrodynamic influences.

Principle of the device is that during its operation suspension flow is crushed by the rotor into a series of small secondary flows, which enter the gap between the rotor and the stator and are grinded into smaller streams, due to structural features of the stator. Flowing through narrow passages created by holes in the rotor and the stator, the processed raw material undergoes mechanical destruction because of particles impact against lateral surfaces of the rotor and the stator openings. In addition, grinding process occurs when the pressure pulsates due to sudden change of flows direction and frictional forces [12].

Sodium hydroxide is the most common alkali. It represents white solid material. NaOH is highly hygroscopic (it actively absorbs water vapor from the air). It is very soluble in water, during which a large amount of heat is released.

Lignosulfonate is a powder, color of which can vary from light shades of brown to rich brown. It mixes well with water. Lignosulfonate is produced at paper mills during interaction of wood with water, as a result of which wood decomposes and new compounds are formed [13].

After processing, dynamic viscosity of suspension was determined using Brookfield viscometer. To determine true viscosity value, the viscometer operates within confidence interval (10-90%), resulting in shear rate range being limited during examination of various CWF samples due to non-constant rheological properties. Thus, quantitative assessment of CWS viscosity variation has been conducted at equal shear rate (70 rpm), which corresponds to the confidence interval of all considered samples.

Evaluation of sedimentation stability has been conducted by settling of CWF sample in a measuring cup for 24 hours, followed by measuring the height of dense deposit.

3 Results

Figure 2 presents obtained results of coal-water fuel viscosity using various methods of its processing.

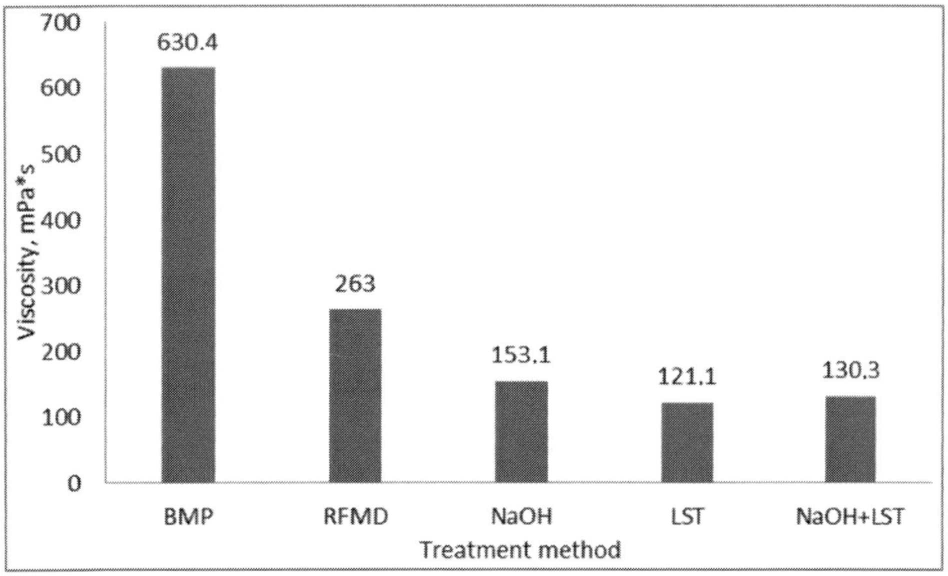

Fig. 2. Viscosity of coal-water fuel depending on treatment method.

As can be seen from Figure 2, maximum effect of suspension viscosity reduction has been obtained using lignosulfonate. Application of mechanical treatment has not showed the expected results, but this type of treatment can be used as an additional stage of CWF processing, since it significantly reduces viscosity of the suspension (more than 2 times).

Figure 3 presents results of coal-water fuel sedimentation stability using various methods of its processing.

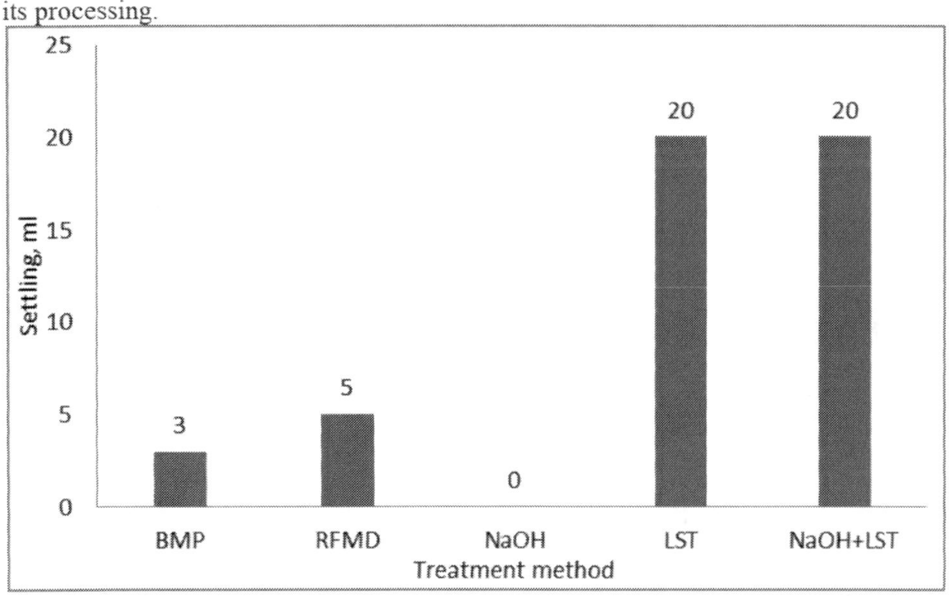

Fig. 3. Sedimentation stability of coal-water fuel depending on treatment method.

It follows from diagram 3 that sodium hydroxide has maximum effect on sedimentation stability. However, despite this, application of NaOH and lignosulfonate mixture causes formation of dense layer (about 20 ml), as in the case of using only lignosulfonate. Hence, it can be concluded that lignosulfonate has a much greater effect on CWF sedimentation stability than sodium hydroxide.

Conclusion

Thus, effect of chemical and mechanical treatment on rheological characteristics of coal-water fuel (viscosity and sedimentation stability) was evaluated. It has been found that maximum decrease of viscosity is observed with lignosulfonate application. However, this additive adversely affects sedimentation stability, which makes its use relevant only if the facility is able to install CWF storage tanks with a stirring mechanism.

Application of mechanical treatment (an additional step in the form of rotary flows modulation device) significantly reduces viscosity of the slurry, which makes it possible to use it in power engineering.

It has been determined that optimal treatment is application of sodium hydroxide, which significantly reduces viscosity of the suspension and does not cause sedimentation.

The research has been accomplished with financial support of the Russian Federation Ministry of Education and Science within implementation of the project No.13.7644.2017/BCH.

References

1. O.S. Dmitrieva, I.N. Madyshev, A.V. Dmitriev, J. Eng. Phys. Thermophys., **90**, 651 (2017)
2. I.N. Madyshev, O.S. Dmitrieva, A.V. Dmitriev, MATEC Web of Conf., **91**, 01019 (2016)
3. A. Uvarov, A. Antonova, A. Vorobjev, MATEC Web of Conf., **37**, 01062 (2015)
4. A.M. Antonova, A.V. Vorobiev, A.S. Orlov, MATEC Web of Conf., **23**, 01069 (2015)
5. O. Marinicheva, Energetika I promyshlennost Rossii, **11**, (2012)
6. A. A. Dekterev, P. V. Osipov, M. Y. Chernetskiy, A. F. Ryzhkov, Solid Fuel Chem., **51**, 17 (2017)
7. G. S. Khodakov, Thermal Eng., **54**, 36 (2007)
8. M. Y. Chernetskiy, V. A. Kuznetsov, A. A. Dekterev, N. A. Abaimov, A. F. Ryzhkov, Thermophys. and Aeromech., **23**, 591 (2016)
9. A. G. Korotkikh. K. V. Slyusarskiy, K. B. Larionov, V. I. Osipov, J. of Phys: Conf. Series, **754**, 052005 (2016)
10. Z. Gao, S. Zhu, M. Zheng, Z. Wu, H. Lü, W. Liu, Int. J. of Coal Sc. and Tech., **2**, 211 (2015)
11. A. Mukherjee, S. Pisupati, Energy and Fuels, **29**, 3675 (2015)
12. K. B. Larionov, A. V. Zenkov, S.A. Yankovsky, A.A. Ditc, *11th International Forum on Strategic Technology*, 7884323, 568 (2017)
13. V. E. Tarabanko, Teh. him., 299 (2008)

MATEC Web of Conferences **141**, 01019 (2017) DOI: 10.1051/matecconf/201714101019
Smart Grids 2017

Analysis of the construction of the high-temperature gas infrared radiator with the use of virtual prototyping

Anton N. Ermolaev [1], and *Olga* V. Khaustova [2,*]

[1]Tyumen industrial University, 625000 Tyumen, Russia
[2]National Research Tomsk Polytechnic University, 634050 Tomsk, Russia

Abstract. Method of virtual prototyping with the following mathematical modeling was used to simulate the heat-mass exchange and combustion during the operation of high-temperature gas infrared radiators, and to find optimal technical solutions for its design. The most authoritative and approved software product Ansys Multiphysics was used. The results of the mathematical modeling of heat and mass transfer in a turbulent reaction medium with combustion reproduce the experimental data produced by a measurement in real operating conditions of the gas-fired infrared heat emitter. The temperature distribution along the height of the ceramic nozzle was established. Obtained results enable estimation of the ignition and combustion zones.

1 Introduction

Two acute tasks of energy resources saving and energy efficiency increasing can be achieved by replacing convective heating systems of production buildings, warehouses, large pavilions, hangars with radiant heating systems based on high-temperature gas infrared radiators [1, 2]. Over many years of operation, high-temperature gas infrared radiators has proven itself as a reliable, energy-efficient and environmentally friendly source of converting fuel energy into thermal energy. The conversion factor of such heating devices is 98% [1], but the fraction of the radiative heat flux transferred to the work zone by radiation is only 58-61% [2]. The rest of the heat energy is accounted for heat losses with outgoing gases (24-30%) and the casing of the radiator (9-18%). The analysis of the high-temperature gas infrared radiators design aimed to improve the effectiveness of its work is an up-to-date practice [2].

Typical model of high-temperature gas infrared radiators consists of a number of complex technical elements (Figure 1) and includes a large number of variable parameters. In the 21st century, the technology of virtual prototyping is increasingly used [3-6]. It allows to analyze all the mechanisms of production at the design stage, look inside the technical process, identify design errors, and make pre-operational evaluation of the

* Corresponding author: olgavguk@tpu.ru

DOI: 10.1051/matecconf/201714101019

efficiency with less labor. In other words, a virtual prototype is a digital twin or a preliminary digital assembly of a technological product at the stage of designing.

Therefore, the method of virtual prototyping with the following mathematical modeling was used to simulate the heat-mass exchange and combustion during the operation of high-temperature gas infrared radiators, and to find optimal technical solutions for its design. The most authoritative and approved software product Ansys Multiphysics was adopted as the research tool [4-6].

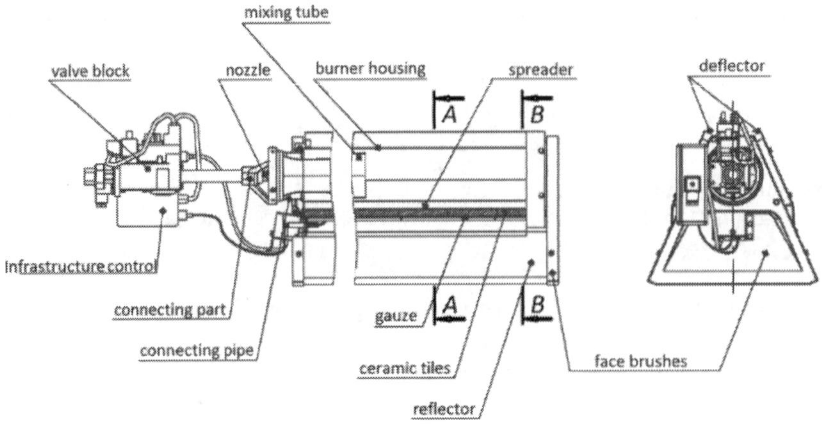

Fig. 1. Typical model of high-temperature gas infrared radiator.

2 Results of the study

Full virtual prototype of high-temperature gas infrared burner was built based on its sketches and complete geometric model (Fig. 1). The construction of the computational grid is performed by ANSYS Workbench, namely, the volume is divided into finite elements - tetrahedrons in the amount of $3 \cdot 10^6$. Boundary conditions are: the material of the radiator construction is 18–8 steel with a thickness of 0.002 m; fuel – natural gas (ρ=0.784 kg/m^3, Q=37.26 MJ/kg); oxidizer – air oxygen; fuel mass flow rate – 0.5 m^3 / hour; air fuel ratio is 10.374 m^3 / m^3. The internal air temperature is assumed to be +20 ° C. Process is in a steady state.

Using ANSYS Fluent, the following processes and phenomena were modeled: combustion of natural gas, turbulence, convective, conductive and radiant heat exchange. To describe the ongoing chemical reaction, a kinetic model of laminar combustion based on a simplified kinetic scheme of the gross reactions of methane oxidation was adopted. Calculation of radiant heat exchange was performed using the Discrete Ordinates radiation model. The validity of the selected mathematical models is confirmed by full-scale tests [1, 2, 6].

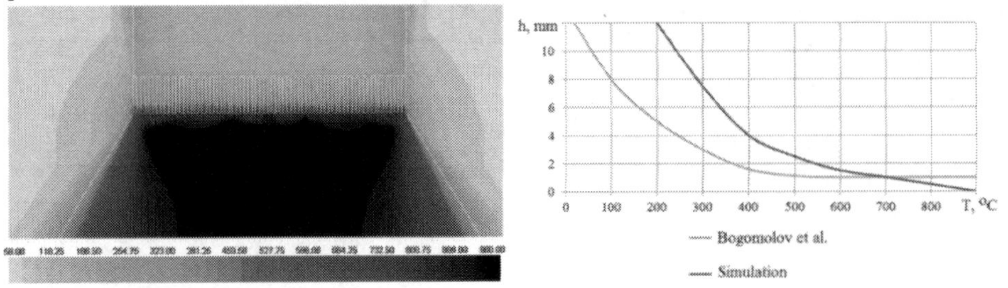

Fig. 2. Temperatures contours (°C) in gas-fired infrared heat emitter (a) and comparison with the experimental data (b) of [1].

The results of the mathematical modeling of heat and mass transfer in a turbulent reaction medium with combustion reproduce the experimental data produced by a measurement in real operating conditions of the gas-fired infrared heat emitter (see Fig. 2). According to the designed virtual prototype, the temperature of the ceramic tiles (see Fig. 1) for all investigated emitters was 900 °C (Fig. 2). In the course of analysis of the designed virtual prototype of high-temperature gas infrared burner, a dependence describing the temperature distribution along the height of the ceramic nozzle was established. The graph in Figure 2 demonstrates a qualitative agreement in mechanisms of combustion initiation and the location of the kinetic combustion region between the virtual prototype of the gas-fired infrared heat emitter and the experiment [1]. The numerical difference in the values of the temperatures at the entrance to the perforation of the ceramic nozzle is explained by the design and regime characteristics of the radiators [1, 2].

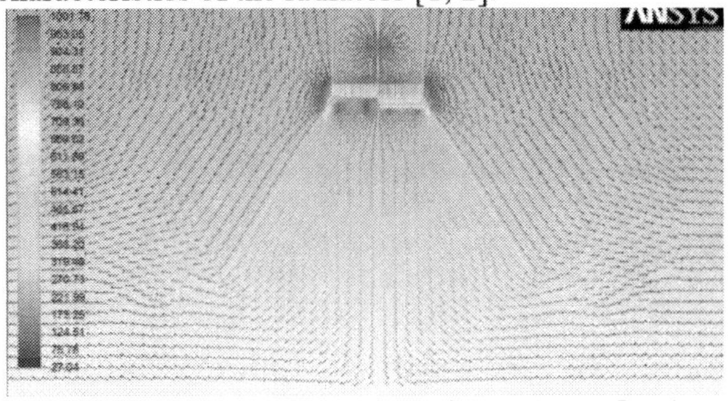

Section A-A (see fig. 1) Section B-B (see fig. 1)

Fig. 3. Vectors colored by temperature (°C) in the longitudinal section of the ceramic tiles of the gas-fired infrared heat emitter.

Detailed temperature fields and vectors in the longitudinal section of the ceramic tiles of the gas-fired infrared heat emitter (Fig. 3) enable estimation of the ignition and combustion zones. The ignition takes place at the outlet of the ceramic tiles, because fuel is premixed with air in mixing tube (see Fig. 1) before entering the holes in the ceramic tile. As a result narrow laminar combustion front moves through holes of ceramic tiles, and has its maximum at the outlet of the cylindrical holes.

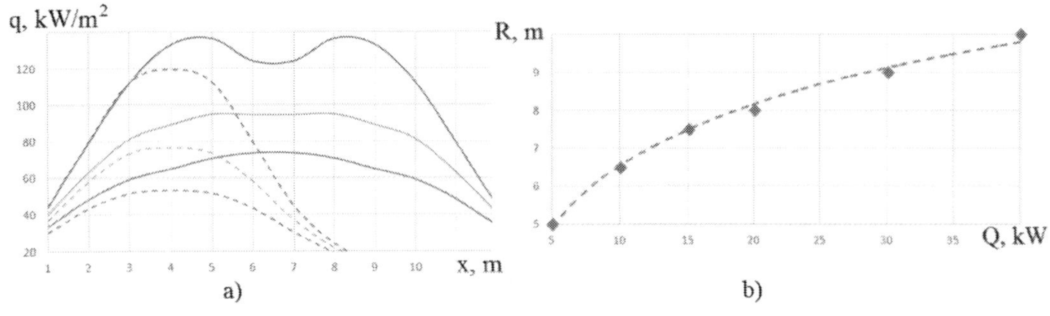

Fig. 4. The distribution of the irradiation intensity (a) and diameter of 10 W/m² isosurface in the working zone.

To verify the fulfillment of thermal comfort conditions, the distribution of the irradiation intensity in the working zone was calculated depending on the parameters studied. The maximum intensity of infrared irradiation of the workplace should not exceed 150 W/m². For the parameters studied, this condition corresponds to the following heights of the virtual prototypes of high-temperature gas infrared burner suspension: 5 kW - 4 m (119.3 W/m²); 10 kW - 6 m (106.1 W/m²); 15 kW - 7 m (116.9 W/m²); 20 kW - 8 m (119.3 W/m²); 30 - 10 kW (114.6 W/m²). For the virtual prototype of high-temperature gas infrared burner with a power of 40 kW this condition is violated (152.8 W/m^2) when placed at a maximum height of 10 m. Dependencies have been established, indicating that the higher the level of high-temperature gas infrared burner suspension results in the more uniform the distribution of local heat fluxes in the working zone and the lower their amplitude (Fig. 4, a). The impact of the high-temperature gas infrared burner thermal power on the diameter of 10 W/m^2 isosurface in the working zone is close to the logarithmic (Fig. 4, b).

3 Short conclusion

Virtual prototype of high-temperature gas infrared burner has been designed to search for a technical solution in accordance with the chosen criterion of optimality and check the efficiency and effectiveness of such solutions with less labor and investment.

Optimization of the design will be achieved by sequential refinement of the virtual prototype of the radiator. With an increase in the degree of detail of virtual prototyping, a more precise structural and functional imitation will become possible. The ideal situation will be reached when the virtual prototype will have the whole set of characteristics of the physical prototype. Thus, virtual prototyping eliminates the need for expensive and labor-intensive physical prototypes.

Using the virtual prototyping it was estimated that the most uniform intensity of irradiation of the working zone is ensured when the high-temperature gas infrared burner is installed of the same power at a distance shown in fig. 4. The value of heat flows for such a heating system organization is 25% higher than when a single high-temperature gas infrared burner located at a greater distance.

References

1. A.I. Bogomolov, D. Ya. Vigdorchik, M.A. Maevsky, *Gas burners of infrared radiation and their application* (Stroiizdat, Moscow, 1967)
2. A. N. Ermolaev, S. A. Khaustov, EPJ Web Conf., **110**, 01038 (2017)
3. A. V. Kazakov, R. B. Tabakaev, P. Y. Novoseltsev, A. V. Astafev, A.V., MATEC Web Conf., **19**, 01014 (2014)
4. S. A. Khaustov, A. S. Zavorin, K. V. Buvakov, L. D. Kudryashova, A. V. Tshelkunova, EPJ Web Conf., **82**, 01041 (2015)
5. S. A. Khaustov, A. S. Zavorin, K. V. Buvakov, V. A. Sheikin, EPJ Web Conf., **82**, 01039 (2015)
6. S. A. Khaustov, Y. A. Belousova, K. V. Buvakov, A. Y. Dolgih, R. N. Kulesh. *Proc. of 11th International Forum on Strategic Technology (IFOST - 2016)*, **2**, 548-551 (2016)
7. *Sanitary rules and regulations 2.2.4.548-96. 2.2.4 Physical factors of the production environment. Hygienic requirements for the microclimate of production facilities* (Information and Publishing Center of the Ministry of Health of Russia, Moscow, 1997)
8. *Code of Regulations 60.13330.2012. Heating, ventilation and air conditioning* (Ministry of Regional Development of Russia, Moscow 2012)

MATEC Web of Conferences **141**, 01020 (2017) DOI: 10.1051/matecconf/201714101020
Smart Grids 2017

Dynamic heating and ventilation of transformer substation buildings with adjustable resistance to heat transfer in windows

Andrey A. Yablokov[1], *Nikolay* N. Smirnov[1,*], *Vladimir* V. Tyutikov[1], and *Vladimir* A. Gorbunov[1]

[1]Ivanovo State Power Engineering University, 153003 Ivanovo, Russia

Abstract. The article considers issues related to providing dynamic heating and ventilation of transformer substation buildings featuring designed digital voltage transformers demanding particular ambient temperatures. We have shown the efficiency of using heat reflecting screens in windows with lower transmission heat losses from the transformer substation buildings concerned.

1 Introduction

As we know, the reserves of fossil fuels such as natural gas, coal and oil are limited and, according to different forecasts, will be exhausted within a few centuries [1, 2].

At electric power substations, it is essential to maintain such indoor environment parameters as will ensure comfortable working conditions for equipment and personnel. Special requirements regarding indoor thermal balance apply in the case of equipment related to relay protection systems, power and measuring transformers. Proper air renewal inside buildings must be provided during the warm season to ensure normal operation of digital transformers featuring a resistive voltage divider [3] and prevent overheating of resistive elements [4].

With regard to substation personnel, we have to note that in buildings where people are con-stantly present it is essential not only to maintain indoor environment parameters, but to reduce fatigue among personnel. In industries characterised by monotonous and stressful work, stable indoor environment conditions lead to increased fatigue among personnel and lower labour efficiency. A dynamic indoor environment system (DIE) [5], which increases capacity for work by stimulating the central nervous system, supplies air with parameters that constantly change with time. A dynamic indoor heating system (DH) also promotes lower levels of fatigue among personnel and better attention to and concentration on their tasks.

Reduction of electric heating costs in electric substation buildings demands reduced transmission heat losses via windows by using heat reflecting screens installed in the windows [6]. It is essential to examine the impact of the techniques related to using

* Corresponding author: nsmirnov@bk.ru

 DOI: 10.1051/matecconf/201714101020

windows with screens as developed by the authors on the indoor environment in transformer substation buildings.

2 Ventilation and heating systems in various types of substation buildings

Heating is not envisaged during the cold season in buildings housing power and measuring transformers (see Fig. 1) and in reactor chambers, due to large heat release.

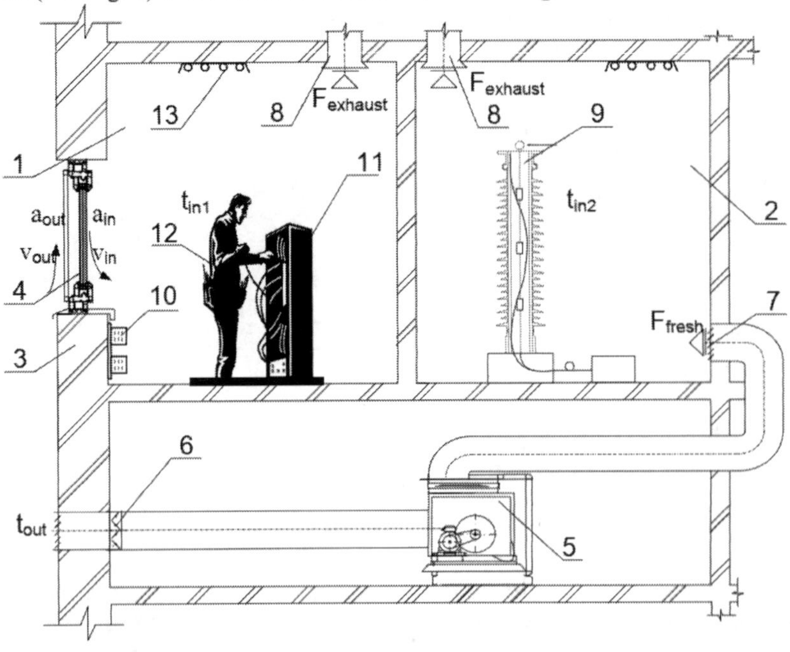

Fig. 1. System for maintaining a technologically appropriate and comfortable indoor environment in substation buildings with adjustable resistance to heat transfer in windows: 1 – heated building; 2 – transformer chamber; 3 – wall; 4 – window with externally mounted panel type heat reflecting screen; 5 – ventilation unit; 6 – adjustable shutters with electric drive; 7 - air inlet screen; 8 – extract duct; 9 - digital transformer; 10 – heating devices (electric heaters); 11 – relay panels; 12 – maintenance personnel; 13 – lighting.

During the warm season, to remove excess heat from working transformers (including those based on resistive voltage dividers), combined extract and input ventilation units are provided, usually with mechanical air input and natural extract air. Our simulation of transformer thermal condition and numerical solution methods [4] enable us to formulate and solve tasks related to choosing optimal transformer design (Fig. 2) and operation modes to ensure the most accurate measurements in the course of limiting the thermal condition of resistive elements, and also to calculate the required air renewal in the building. This approach enables us to develop technological measures aimed at increasing the reliability and prolonging the service life of transformers featuring resistive voltage dividers.

A verified mathematical digital transformer simulation with COMSOL Multiphysics software was used to study the thermal field of a digital transformer featuring a resistive voltage divider to determine the location of resistive elements that results in their minimal self heating.

Our studies showed that moving the resistive divider to the inner insulator wall permits reduction of resistive element temperature from 81 °C to 54 °C, with a further reduction by 3.4 °C if resistive elements are positioned in a spiral.

DOI: 10.1051/matecconf/201714101020

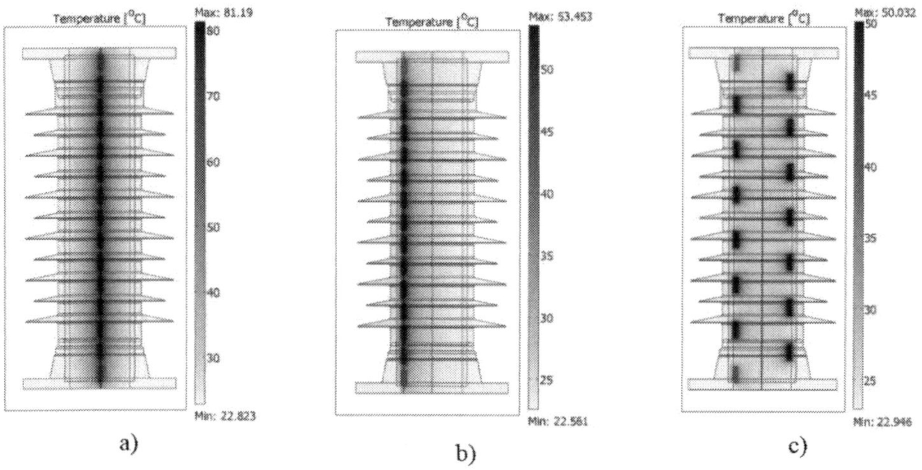

Fig. 2. Comparing thermal fields in a transformer featuring a resistive divider with axial (a), edge (b) and spiral (c)positioning of resistive elements.

Physical diagnostics of the thermal condition of digital voltage and current transformers (Fig. 1, item 9), involved in laboratory and commercial recording of electric power consumption also create a better understanding of the role of such measuring equipment in the thermal balance of electric substation buildings.

Heated premises (for example, relay panel facilities, workshops etc.) are provided with heating devices such as electric heaters, with automatic control from temperature sensors installed on the premises, and by means of an algorithm maintaining dynamic heating mode. Dynamic heating improves work capacity and attention among personnel engaged in intense and monotonous work during a work shift. Both normal (constant) and emergency ventilation of the heated premises are envisaged.

3 Developing techniques of using heat reflecting screens in windows. Results of testing these screens in windows

Staff from ISPEU and the National Applied Sciences Institute in Strasbourg (INSA de Strasbourg) have developed and patented windows designed with roll, jalousie and panel type heat reflecting screens made of metal [6, 7]. These significantly reduce transmission heat losses. The use of such screens is convenient at night, or in the absence of people. Fig. 1 shows an example of a panel type heat reflecting screen mounted on the outer surface of a window (item 4) in a heated building where required air parameters are maintained with the help of a heating and ventilating system. The use of such screens not only reduces transmission heat losses, but also permits lower ambient during setback heating mode [6].

In the course of mathematical simulation and physical experiments, it was proved [7] that in windows with heat reflecting screens not only is two-position control of resistance to heat transfer R_0 provided by moving the screens, but, as is evident from simulation data (Fig. 3, a), the given (and thermal) resistance to heat transfer depends to an even greater degree on the difference between indoor and outdoor air temperature, Δt. Moreover, this difference becomes more pronounced as the number of screens is increased. The value of transmission heat losses through windows is also a non-linear value (see Fig. 3, item 6).

DOI: 10.1051/matecconf/201714101020

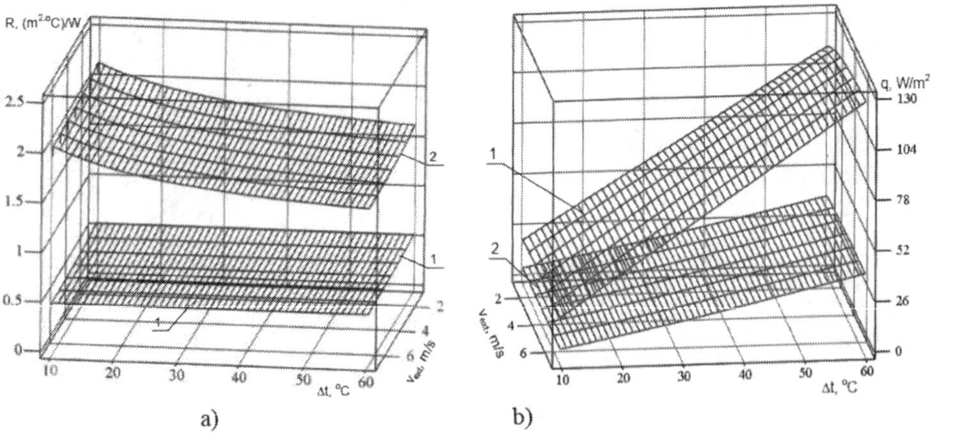

Fig. 3. Relation between given resistance to heat transfer *R* (a) and specific heat losses *q* (b) in the transparent area of a window according to the formula 4M1x10x4M1x10x4M1 in the event of natural convection on the inner surface of the window: 1 – without screens (control); 2 – with 3 metal screens mounted on the outer surface.

4 Mathematical simulation of dynamic heating in buildings with windows featuring heat reflecting screens

We developed a mathematical simulation of dynamic heating in an electric substation build-ing with adjustable window resistance to heat transfer, taking into account non-linear dependence of resistance to heat transfer in transparent structures on indoor and outdoor air parameters and adjustable window design during a 24-hour period. We also determined the energy efficiency of using heat reflecting screens with windows and further reduction of ambient temperature during non-working hours (for example, in production facilities and workshops used only during day shifts).

Fig. 4 and 5 show the results of heat exchange simulation through windows at an industrial facility in Moscow (for January conditions), with screens, lowering of indoor air temperature during non-working hours and use of ACS (air conditioning system) to maintain a dynamic indoor environment.

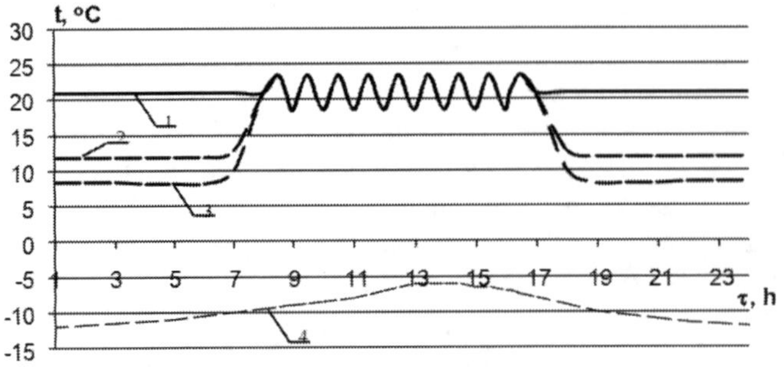

Fig. 4. Changes in indoor and outdoor air temperature over a 24-hour period: 1 – control indoor air temperature ($t_{stb} = \bar{t}_{die} = 21$ °C); 2 – with temperature lowered to $t_{stb} = 12$ °C; 3 – same with t_{stb} lowered to minimum possible temperature assuming prevention of condensation; 4 – outdoor air temperature (*stb – setback mode; die – dynamic indoor environment*).

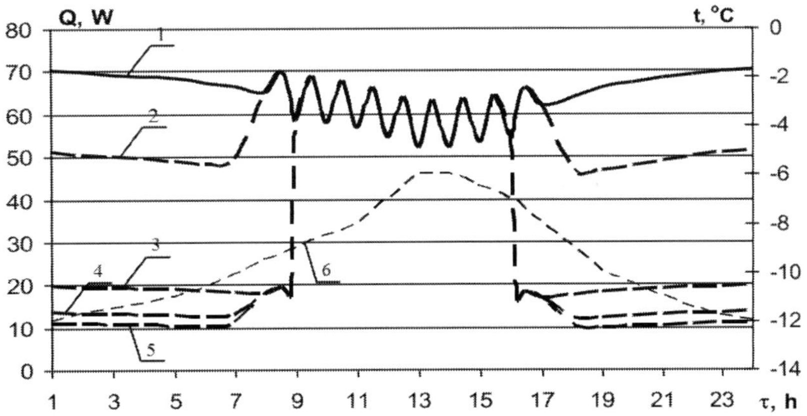

Fig. 5. Changing heat losses Q through 1 m^2 of window surface over 24 hours: 1 – control; 2 – with indoor air temperature reduced to t_{stb}=12 °C; 3 – using screens during night time; 4 – with screens and indoor air temperature lowered to t_{stb}=12 °C during night time; 5 - using screens at minimal indoor air temperature $t_{stb.\ min}$ during night time; 6 – outdoor air temperature.

On the basis of our simulation (Fig. 4 and 5) we may conclude that a major contribution to reducing heat losses through window structures is achieved by using heat reflecting screens (losses reduced 3.5 times). Additional heat loss reduction in setback heating mode (with the use of screens) permitted a fivefold reduction in transfer losses. Heat loss changes during a whole day are also related to fluctuating outdoor air temperatures.

5 Conclusion

Reducing transmission heat losses through the use of window screens and reduced air temperature during non-working hours significantly increases energy efficiency in the functioning of systems intended to maintain a dynamic indoor environment, with a 30-40% reduction in annual heating costs. Using a simulation of the thermal condition of a digital transformer enables us to more accurately determine the required air renewal in electric substation buildings.

This research was carried out at Ivanovo State Power Engineering University with the help of a grant from the Russian Scientific Foundation (Project No. 17-79-10455).

References

1. R. Tabakaev, I. Shanenkov, A. Kazakov, A. Zavorin, J. Anal. Appl. Pyrolysis, **124** (2017), DOI: 10.1016/j.jaap.2017.02.016
2. M. Gaydabrus, I. Razov, R. Tabakaev, V. Lebed, MATEC Web Conf., **110** (2017), DOI: 10.1051/matecconf/201711001030
3. V.D. Lebedev, A.A. Yablokov, IOP Conference Series: MSE, **177** (2017)
4. V. Lebedev, V. Zhukov, A. Yablokov, IOP Conference Series: MSE, **93** (2015)
5. V.V. Lovtsov, Yu.N. Khomutetsky, *Systems of conditioning of dynamic micro-climate of premises* (Stroiizdat, Leningrad, 1991)
6. V.M. Zakharov, V.V. Tyutikov, N.N. Smirnov, B. Flament, IOP Conference Series: MSE, **93** (2015), DOI: 10.1088/1757-899X/93/1/012015
7. N. Smirnov, V. Tyutikov, V. Zakharov, MATEC Web Conf., **110** (2017), DOI: 10.1051/matecconf/201711001096

Numerical modeling of gas infrared heating of production buildings

Anton N. Ermolaev [1,*]

[1]Tyumen industrial University, 625000 Tyumen, Russia

Abstract. The paper presents the method of simulating the infrared heating of a ventilated room based on the finite element analysis (FEM) with application of the engineering software. Approbation of the method was carried out by studying the influence of various parameters of infrared gas radiators on the temperature distribution in the volume of industrial facility. Analysis of the obtained data indicates that high-temperature gas-fired infrared heating system produces significant thermal radiation in the upper zone of the computational domain by the heated surfaces with simultaneous flow of convective heat and mass transfer processes. Temperature difference between the working area and the ceiling space varies from 17 ° C to 65.2 ° C. The obtained results confirm the applicability of the engineering software ANSYS Multiphysics for modeling the combustion process and the heat and mass transfer in solving problems of designing and optimizing high-temperature gas infrared heating systems.

1 Introduction

Designing a gas infrared heating system is a complex engineering practice. It includes solving the thermal balance of a radiant heating system, selecting heating devices by type of execution and heat output, determining the optimal locations for gas infrared radiators in the room, selecting regulating equipment, satisfaction of the received heat load with sanitary and hygienic requirements [1, 2].

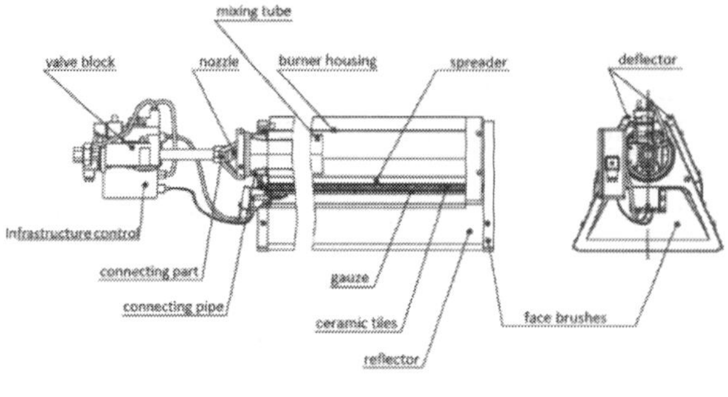

Fig. 1. Scheme of a typical high-temperature gas infrared radiator.

* Corresponding author: ermolaevanton03@gmail.com

The difficulty is in the fact that there is currently no single standard methodology for the design and calculation of gas infrared heating systems. The current normative and technical documentation provides only regulation of indoor air temperature, taking into account heat losses through the enclosing structures and heat consumption for air heating [1–4]. However, it does not fully account all the specifics of radiation heating systems. For example, an excessive number of gas infrared radiators (Fig. 1) and their incorrect placement lead to increased operational costs and a violation of the irradiation requirements. "Despite the fact that high-temperature gas infrared radiators have high energy efficiency, during their operation for heating large-scale production buildings a number of drawbacks was discovered: upper zone was overheated; smut appeared on walling surface resulting in wall subsequent destruction" [5]. Listed drawbacks are caused by energy losses in stack (24-30 %) and by high-temperature gas infrared radiator's casing (9-18 %).

In the 21 century, the best methods of mathematical modeling became available to a wide range of researchers in the form of commercial software products. Among the software products, the most common, authoritative and proven tool is Ansys Multiphysics, which uses approved mathematical models and discretization of the computational domain using the finite element method to solve the problems of heat and mass exchange and combustion modeling.

The gas infrared heating system includes a lot of variable parameters: the geometric dimensions, shapes and materials of all elements of the system, the optical properties of the environment, the location of the emitters and their design. Therefore, in order to work out the methodology for designing the gas infrared heating system, numerical modeling was carried out with changing the boundary conditions.

2 Computational model

The differential equations of computational model with the following boundary conditions were solved. Natural gas with the density 0.784 kg/m3 and calorific value 37.26 MJ/kg was set as the fuel. Oxidizer was air with oxygen mass fraction of 0.23. Gas-fired infrared radiators with thermal capacity of 5, 10, 15, 20, 30, 40 kW were studied. Thermal capacity of infrared radiators was defined by varying the number of ceramic tiles (see fig. 1), 1.5 kW each; fuel consumption per tile was 0,5 m3/hour. Wall material was sandwich panels with polyurethane foam insulation. The material of the light opening is glass with the thickness of 0.27 m. The internal emissivity of the inner enclosing surfaces is assumed to be $\varepsilon = 0.8$. On the walls that bound the calculated region, heat exchange conditions of the third kind are given. The outside temperature was minus 35 °C. The ventilation was organized as follows: air supply in the lower zone of the room, and air intake in the upper zone. The most common gas infrared heating system layouts were considered: industrial facility with dimensions of 60 m x 24 m x 12 m with high-temperature radiators of thermal capacities of 5, 10, 15, 20, 30, 40 kW at hanging heights from 4 up to 10 meters. We carried out a series of numerical experiments to obtain temperature fields formed in the area above high-temperature gas-fired infrared emitter's installation.

The computational domain of high-temperature gas infrared radiator is a conjugate gas-dynamic zone with double walls at the points of contact. With the help of a grid generator, the computational volume is divided into tetrahedral finite elements. Trial calculations using grids with different topologies and the number of cells showed that the mesh adopted in this work, containing about $3 \cdot 10^5$ control volumes, provides a low sensitivity of the results to further grinding the mesh. The finite element grid of the high-temperature gas infrared heating system is shown on Fig. 2. A standard mathematical k-ε model is chosen to model the turbulence described in paper [6]. Calculation of radiant heat exchange was performed using the Discrete Ordinates radiation model described in paper [5]. To describe the ongoing chemical reaction, a kinetic model of laminar combustion was adopted. The reliability of the selected mathematical models is confirmed by field trials [5 – 8].

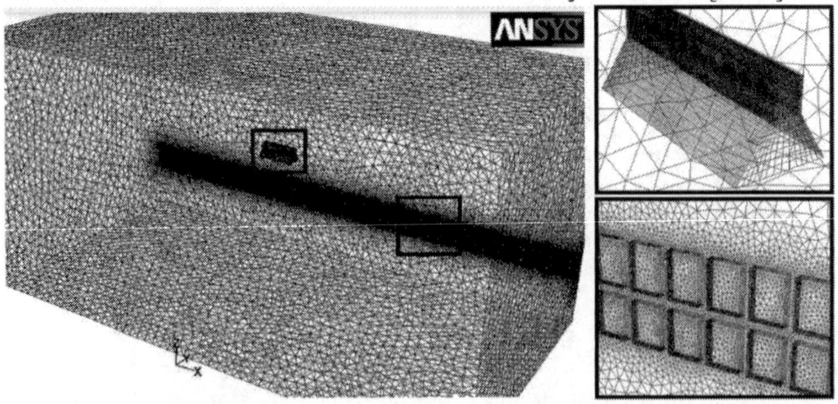

Fig. 2. Computational mesh of high-temperature gas-fired infrared heating system.

3 Results of the study

Analysis of the obtained data indicates that high-temperature gas-fired infrared heating system produces significant thermal radiation in the upper zone of the computational domain by the heated surfaces with simultaneous flow of convective heat and mass transfer processes. According the results of the simulation (Figure 3), the temperature difference between the working area and the ceiling space varies from 17 ° C (the gas infrared radiators were placed at a height of 4 m) to 65.2 ° C (the gas infrared radiators were placed at a height of 10 m).

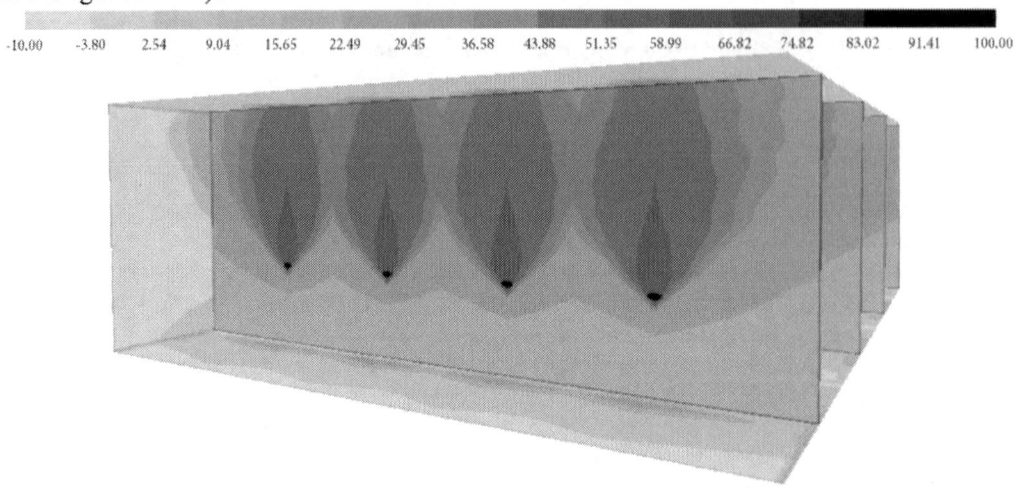

Fig. 3. Temperature distribution above high-temperature gas-fired infrared emitters.

The maximum value of the temperature was observed above the gas infrared radiators, then the temperature is decreased inversely proportional to the square of the distance from the radiator. Isolines of temperature have a V-shaped closed contour (fig. 3), increasing the curvature with temperature.

4 Conclusion

The results of research and mathematical modeling were:

1. Effective use of infrared heating systems is a multiparameter task that requires the effective development of formalized optimization procedures.
2. We have developed method of simulating the infrared heating of a ventilated room based on the finite element analysis (FEM) with application of the engineering software.
3. Approbation of the method was carried out by studying the influence of various parameters of infrared gas radiators on the temperature distribution in the volume of industrial facility.
4. The obtained results confirm the applicability of the engineering software ANSYS Multiphysics for modeling the combustion process and the heat and mass transfer in solving problems of designing and optimizing high-temperature gas infrared heating systems. It should be noted that approaches similar to the one proposed have not previously been applied to the analysis of the microclimate of industrial facilities.

References

1. *Sanitary rules and regulations 2.2.4.548-96. 2.2.4 Physical factors of the production environment. Hygienic requirements for the microclimate of production facilities* (Information and Publishing Center of the Ministry of Health of Russia, Moscow 1997)
2. *Code of Regulations 60.13330.2012. Heating, ventilation and air conditioning* (Ministry of Regional Development of Russia, Moscow 2012)
3. A. V. Kazakov, R. B. Tabakaev, P. Y. Novoseltsev, A. V. Astafev, A.V., MATEC Web Conf., **19**, 01014 (2014)
4. A.I. Bogomolov, D. Ya. Vigdorchik, M.A. Maevsky, *Gas burners of infrared radiation and their application* (Stroiizdat, Moscow, 1967)
5. A. N. Ermolaev, S. A. Khaustov, EPJ Web Conf., **110**, 01038 (2017)
6. S. A. Khaustov, A. S. Zavorin, K. V. Buvakov, L. D. Kudryashova, A. V. Tshelkunova, EPJ Web Conf., **82**, 01041 (2015)
7. S. A. Khaustov, A. S. Zavorin, K. V. Buvakov, V. A. Sheikin, EPJ Web Conf., **82**, 01039 (2015)
8. S. A. Khaustov, Y. A. Belousova, K. V. Buvakov, A. Y. Dolgih, R. N. Kulesh. *Proc. of 11th International Forum on Strategic Technology (IFOST - 2016)*, **2**, 548-551 (2016)

MATEC Web of Conferences **141**, 01022 (2017) DOI: 10.1051/matecconf/201714101022
Smart Grids 2017

Increasing the thermal efficiency of boiler plant

Evgeniy Uyanchinov[1,*], and *Igor* Razov[2]

[1]National Research Tomsk Polytechnic University, 634050 Tomsk, Russia
[2]Tyumen industrial University, 625000 Tyumen, Russia

Abstract. The thermal efficiency increase of boiler plant is actual task of scientific and technical researches. The optimization of boiler operating conditions is task complex, which determine by most probable average load of boiler, operating time and characteristics of the auxiliary equipment. The work purpose – the determination of thermodynamic efficiency increase ways for boiler plant with a gas-tube boiler. The tasks, solved at the research are the calculation of heat and fuel demand, the exergetic analysis of boilerhouse and heat network equipment, the determination of hydraulic losses and exergy losses due to restriction. The calculation was shown that the exergy destruction can be reduced by 2.39% due to excess air reducing to 10%; in addition the oxygen enrichment of air can be used that leads to reducing of the exergy destruction rate. The processes of carbon deposition from the side of flame and processes of scale formation on the water side leads to about 4.58% losses of fuel energy at gas-tube boiler. It was shown that the exergy losses may be reduced by 2.31% due to stack gases temperature reducing to 148 °C.

1 Introduction

One of the most important actions to efficiency increase of boiler plant is correct choice of capacity and boiler amount for projected heat supply source [1–3]. The optimization of boiler operating conditions is task complex, which determine by most probable average load of boiler, operating time and characteristics of the auxiliary equipment [4–7]. The work purpose – the determination of thermodynamic efficiency increase ways for boiler plant with a gas-tube boiler. The tasks, solved at the research are the calculation of heat and fuel demand, the exergetic analysis of boilerhouse and heat network equipment, the determination of hydraulic losses and exergy losses due to restriction. The work scientific novelty is applying of modern exergetic analysis methods to new object of research – gas-tube boiler and boilerhouses, equipped this boilers. The work novelty – the known data improvement and addition about research object and topic.

* Corresponding author: uyanchinov9576@gmail.com

2 Method of the research

The installation variants of boilers with various output were analyzed for determination of boilers amount, installed in the boiler house. The analysis was carried out by developed software package, which allows calculate the technical and economic characteristics of the boiler plant. The input parameters are the fuel and air energy and exergy, the output parameters are the coolant energy and exergy. The identification of exergy destruction allows find the thermal efficiency increase ways. The first propositions about exergy using for economic evaluation of efficiency were proposed by A. Andryushchenko [8, 9], Z. Rant [10, 11], J. Shargut [12, 13]. The influence of air combustion temperature and excess air, the influence of heat losses renewal at the steam generator and influence of stack effluents temperature to the exergy destruction at combustion chamber, steam generator and funnel can be modelled by modern software products.

3 Results and discussion

The calculations (Table 1) were shown that the thermal efficiency, the exergy efficiency and the boiler energy are accordingly 74.16%, 31.55% and 0.43%. The thermal efficiency is sufficiently low due to excessive heat funnel and steam generator losses, just as conservation index has a low quality due to the very low coolant exergy/energy ratio, which does not exceed 0.5. It was established that stack effluents have 22.59% of fuel energy due to high temperature of boiler outlet, although only a part of this loss can be recovered [14].

Table 1. The initial data for economic evaluation of operating.

Parameter	Value	Unit of measurement
Natural gas	4.12 to 7.29	rub/m^3
Cost of energy unit	0.0003	rub/kJ
Cost of exergy unit	0.0003	rub/kJ
Electricity	4.5	rub/(kW·hour)
Operating hours	8	thousand hour/year

It was found that the loss with heat of stack gases, which is the main source of energy losses and leads to 82.3% of total heat losses due to very high stack effluents temperature at funnel, is only 8.7% of total exergy destruction.

The operating costs during operation of boiler house was evaluated and aspects of fuel exergy economy, such as fuel cost, boiler water preparation cost, boiler water pumping cost and cost of air supplying for combustion, were identified in this work. The thermal and economic analysis of gas-tube boiler, considered the current and optimal operating costs, was made. It was established that the heat and exergy efficiency of researched boiler was not optimal due to excessive losses at combustion chamber, steam generator and funnel. The operating costs were also high due to fast exergy destruction. The approach of intrinsically value, which corresponds to current boiler operation, and marginal number of cost, which corresponds to optimal operation, lead to annualized savings at the 5,561,214 million rubles. Also, the modelling has shown that the optimal values of air combustion temperature, excess air and stack effluents temperature are accordingly 135 °C, 10% and 150 °C. The optimization of air combustion temperature, excess air, recuperation in convectional heating surfaces and temperature of stack gases was made to increase boiler efficiency.

The result of parametric optimization was the reducing the required fuel exergy during the boiler operation (5,648.2 kW) by 10.66% in comparison with the initial state

(6,322.3 kW). The losses with thermodynamic irreversibility were also reduced by 3.9% and 1.3 % at the combustion chamber and funnel accordingly.

The comparison of researched and optimized boilers is shown in the table 2. The optimization enables increasing the thermal efficiency by 4.91% that lead to increase of exergy efficiency by 3.62%, the economy of fuel costs and electricity amounts to 10.66% and 10.9% accordingly. Thus, the annualized savings amounts to 5561214 million rubles those equivalents to 11.8% increase of net efficiency.

Table 2. The parametric optimization based on thermoeconomic theory.

Parameter	Unit of measurement	Before optimization	After optimization
Efficiency	%	74.16	77.78
Exergy efficiency	%	31.55	36.25
Fuel costs	million rubles/gigacalorie	790.76	706.61
Boiler water preparation cost	rubles /gigacalorie	302.064	302.064
Boiler water pumping cost	rubles /gigacalorie	8.9	8.9
Cost of air supplying	rubles /gigacalorie	23.89	20.17
Total of operating costs	rubles /gigacalorie	1125.72	1037.80

4 Conclusions

It was established that the exergy destruction can be reduced by 2.39% due to excess air reducing to 10%. In addition, the rate of exergy destruction can be reduced by oxygen enrichment of air. The steam generator has a low energy transfer to the final product due to a very low coolant exergy/energy ratio. Consequently, only small part of the exergy destruction may be removable. It was established that the processes of carbon deposition from the side of flame and processes of scale formation on the water side is cause of heat energy loss to 4.58% at gas-tube boiler. And then, exergy losses may be reduced by 2.31% due to stack gases temperature reducing to 148 °C. The further decrease of stack gases temperature does not allowed; because it leads to low-temperature corrosion in the boiler pipes and funnel lining. The modelling shows that exergy destruction of steam generator may be additionally reduced by cleaning and proper maintenance.

References

1. E. L. Paley, *The normative requirements and practical recommendations at the boilerhouses design. Reference and practical book* (Peter, St. Petersburg, 2014)
2. R. B. Tabakaev, A. V. Kazakov, A. S. Zavorin, Solid Fuel Chemistry **49**, 267 (2015)
3. S. A. Khaustov, A. S. Zavorin, K. V. Buvakov, L. D. Kudryashova, A. V. Tshelkunova, EPJ Web Conf., **82**, 01041 (2015)
4. S. A. Khaustov, A. S. Zavorin, K. V. Buvakov, V. A. Sheikin, EPJ Web Conf., **82**, 01039 (2015)
5. Khaustov S. A., Belousova Y. A., Buvakov K. V., Dolgih A. Y., Kulesh R. N. *Proc. of 11th International Forum on Strategic Technology (IFOST - 2016)*, **2**, 548 (2016)
6. A. V. Kazakov, R. B. Tabakaev, P. Y. Novoseltsev, A. V. Astafev, A.V., MATEC Web Conf., **19**, 01014 (2014)
7. Sannikov D. I. About amount of the boilers for designed boiler plant [Electronic source] // RosTeplo – Access: http://www.rosteplo.ru/Tech_stat/stat_shablon.php?id=2921

8. A. I. Andryushchenko, *The technical working capacity of thermodynamical systems* (Saratov institute of Automobile and Road, Saratov, 1956)
9. A. I. Andryushchenko. *The thermodynamical calculations of optimal parameters on thermal stations* (Gosenergoizdat, Moscow, 1963)
10. Z. Rant, Forsch. Ing.–Wes., **22**, 212 (1956)
11. Z. Rant, BWK, **12**, 297 (1960)
12. J. Szargut, Archiwun Hutnictwa, **6**, 23 (1961)
13. J. Szargut, Energetyka Przemyslowa, **9**, 73 (1961)
14. N. Tobish., M. Zafar., M. Rustam., S. H. Javed., N. Feroze., M. Kazimi., Jurnal of Pakistan Institute of Chemical Engineers., (2014)

MATEC Web of Conferences **141**, 01023 (2017) DOI: 10.1051/matecconf/201714101023
Smart Grids 2017

Hydraulic resistance of thermal deaerators of thermal power stations (TPS) with jet-film contact devices

Ilnur N. Madyshev[1*], *Oksana* S. Dmitrieva[2], and *Andrey* V. Dmitriev[2]

[1]Kazan National Research Technological University, 420016 Kazan, Russia
[2]Kazan State Power Engineering University, 420066 Kazan, Russia

Abstract. One of the methods of solving the problem of improvement of heat-mass efficiency of thermal deaerators of TPSs is application of jet-film contact devices for interaction of vapor and liquid flows. The design of these devices has been studied. Estimate calculations of hydraulic resistance have been carried out. We have obtained the equation for calculation of resistance coefficient, depending on the height of element of contact device.

1 Introduction

Thermal deaeration of heat transfer water agent is one of the most important activities, ensuring water-chemistry conditions for power facilities [1]. At the same time, reliability and operation efficiency of power asset depends on the efficiency of water deaeration processes. Reliability is due to the intensity of corrosion processes of steel pipelines and equipment, and the efficiency is related to the presence of non-condensable gases in the water vapor, which hinder the heat transfer during condensation [2].

A possible way of solving the problem of improvement of heat-mass efficiency of thermal deaerators of TPSs is application of jet-film contact devices for interaction of vapor and liquid flows. Currently known designs of jet bubbling, jet and film deaerators are of large dimensions and very metal-intensive. This is due to the relatively low intensity and efficiency of heat-mass transfer processes, occurring in these devices. Besides, many designs of contact devices have low throughput capacity, while operation within area of high vapor velocities leads to an intensive liquid removal and the "flooding" of deaerator column.

2 Description of jet-film contact devices

The aforementioned shortcomings were eliminated in the jet-film contact devices [3], developed by the authors. The contact devise (Fig. 1) consists of parallel square drain cups *1* with vertical walls, required for maintaining the liquid level inside of them. Vertical perforated baffles *2* are the supports for drain cups *1* and have slotted holes for the

* Corresponding author: ilnyr_91@mail.ru

installation of appropriate drain cups *1*. The drain cups *1* are open from the top end, but recurved leaves *3* in the form of circular segments are executed in the bottom of cups. These segments are required for the liquid distribution over the surface of vertical perforated baffles *2*.

The proposed jet-film contact device operates as follows. The liquid, through the plurality of recurved leaves *3*, performed in the bottom of drain cups *1*, is dispersed in the form of jets to the below located vertical perforated baffles *2*. Thus, the liquid level in these drain cups *1*, is maintained by means of vertical walls of drain cups. The drain cups *1* are arranged horizontally chequerwise, forming a tray. The below tray has an offset of drain cups, forming vertical chequerwise arrangement of them. For this reason, vapor, coming to underneath the plate, acquires a zigzag nature of movement.

Fig. 1. Jet-film contact device: *1* – drain cup; *2* – baffles; *3* – recurved leaves.

When liquid jets move along the surface of vertical baffles *2*, the liquid is distributed with formation of a stable film flow. Thus, the coming down liquid film interacts with an ascending vapor flow. After that, the formed film, colliding with liquid surface inside of drain cups *1*, collapses. Thus, there is a developed, continuously renewing surface of contact of phases, which is determined by the presence of relatively small vapor bubbles in the liquid layer and droplets, departing from the surface. The presence of holes leads to alignment of the profile of concentrations in the coming down liquid film and its more uniform distribution over the surface of baffles. Besides, metal consumption of the proposed design reduces. When distances between the drain cups, arranged at the same level, are equal to width of a drain cup, it provides stable, uniform vapor flow, which leads to reduction of hydraulic resistance of jet-film contact device. Thus, organization of original interaction between vapor and liquid allows stimulating heat-mass transfer processes both in liquid and in vapor phases at relatively simple apparatus design.

3 Analysis of the results of numerical calculations

The purpose of numerical studies is to determine the hydraulic resistance of proposed contact devices. For that purpose, all-gas (air) flows with a temperature of 20°C through the layer of contact elements, consisting of 3 stages, were computer simulated via ANSYS Fluent software application. Width of drain cup has been taken equal to 100 mm. During numerical studies, the height of drain cup has been varied from 10 up to 30 mm, as well as gas velocity in the cross section of apparatus – from 1 up to 10 m/s. As the main model, we

 DOI: 10.1051/matecconf/201714101023

chose a modification of the two-parametric turbulence model k-ω SST-model, showing satisfactory agreement with experimental data, obtained in earlier studies [4, 5].

The resistance of dry layer of contact elements can be determined by Darcy-Welsbach modified equation:

$$\Delta p = \xi \frac{H}{b} \frac{\rho_G W_G^2}{2}, \quad (1)$$

where ξ – resistance coefficient; H – height of contact elements layer, m; b – width of drain cup, m; ρ_G – gas density, кg/m³; W_G – gas velocity in the cross section of apparatus, m/s.

Height of contact stage is related to height of drain cup by the following ratio:

$$h_{cm} = b + h_1, \quad (2)$$

где h_1 – высота сливного стакана, м.

Fig. 2 shows the dependence of hydraulic resistance of jet-film contact devices from gas flow velocity in the cross section of apparatus. The diagram shows that increasing the height of drain cup from 10 up to 30 mm leads to decrease of hydraulic resistance by 1.43–1.46 times.

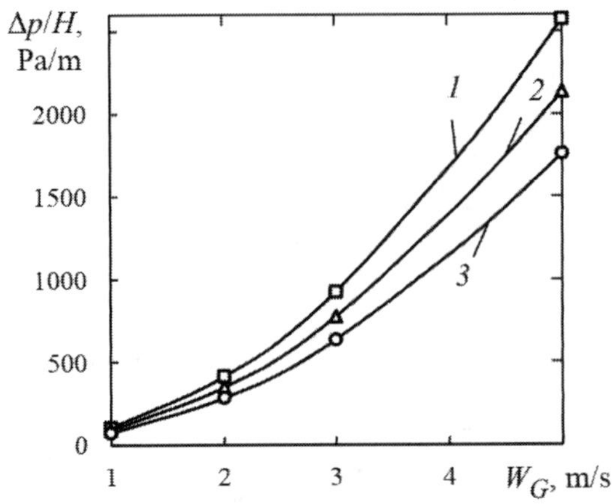

Fig. 2. Specific hydraulic resistance of jet-film contact devices, depending on gas velocity in the cross section of apparatus at different height of drain cup h_1, mm: *1* – 10; *2* – 20; *3* – 30.

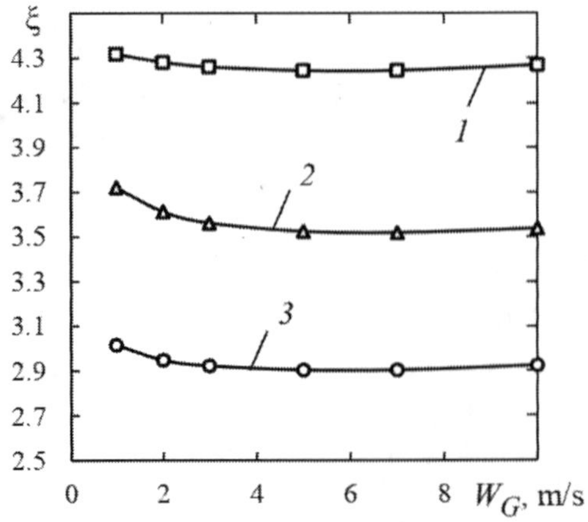

Fig. 3. Dependence of resistance coefficient of jet-film contact devices from gas velocity in the cross section of apparatus at different height of drain cup h_1, mm: *1* – 10; *2* – 20; *3* – 30.

Analyzing the data, shown in Fig. 3, it can be concluded that resistance coefficient is practically independent from gas velocity in the cross section of apparatus, as its change under the studied conditions does not exceed 5.7%. Processing of numerical simulation results allowed us to obtain dependence of resistance coefficient from height of drain cup in the following form:

$$\xi = -67.134 h_1 + 4.9508, \quad (3)$$

approximation certainty factor of which is equal to 0.9992.

4 Conclusion

On the basis of numerical calculations we can draw the following conclusion: a equation (3) calculates the resistance coefficient with accuracy sufficient for practice, and it can be used to predict the hydraulic resistance of thermal deaerators of TPS with jet-film contact devices. Thus, it can be said that application of proposed contact devices inside of deaerators of industrial TPSs is promising, as they can improve the efficiency of water thermal deaeration process at low energy costs.

The reported study was funded by RFBR, according to the research project No. 16-38-60081 mol_a_dk.

References

1. A.A. Abramovskikh, U.A. Ulyanova, S.V. Lavrinenko, A.A. Matveeva, D. Malishev, MATEC Web of Conferences, **91**, 01002 (2017)
2. V.I. Sharapov, E.V. Kudryavtseva, Power Tech. and Eng., **50**, 2, 204 (2016)
3. O.S. Dmitrieva, A.V. Dmitriev, L.V. Kruglov, Procedia Engineering, **150**, 753 (2016)
4. O.S. Dmitrieva, I.N. Madyshev, A.V. Dmitriev, J. Eng. Phys. Thermophys., **90**, 3, 651 (2017)
5. O.S. Dmitrieva, A.V. Dmitriev, I.N. Madyshev, A.N. Nikolaev, Chemical and Petroleum Engineering, **53** (1-2), 130 (2017)

MATEC Web of Conferences **141**, 01024 (2017) DOI: 10.1051/matecconf/201714101024
Smart Grids 2017

Technical and economic analysis of the combined-cycle plant with back-pressure steam turbine

Aleksandra Antonova[1,*], *Aleksandr* Vorobiev[1] , and *Artem* Uvarov[1]

[1]National Research Tomsk Polytechnic University, 634050 Tomsk, Russia

Abstract. The cogeneration combined-cycle gas turbine plant with back-pressure steam turbine is considered. A simple circuit of the plant increases its reliability, and the use of optimal temperatures of the heating and return water allows obtaining the highest thermal efficiency of the plant taking into account the ratio of its electric and heat power. The technical and economic efficiency of the combined-cycle plant with back-pressure steam turbine is studied to estimate the capital investment. The parameters and conditions to maximize the net present value growth relative to the base version are determined.

1 Introduction

One of the most effective energy saving technologies is the improvement of existing heat and mass exchange units [1-4]. Combined-cycle technologies have occupied a leading position in heat power engineering due to high heat efficiency and environmental friendliness. Generating equipment in domestic thermal power plants and in the world energy is mainly implemented through combined-cycle power plants (CCPP) [5]. In the most efficient binary-cycle CCPP with bottoming cycle and is called a combined cycle gas turbine (CCGT) plant. The exhaust heat of gas turbine (GT) is recovered in a heat recovery steam generator (HRSG) for generation of superheated steam, which then flows to the steam bottoming cycle. The efficiency of these CCGTs based on high-temperature GT reaches 60% and higher [6, 7]. Higher levels of heat efficiency are obtained in cogeneration combined-cycle plants supplied with different cycles and circuits. One of the versions of this plant is CCGT with a back-pressure steam turbine (CCGT-BP) [8].

2 CCGT with a back-pressure steam turbine

In CCGT-BP, the exhaust GT gas heat is recovered in HRSG in two sections of the heating cycle: upon supply of the superheated steam (hot water, steam generation, superheating) and during heating the return water in the HRSG back-end surfaces. The heat generated by the exhaust steam is transferred to the steam or to the water supplied directly to the network.

* Corresponding author: anton@tpu.ru

DOI: 10.1051/matecconf/201714101024

CCGT-BP is very simple in its design since no several steam pressure levels and reheating are required. Fig. 1 shows a CCGT-BP schematic diagram to supply heat to heat and hot water consumers and for household needs.

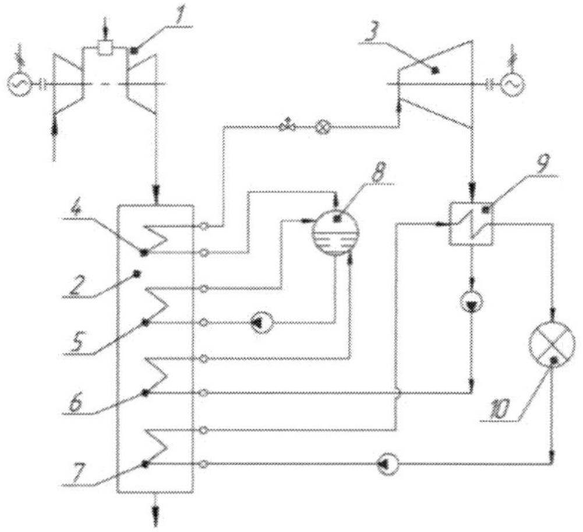

Fig. 1. Schematic diagram of CCGT with a back-pressure steam turbine: GT (1); HRSG (2); steam turbine (3); steam superheater (4); evaporator (5); economizer (6); GWNH (7); drum (8); network heater (NH) (9); heat consumer (10).

The efficiency of HRSG and that of the whole plant can be increased through deep heat recovery due to reduced exhaust gas temperature during transmission of heat to the return heating water in the gas-water network heater (GWNH). The temperature chart of the heat supply system is crucial for the performance of the CCGT-BP steam section, since the temperature of the heating water affects the pressure of the turbine exhaust steam; its value is close to the atmospheric pressure or exceeds it.

3 Thermal efficiency of CCGT-BP

The thermal efficiency of CCGT-BP can be estimated using the so-called total efficiency of CCGT, which shows the portion of the fuel heat utilized in the plant.

Utilized heat is the sum of the CCGT electric power N_E, and heat power transferred to heat consumers Q_{HC}. The total efficiency of CCGT

$$\eta_{CCGT} = \frac{N_E + Q_{HC}}{Q_{CC}}, \tag{1}$$

where Q_{CC} is the fuel heat input to the working fluid in the GT combustion chamber.

In turn, the CCGT electric power is the sum of two summands

$$N_E = N_E^{GT} + N_E^{ST}, \tag{2}$$

where $N_E{}^{GT}$ and $N_E{}^{ST}$ are the electric power of gas and steam turbine.

Transform the formula of the CCGT total efficiency

$$\eta_{CCGT} = \frac{N_E + Q_{HC}}{Q_{CC}} = \frac{N_E^{GT} + N_E^{ST} + Q_{HC}}{Q_{CC}} = \frac{N_E^{GT}}{Q_{CC}} + \frac{N_E^{GT} + Q_{HC}}{Q_{CC}} \tag{3}$$

Transforming expression (3), we obtain

DOI: 10.1051/matecconf/201714101024

$$\eta_{CCGT} = \frac{N_E^{GT}}{Q_{CC}} + \frac{N_E^{ST} + Q_{HC}}{Q_{CC}} = \eta_{GT} + \frac{Q_{CC} \cdot (1 - \eta_{GT}) \cdot \eta_{HRSG}}{Q_{CC}} = \eta_{GT} + (1 - \eta_{GT}) \cdot \eta_{HRSG} \quad (4)$$

Thus, the CCGT total efficiency is

$$\eta_{CCGT} = \eta_{GT} + (1 - \eta_{GT}) \cdot \eta_{HRSG} \quad (5)$$

This formula lacks the efficiency of the steam turbine or any other factor that indicates its heat efficiency. Hence, the effectiveness of the combined-cycle plant with a back-pressure turbine does not depend on the steam cycle (SC) efficiency. This can be attributed to the fact that SC works without cycle losses: the exhaust steam condensation heat is fully utilized to cover the consumers' heat demands.

In addition to the total efficiency, the most important characteristic of the CCGT-BP heat efficiency is the power ratio N_E^{ST}/Q_{HC}.

The aim is to study the effect of a number of factors on CCGT-BP heat efficiency. These factors are as follows: temperature of the heating water t_{hw}, temperature of the return water t_{rw}, initial steam pressure p_0, initial steam temperature t_0.

In accordance with the recommendations [9], a number of variables are taken: temperature difference at the evaporator of 5 °C, subcooling to the saturation temperature in network heater of 5 °C, temperature difference at the GWNH of 5 °C, initial steam temperature t_0.

The design conditions for CCGT-BP imply its operation at ambient temperature of –40 °C (temperature specified for heating system design in Tomsk). The key GT characteristics are calculated with regard to its energy parameters; air heating to 3.4 °C through air recirculation at the compressor outlet is considered. The Siemens SGT5-4000F gas turbine was used as the CCGT-BP basis. The GT performance parameters at external temperature of –40 °C are as follows: GT load is 100%, electrical efficiency is 35.37 %, electric power is 247.6 MW, exhaust gas temperature is 594.8 °C, exhaust gas flow is 644.9 kg/s.

The model was made in "MathCad" software and tested in the special "Thermoflex" complex.

In this study, the total capital investment cannot be taken as a criterion for the solution choice since the electric power of the steam turbine varies considerably. Therefore, to choose the best option, a net present value or net value gain relative to the base case should be used [10].

4 Results of the design efficiency of CCGT-BP

4.1 Effect of heating water temperature

The results of calculations for the CCGT-BP steam section performed at different heating water temperature are shown in Fig. 2. The return water temperature is 60 °C; the initial parameters of steam: 5.0 MPa and 540 °C.

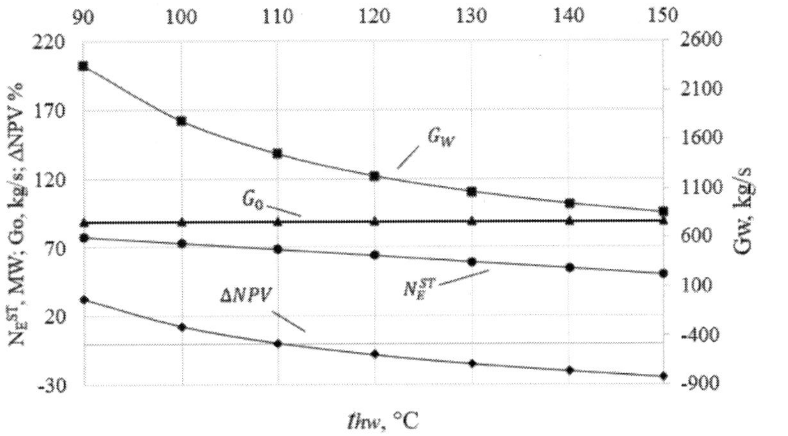

Fig. 2. Variation in CCGT-BP parameters against the temperature of the heating water.

4.2 Effect of return water temperature

The results of the calculations carried out for different values of the return water temperature are shown in Fig. 3. The heating water temperature is 110 °C; the initial parameters of steam: 5.0 MPa and 540 °C.

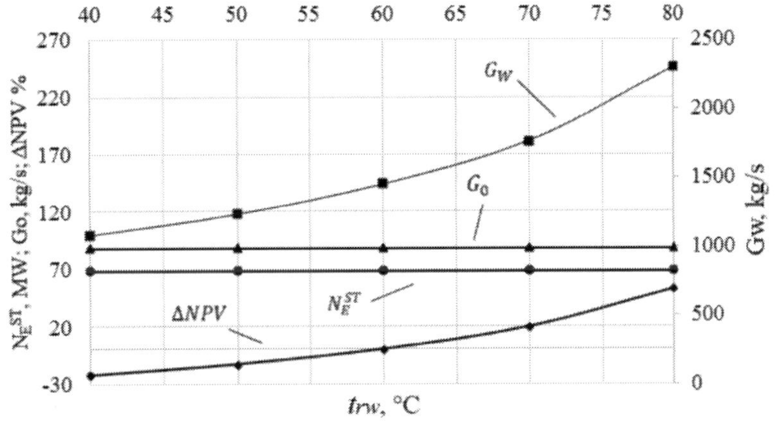

Fig. 3. Variation in CCGT-BP parameters against the temperature of the return water.

4.3 Effect of initial parameters of steam on CCGT-BP efficiency

The calculation results for different values of initial parameters of steam are shown in Fig. 4. The heating water temperature is 110 °C, and the return water temperature is 60 °C.

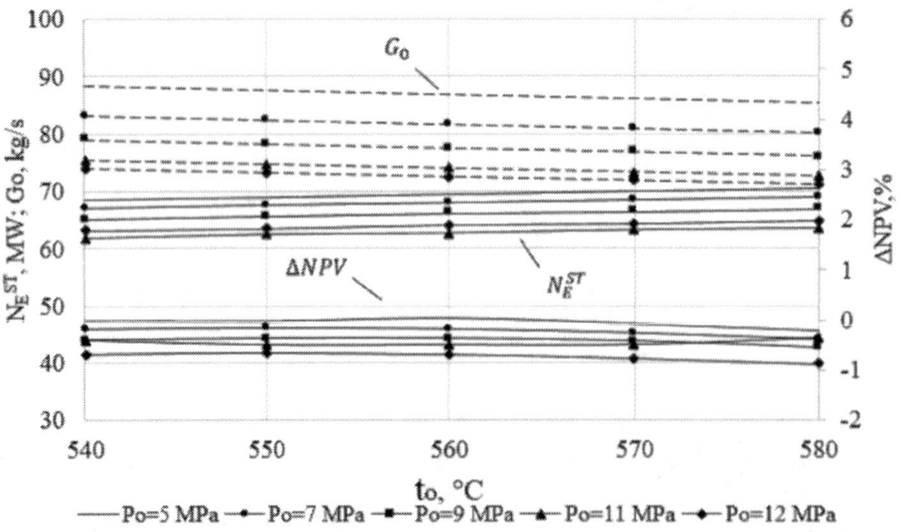

Fig. 4. Variation in CCGT-BP parameters against initial parameters of steam.

As can be seen from Fig. 4, the initial steam parameters have a negligible effect on the CCGT-BP basic parameters.

Taking into account the impact of ecological benefits of the installation can be using the software complex [11].

4 Conclusions

The efficient performance of the steam section does not affect the CCGT-BP heat efficiency since bottoming cycle runs without cycle losses. In addition, simple circuit of the plant due to single-loop bottoming cycle and lack of steam reheating increases the CCGT-BP reliability.

The values of the CCGT-BP total efficiency and that of HRSG do not depend on temperature of the heating water since useful power of HRSG does not change due to constant temperature of gases at its inlet and outlet. Reduced temperature of the heating water results in increased efficiency of the investment in the project as a result of increased electric power of the steam turbine unit and increased gain in the net present value ΔNPV.

When the return water temperature grows from 40 °C to 80 °C, the HRSG efficiency drops by 7 %, which reduces the amount of the delivered heat by 8.5 % and reduces the total value of the CCGT-BP efficiency by 4.3 %. In addition, as temperature grows, the net present value ΔNPV changes by 74.8 %.

The growth of the initial temperature in that range (by 7.4 %) leads to a decrease in the live steam flow. At increased temperature, the electric power slightly increases regardless of the initial pressure level, and the electric power gain is 1.6–1.9 MW. The value of ΔNPV relative to the base version is negligibly negative in all the cases, and it does not virtually change.

At increased initial steam pressure, the parameter of the heat transferred to each surface of HRSG, except for GWNH, decreases. As a result, the steam flow to the turbine reduces and the electric power decreases. At increased pressure, the total capital investment remains virtually unchanged: at pressure increased in the above range, it reduces by 0.73 %–0.75 %. The NPV growth is not observed.

Based on the above, it should be noted that the technical and economic efficiency of CCGT-BP largely depends on the temperature chart of the heating system, i.e., on temperatures of the heating and return water. Variation in the initial parameters of steam has a negligible impact on the CCGT-BP performance parameters.

References

1. A.V. Dmitriev, O.S. Dmitrieva, I.N. Madyshev, Therm. Eng., **63** (9), 674 (2016), DOI: 10.1134/S0040601516080036
2. S. Lavrinenko, P. Polikarpov, A. Matveeva, V. Martyshev, MATEC Web of Conf., **110**, 01049 (2017), DOI: 10.1051/matecconf/201711001049
3. D. Gvozdyakov, V. Gubin, S. Shvab, A. Tanishev, MATEC Web of Conf., **72**, 01037 (2016)
4. E. Pavelyev, S. Lavrinenko, MATEC Web of Conf., **72**, 01085 (2016), DOI: 10.1051/matecconf/20167201085
5. G. Olkhovsky, V. Trushechkin, Thermal power plants, **1**, 2 (2013)
6. A.M. Antonova, A.V. Vorobiev, A.S. Orlov, MATEC Web of Conf., **23**, 01069 (2015), DOI: 10.1051/matecconf/20152301069
7. A. Uvarov, A. Antonova, A. Vorobiev, MATEC Web of Conf., **37**, 01062 (2015), DOI: 10.1051/matecconf/20153701062
8. A. Gushchin, V. Torzhkov, Turbiny i dizeli, **7**, 16 (2007)
9. A. Trukhniy, *Parogazovyye ustanovki elektrostantsiy* (MEI, Moscow, 2013)
10. A. Devyanin, *Optimization of parameters of the cycle for three-loop combined-cycle plants* (Abstract of diss., Moscow, 2009)
11. A.V. Vorobiev, E.M. Dutova, A.S. Latyshev, MATEC Web of Conf., **72**, 01125 (2016), DOI: 10.1051/matecconf/20167201125

 DOI: 10.1051/matecconf/201714101025

Decision working medium for lower cycle CCGT trinary type

Sviatoslav Tsibulskiy[1,*], *Nikolay* Galashov[1], *Al'bina* Bannova[1], *Denis* Melnikov[1], and *Alexander* Kiselev[1]

[1]National Research Tomsk Polytechnic University, 634050 Tomsk, Russia

Abstract. The object of the study are low-boiling substances, which can be used as working fluid in the Organic Rankine cycle (ORC) combine cycle gas turbine (CCGT) trinary type. Purpose of research is search for substances with the best thermodynamic, thermophysical and ecological properties. The calculation algorithm represents a mathematical calculation of CCGT in which the lower cycle, operating on low-boiling fluid and utilizes low-potential heat, as an air-cooling device On the basis of available scientific and practical facts on the use of working fluids the organic Rankine cycle for analysis were selected ozone-safe refrigerant: butane, pentane, R236ea, R236fa, R123, R245ca, R245fa, R365mfc, RC318. For these substances were found and tabulated critical pressure and temperature, condensation pressure, density and dynamic viscosity at a given condensation temperature. Studies have shown that the highest thermal net efficiency is achieved when using natural Freon pentane and synthetic R365mfc. Net efficiency CCGT strongly depends on the condensation temperature of the substance. For example, when using pentane when the condensation temperature decreases on 10 °C net efficiency increases about 0.71 %.

1 Introduction

The most perfect are recycling three-circuit CCGT with intermediate steam superheat on the based the most powerful gas turbine unit (GTU) General Electric GE 9HA.02 with nominal electric power 544 MW and net efficiency 43.9 % can achieve summary net efficiency 63.5 % when arrangement 1GTU+1STU (steam turbine unit) or 63.7 % when arrangement 2GTU+1STU [1]. At the same time, all the possibilities of improving the CCGT equipment were used, and when consume water steam as a working fluid in the steam turbine plant, further increase in the efficiency of the CCGT is practically impossible. This is due to the fact that the heat removal temperature in the condenser during the operation of the STU with water vapor is limited by the freezing point of water – 0 °C, due to economic conditions, is usually not lower 15 °C.

* Corresponding author: s.tzibulsky@yandex.ru

DOI: 10.1051/matecconf/201714101025

In the north regions, where the air temperature for a long time is lower 0 °C, it is possible to lower the heat removal temperature in the cycle using instead of water air capacitors (AC), and as a working fluid to use substances, that do not freeze at low temperatures, which are used in the organic Rankine cycle (ORC). But these substances are destroyed at temperatures above 400 °C. Therefore, a CCGT of a trinary type was proposed on the basis of three cycles: higher – Brighton gas turbine cycle, average – Rankine cycle steam turbine and lower – ORC [2-6]. The use of low-boiling substances (LBS) in ORC makes it possible to reliably remove heat in the AC at the temperature below 0 °C [7], allows downsizing and installation costs of the turbine and condenser, and also makes it possible to significantly reduce water consumption for technological purposes.

The ecological purity of Freon is determined by the potential for ozone depletion (ODP) and the potential of global warming GWP.

2 Selection of substances and analysis based on the REFPROP

In this paper, for the analysis and selection of the LBS for the ORC use the base "REFPROP" [8], which allows us to calculate the thermodynamic and thermophysical properties of most substances used in production processes.

At present, low-boiling substances are divided into 4 groups: ozone-depleting chlorofluorocarbons and hydrochlorofluorocarbons, transient hydrofluorocarbons, ozone-friendly hydrofluorocarbons and natural.

For research based on the analysis of available publications [9–10] and the requirements for working fluids the following Freon were chosen: pentane, butane, R365mfc, RC318, R236ea, R236fa, R123, R245ca, R245fa. These Freon have a low potential for ozone depletion. The remaining Freon are not flammable, but they have a significant GWP. All the selected Freon have a positive slope of the saturated vapor boundary curve, so the process in the turbine will pass in the superheated steam and after the turbine it is required to install the desuperheater.

For selected substances in the table 1–2 are given the thermophysical and thermodynamic parameters determined on the basis of the "REFPROP", where designation of parameters following: M – molecular mass; t_{TP} – temperature of triple point; t_0 – temperature boiling point at atmospheric pressure 98 kPa; t_{CR} – critical temperature; P_{CR} – critical pressure; ρ_{CR} – critical density; ξ – the value of the derivative dT/dS on the saturated vapor line; GWP – global warming potential.

Table 1. Properties of ozone-safe refrigerant.

Refrigerant	M, g/mol	t_{TP}, °C	t_0, °C	ρ_{CR}, mol/l	ξ, J/(kg·K^2)	GWP
R236ea	152	-	6.5	3.70	1.39	1200
R236fa	152	-93	-1	3.70	0.76	8000
R123	153	-107	28	3.60	2.76	90
R245ca	134	-	25	3.91	1.38	610
R245fa	134	-102	15	3.85	0.19	950
R365mfc	148	-34	40	3.20	0.86	<1500
RC318	200	-40	-7	3.10	1.12	9100
butane	58	-138	-1	3.92	1.03	3
pentane	72	-130	36	3.22	1.51	11

The lowest value of the global warming potential is possessed by natural Freon: pentane and butane, further along the increasing go: R123, R245ca, R245fa, R236ea, R365mfc, R236fa and RC318. All considered Freon have a triple point temperature lower -30°C, which determines their reliable operation in the winter and the possibility of heat removal in the air condenser at very low temperatures. All Freon have ξ close to 0, which ensures their operation in the turbine without loss of moisture.

High temperature of boiling points (t_0) at atmospheric pressure have following Freon (in decreasing order): pentane, R365mfc, R123, R245ca. This indicator characterizes below what temperature in the condenser there is a vacuum.

A number of operational parameters for selected substances are given in the table 2. P_{CR} and t_{CR} – critical properties steam of Freon; P_C – condensation pressure at temperature 15 °C; ρ, λ – density and dynamic viscosity of liquid at temperature 15 °C.

Table 2. Operational characteristics of refrigerant.

Refrigerant	P_{CR}, MPa	t_{CR}, °C	P_C, MPa	ρ, kg/m^3	λ, mPa·s	$\rho\cdot\lambda\cdot10^5$
R236ea	3.50	139.29	0.143	1455.78	440.90	6.42
R236fa	3.20	124.85	0.101	1392.75	324.29	4.52
R123	3.66	183.68	0.062	1489.18	469.61	6.99
R245ca	3.925	174.42	0.0676	1411.07	621.54	8.77
R245fa	3.651	154.01	0.1007	1365.27	464.27	6.34
R365mfc	3.266	186.85	0.0372	1278.10	463.22	5.92
RC318	2.777	115.23	0.224	1537.76	420.39	6.46
butane	3.769	152.00	0.18	584.25	174.39	1.02
pentane	3.370	196.55	0.05	630.56	237.45	1.50

The highest values of critical temperatures are pentane, R365mfc, R123.

The lowest values of viscosity-density characteristics have (with increasing): butane, propane, R236fa, R365mfc, R245fa, R236ea, RC318, R123, R245ca.

3 Investigation of the thermal scheme of a combined-cycle plant

For research on the selection of Freon for the ORC was developed the basic thermal scheme of a combined-cycle plant of the trinary type (fig.1).

The installation works as follows. In the GTU cycle, compressed air in the compressor enters the combustion chamber, where fuel is supplied (gas). Combustion products are sent to the gas turbine, the flow passes through the part where the commit operation. Then the combustion products are sent to the waste heat boiler, where heat is given to water and Freon. Steam-water mixture is formed in the evaporator of the waste heat boiler. Saturated steam separates from the mixture in the blow case. Steam after overheating in a superheater with temperature t_0 enters the high-pressure cylinder (HPC) of the steam turbine (ST) and performs work on the shaft. The steam after the HPC is enters to the condenser -evaporator (CE), where it transfers heat to the Freon and condenses to the saturation state of the liquid. The water condensate is fed by a feed pump (FP1) through the economizer (EC1) into the drum and the steam-water cycle closes. Freon steam from CE enters in low pressure cylinder (CLP) turbine, performs work and, through a desuperheater (DH), is condensed into an air condenser (AC). LBS condensate by condensate pump (CP) is fed through a DH to heat its own stream of superheated steam into a regenerative heater (RH). In the RH receives Freon steam from the regenerative bleeding off LPC and heats the condensate to the required temperature 60 °C. Heated condensate with a feed pump (FP2) is pumped through an economizer (EC2), where it is heated by combustion products. The condensate is fed into the CE of ORC and the cycle is closed. Exhaust gas are diverted to the chimney. The work received through turbines GTU and STU through the shaft is transmitted to the power generator, where electricity is generated.

DOI: 10.1051/matecconf/201714101025

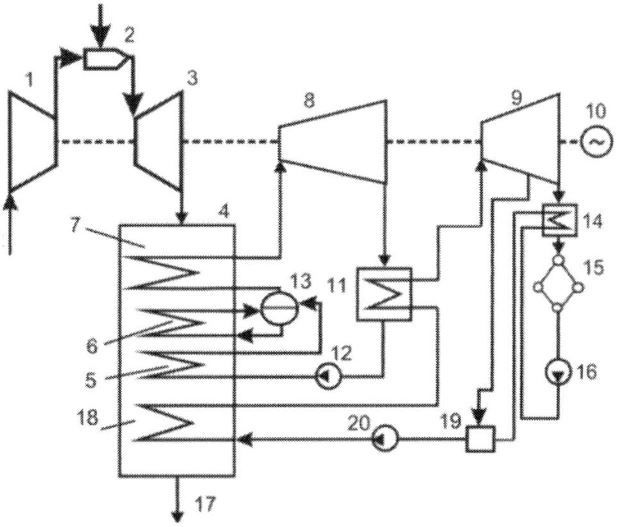

Fig. 1 – The basic thermal scheme of a combined-cycle gas turbine of the trinary type: 1 – basic air compressor (AC), 2 – combustion chamber (CC), 3 – gas turbine (GT), 4 – heat-recovery steam generator (HRSG), 5 – water economizer (EC1), 6 – evaporator (E), 7 – superheater (SH), 8, 9 – high and lower pressure cylinders HPC and LPC, 10 – electric generator (EG), 11 – condenser-evaporator (CE), 12 – feed pump (FP1), 13 – blow case (B), 14 – desuperheater (DH), 15 – air condenser (AC), 16 – Freon condensate pump (CP), 17 – exhaust gases, 18 – Freon economizer (EC2), 19 – regenerative heater (RH), 20 – Freon feed pump (FP2).

4 Analysis of the results of the algorithm for calculating the thermal diagram on the bases REFPROP

Based on this diagram (fig. 1) was developed mathematical algorithm for calculating the performance of CCGT of a trinary type. The algorithm is implemented by the program in an Excel spreadsheet package. Calculations of the properties of all substances in the program are performed on the basis of the functions of the database "REFPROP" [4].

The calculations are based on the following data.

Cycle GTU based on the turbine GE 9HA.02: electricity power 544 MW; net efficiency 43.9 %; gas flow rate and temperature at the entrance to the HRSG 986.89 kg/s и 636 °C; temperature of the Freon inlet in the economizer 60 °C, temperature of exhaust gases 100 °C.

Steam-water cycle: temperature and pressure at the inlet to the HPC 610 °C and 18 MPa, pressure of water steam after HPC 0.2 MPa.

ORC: Freon saturation temperature into CE 109.9 °C; Freon condensing temperature AC -20 ÷ +20 °C.

The results obtained from the studies of efficiency gross and net is dependent on the condensation temperature and Freon type in the fig. 2.

In fig. 2 that the efficiency of the CCGT depends essentially on the condensation temperature and the type of Freon. The type of Freon strongly affects the efficiency of the CCGT. This is due to the cost of own needs due to the increased density, viscosity and pressure. Of the Freon considered according to the efficiency of CCGT net, the best is pentane, in second place is R365mfc, followed by R123 and R245ca. For pentane, a decrease in the condensation temperature by 10 °C increases the efficiency of gross and net 0.71%.

The conducted studies show that when ORC on pentane or R365mfc CCGT is used, the trinary type at a condensation temperature below 0 °C will surpass the net efficiency of a three-circuit CCGT running on water.

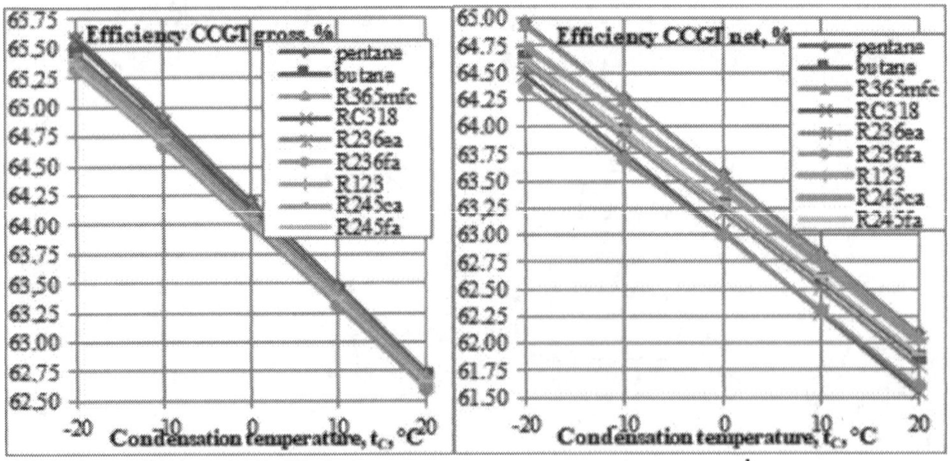

a b

Fig. 2. Dependence efficiency CCGT gross (a) and net (b) on the condensation temperature of Freon in AC.

References

1. URL: https://www.gepower.com/gas/gas-turbines/9ha.
2. N.N. Galashov, S. A. Tsybulsky, Power Technology and Engineering, **48**, 6 (2015), DOI: 10.1007/s10749-015-0555-9
3. K. Larionov, D. Gvozdjakov, A. Zenkov, V. Zaytsev, EPJ Web of Conferences, **110**, 01034 (2016), DOI: 10.1051/epjconf/201611001034
4. O. Romashova, L. Belyaev, A. Tubolev, E. Skrebatun, MATEC Web of Conferences, **91**, 01009 (2016) DOI: 10.1051/matecconf/20179101009
5. O. Romashova, A. Minor, V. Martyshev, A. Tubolev, A. Katalevskaya, MATEC Web of Conferences, **110**, 01054 (2017), DOI: 10.1051/matecconf/201711001054
6. S. Yankovskiy, A. Matveev, A. Tolokolnikov, A. Zenkov, MATEC Web of Conferences, **92**, 01036 (2016), DOI: 10.1051/matecconf/20179201036
7. N.N. Galashov, S.A. Tsibul'ski, Power Technology and Engineering, **49**, 5 (2016), DOI: 10.1007/s10749-016-0629-3
8. E.W. Lemmon, M.L. Huber, M.O. McLinden, Reference fluid thermodynamic and transport properties-REFPROP, standard reference database 23, version 8.0, National Institute of Standard and Technology (2007)
9. B.N. Maksimov, V.G. Barabanov, I.L. Serushkin, *Industrial fluororganic products: reference edition* (St. Petersburg: Chemistry, 1996)
10. Freons: types and properties. URL: http://www.newchemistry.ru/letter.php?n_id=817

Influence of $Cu(NO_3)_2$ initiation additive in two-stage mode conditions of coal pyrolytic decomposition

Kirill Larionov[1,*], *Ilya* Mishakov[1,2], *Konstantin* Slyusarskiy[1], *Darya* Bolgova[1], and *Sergey* Lavrinenko[1]

[1]National Research Tomsk Polytechnic University, 634050 Tomsk, Russia
[2]Boreskov Institute of Catalysis SB RAS, 630090 Novosibirsk, Russia

Abstract. Two-stage process (pyrolysis and oxidation) of brown coal sample with $Cu(NO_3)_2$ additive pyrolytic decomposition was studied. Additive was introduced by using capillary wetness impregnation method with 5% mass concentration. Sample reactivity was studied by thermogravimetric analysis with staged gaseous medium supply (argon and air) at heating rate 10 °C/min and intermediate isothermal soaking. The initiative additive introduction was found to significantly reduce volatile release temperature and accelerate thermal decomposition of sample. Mass-spectral analysis results reveal that significant difference in process characteristics is connected to volatile matter release stage which is initiated by nitrous oxide produced during copper nitrate decomposition.

1 Introduction

Catalytic fuel combustion is promising technology which allows to sustain low-temperature combustion regime with minimal antropogenic impact [1]. The oxides of rare-earth, transitional and alkaline metals are known to be the most widespread among catalytic agents. The main effect appearing during application of such promoters is reducing of initial temperature of volatile matter release as well as shifting fuel oxidation into low-temperature area [2-4] which has a favorable effect on the efficiency of the energy equipment [5-8]. Scientific studies reported that activation of metal-based catalytic additives during fuel oxidation appears to be most effective only at high temperatures (above 500 °C) [9-11]. In our previous studies [12] it was found that additives of metal oxide precursors in form of mineral salts are able to provide sufficient initiative impact on coal oxidation process. During discussion on possible mechanism of fuel and initiative additive interaction the hypothesis about composite observation effect was proposed. The first one is connected to release of oxide gaseous phase in the result of additive decomposition. The second one is connected to formation of disperse non-stoichiometric metal oxide which plays role of catalyst and accelerates oxidation process. To prove composite effect of initiated combustion with salts and their bases the process was proposed to be divided into two parts: volatile matter release during additive decomposition and

* Corresponding author: laryk070@gmail.com

further carbon oxidation at the presence of formed non-stoichiometric metal oxide. In current study, the results of research on kinetic characteristics of two-stage thermal decomposition of modified by $Cu(NO_3)_2$ additive brown coal were presented.

2 Experimental setup and study technique

The brown coal of Borodinskoe deposit from Cansk-Achinsk coal-basin was used as initial fuel. Rough-disperse samples (d=5:10 mm) was taken and ground using ball mill at following conditions: milling bodies to material mass ratio was 1:1, milling time was 9 hours. Later samples were undergo fractionation. Physical properties of obtained sample were defined using known methodic and given in table 1.

Table 1. Physical properties of studied sample

Index	Unit of measure	Value
Ash content	wt.%	5.2
Moisture content		3.0
Volatile compounds		38.8
Carbon		55.0
Humidity ratio	ml/g	3.2

As it can be seen from table 1, studied sample has a high content of volatile substances and a low content of the mineral residue, which is explained by its stage of metamorphism [13, 14].

The initiating additive $Cu(NO_3)_2$ was dissolved in a water-alcohol solution in a volume ratio H_2O / C_2H_5OH = 50:50 and applied to the obtained sample by capillary impregnation method with respect to the moisture capacity. The content of the additive into the modified sample was 5 wt.%. Resulting sample was dried into a drying oven at 105 °C temperature for 20 hours.

For comparative analysis, a reference sample with no initiating additive was prepared and subjected to exactly same treatment procedures.

The study of the stepwise process of thermal decomposition of coal was carried out using a synchronous thermal analyzer STA 449 C Jupiter (Netzsch, Germany). The experimental regime is schematically shown in figure 1.

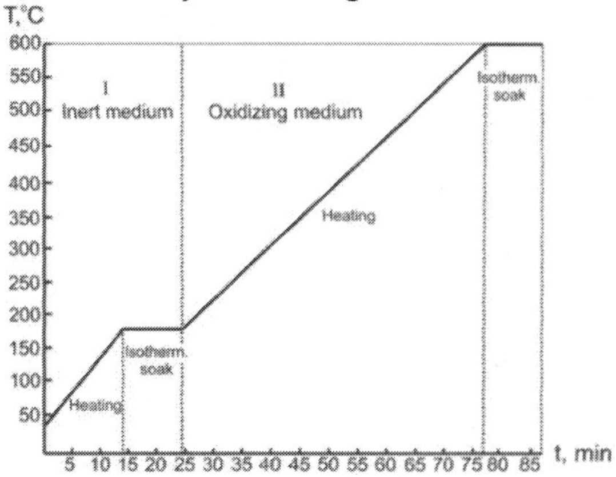

Fig. 1. Scheme of the experimental regime of synchronous thermal analyzer.

Stage I was characterized by decomposition of copper nitrate and the active release of volatiles from coal (pyrolysis). It was carried out in an inert atmosphere (argon) with a 60 ml/min gas flow rate at 10 °C/min heating rate in temperature range from 40 °C to 180 °C (it is temperature corresponding to the decomposition of copper nitrate) and was followed by isothermal soaking for 10 minutes. Stage II includes heating with the subsequent oxidation of the samples in stream of air (60 ml / min) and argon (10 ml / min) with same heating rate (10 °C/min) to 600 °C. After reaching the maximum temperature, the samples were kept in isothermal soaking mode for 10 minutes. All experiments were carried out under ambient pressure. The sample weight was about 8 mg.

To determine the qualitative composition of the gaseous products (NO_x m / z = 30) of sample pyrolysis and oxidation at the outlet of thermal analyzer the mass spectrometric analysis was performed using a QMS 403 D Aeolos (Netzsch, Germany) interfacing quadrupole mass spectrometer.

3 Experimental results

The TG and DTG-curves for pyrolysis (figure 2a) and oxidation (figure 2b) stages are given in figure 2.

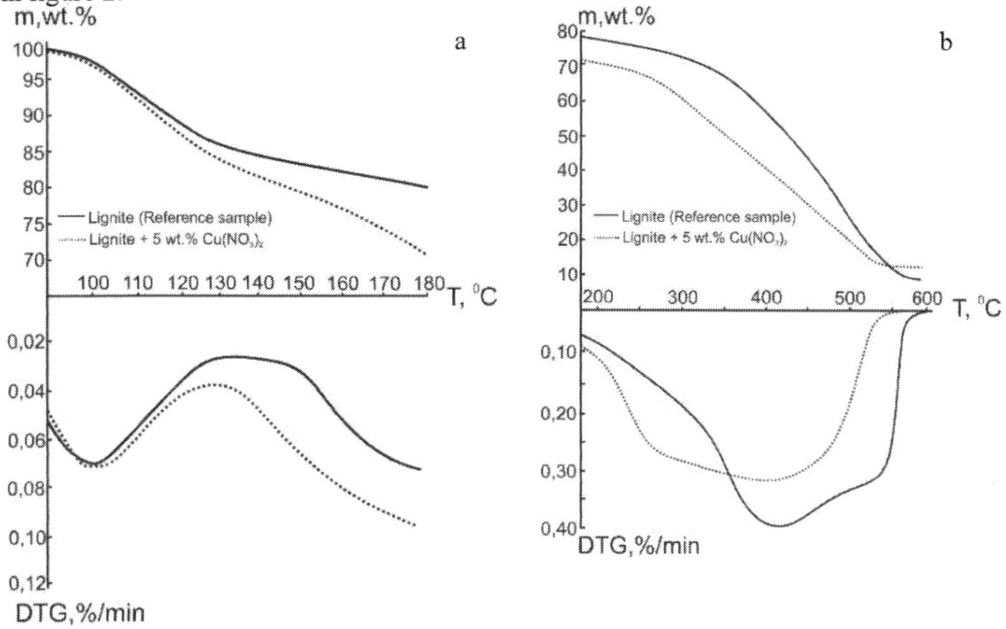

Fig. 2. TG and DTG-curves of pyrolysis (a) and oxidation (b) processes for studied samples.

From the obtained results it could be seen that the introduction of $Cu(NO_3)_2$ initiating additive into the coal sample resulted into decreasing og the initial temperature of volatile matter release (figure 2a) and an increasing of the average reaction rate of oxidation process (figure 2b). Comparing the obtained results of mass-spectrometric analysis of NO_x emission (m/z=30) (figure 3), it can be seen that the main effect of the initiating additive appears during pyrolysis process. It is accompanied by the release of nitric oxide as a result of the copper nitrate decomposition in the temperature range of 140-180 ° C (figure 3a)

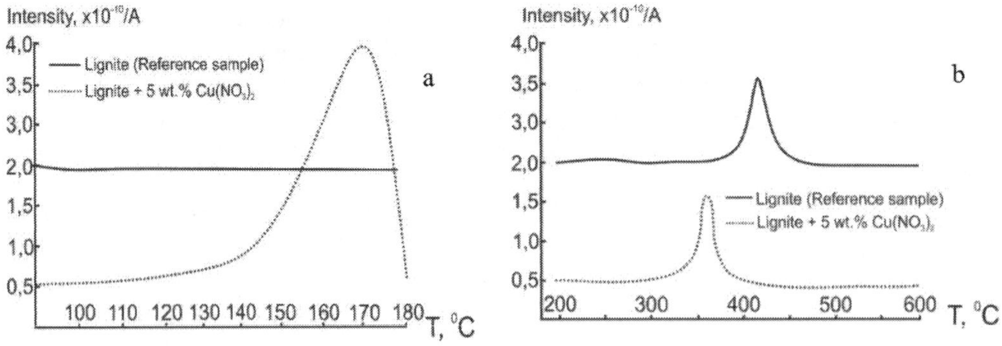

Fig. 3. MS-profiles of NO_x release for pyrolysis (a) and oxidation (b) processes.

The second peak of NO_x release is less intense and connected to formation of nitrogen oxides in the flame regime due to the involvement of molecular nitrogen N_2 in the combustion process.

Considering the possible mechanism of initiating additive action during coal pyrolysis and oxidation it is worth mentioning a number of key chemical reaction equations:

$$Cu(NO_3)_2 \xrightarrow{T,°C} CuO + 2NO_2 + O_2 \tag{1}$$

$$C + NO_2 = CO + NO \tag{2}$$

$$2CO + 2NO_2 = 2CO_2 + N_2 \tag{3}$$

$$NO + \frac{1}{2}O_2 \rightleftarrows NO_2 \tag{4}$$

$$C + O_2 \xrightarrow{CuO_x} CO_2 \tag{5}$$

For the first stage (figure 2a), heating of copper nitrate (up to 180 ° C and above) promotes its decomposition according to equation (1). At the same time, according to the MS analysis data, decomposition of the introduced $Cu(NO_3)_2$ salt begins at lower temperature (140 °C). Nitric oxide decomposition produces NO and NO_2 nitrogen oxides. The NO_2 acts as a strong oxidant by promoting the combustion of carbon and volatile compounds (equations 2 and 3). The interaction of coal with nitrogen dioxide is accompanied by the formation of NO, which is easily oxidized by air oxygen according to reversible reaction (4).

Upon completion of the $Cu(NO_3)_2$ decomposition process, a dispersed non-stoichiometric copper oxide CuO_x is formed. The high catalytic activity of copper oxide for organic substrate and fuel complete oxidation of is well known. It makes it possible to consider the further participation of the formed CuO_x particles as a catalyst for brown coal oxidation acceleration (equation 5).

4 Conclusion

The study on the two-stage (pyrolysis and oxidation) thermal decomposition of the brown coal modified by the copper nitrate additive revealed that the main effect of changing process temperature regime is concentrated in the temperature range of volatile substance sublimation during active decomposition of copper nitrate. The initiation of volatile matter sublimation occurs as a result of the released NO_x oxidizing properties. In turn, it was found that the resulting non-stoichiometric copper oxide also affects the subsequent process of sample carbon residue oxidation.

The work was carried out with financial support of Minobrnauki RF in framework of project No. 13.7644.2017/BCH.

References

1. D.V. Gvozdyakov, V.E. Gubin EPJ Web Conferences, **82**, 01026, (2015)
2. A.N. Kozlov, D.A. Svishchev, G.I. Khudiakova, A.F. Ryzhkov, Solid Fuel Chem., **51**, 205 (2017)
3. A.A. Dekterev, P.V. Osipov, M.Y. Chernetskiy, A.F. Ryzhkov, Solid Fuel Chem., **51**, 17 (2017)
4. M.Y. Chernetskiy, V.A. Kuznetsov, A.A. Dekterev, N.A. Abaimov, A.F. Ryzhkov, Thermophys. Aeromech., **23**, 591 (2016)
5. N.N. Galashov, S.A. Tsibulskii, Power Techn. Eng., **48**, 459 (2015)
6. A.I. Gabdullina, N.N. Galashov, S.A. Tsibulskiy, I.A. Asanov, A.S. Kiselev, MATEC Web Conferences, **91**, 01004 (2016)
7. O.Y. Romashova, D.S. Luzhkovoj, A.A. Tubolev, M.A. Vagner, MATEC Web Conferences, **110**, 01073 (2017)
8. D.S. Luzhkovoy, O.Y. Romashova, A.A. Tubolev,Y.V. Kobenko, E.S. Riabova, MATEC Web of Conferences, **92**, 01054 (2016)
9. G.V. Kuznetsov, A.E. Nee, Therm. and Aerom., **23,** 393 (2016)
10. N.V. Baranovskii, A.E. Ni, J. Eng. Phys. Therm. **87**, 1438 (2016)
11. A.E. Nurpeiis, E.G. Orlova, K.O. Ponomarev, MATEC Web Conf. **110**, 01062 (2017).
12. K.B. Larionov, I.V. Mishakov, A.A. Gromov, A.V Zenkov, MATEC Web Conf., **91**, 01007 (2016)
13. S.A. Yankovskiy, A.S. Matveev, A.A. Tolokolnikov, A.V. Zenkov, Matec Web Conf., **92**, 01036 (2016)
14. S.A. Yankovsky, A.A.Tolokolnikov, A.A Matveev, Y.V. Marysheva, Matec Web Conf., **110**, 01089 (2017)

MATEC Web of Conferences **141**, 01027 (2017) DOI: 10.1051/matecconf/201714101027
Smart Grids 2017

Potential of composite fuel application based on low-grade coal and wastes of forest and flour industry

Stanislav Jankovsky[1,*], *Anton* Tolokolnikov[1], and *Konstantin* Slyusarskiy[1]

[1]National Research Tomsk polytechnic university, 634050 Tomsk, Russia

Abstract. Composite fuels based on low-grade brown coal of 3B-grade mixed with forest and flour industry wastes were studied. Three compositions with similar weight content of forest industry waste and low-grade coal as well as flour industry waste in concentration range from 0 to 15 wt. %. The chemical properties and ignition characteristics of samples were studied by means of TG-analysis at air medium and 10 K/min heating rate in temperature range 50-600 °C. The samples with 15 wt. % content of flour waste were found to have higher burnout degree (up to 21.17 %) and mechanical endurance while having lower heating value (up to 6 %) and higher ignition temperatures (up to 24 °C) compared to samples with no flour industry wastes.

1 Introduction

Production of electric energy is known to annually increase [1] as a result of growing population and development of industrial capacities. The analysis of the last 15 years revealed a decrease in the share of electricity generated by nuclear power and petroleum products [2]. However, slight increase was observed for hydro energy and renewable sources. At the same time, the power produced by gas-fired thermal power plants has significantly increased while coal power plants have shown the maximum increase up to 46% [1]. In the Russian Federation, more than 26% of power plants operate on coal [3, 4]. For majority of those thermal power plants, coal is imported fuel. The remoteness from coal deposits increases the cost of energy resources by 2 or more times for final consumer. When this type of fuel is burned, significant emissions of polluting anthropogenic substances occur, such as oxides of sulfur, carbon, nitrogen, fly ash and other [5, 6]. Modern methods of harmful emission reducing are effective but with improper and inadequate maintenance, the efficiency of the filters used is lowered by more than 50% [7]. It leads to significant increasing of heat and power generation costs [7].

Possible solution for this problem may be a less expensive and more environmentally friendly composite fuel based on coal and sawmill waste (fine pine wood) [7]. The effectiveness of such fuels is achieved due to the carbon-neutrality of wood. It is the most common substance on the planet and its processing creates waste at colossal volumes. There is a

* Corresponding author: jankovsky@tpu.ru

problem of storage and subsequent transportation of such composite fuels [7]. The production of fuel briquettes is often carried out by thermal method with the addition of binders.

But they significantly increase the cost of final product. Reducing cost of such fuels can be achieved by adding flour waste to improve the strength characteristics of coal and wood based fuel mixtures. The relevance of study in this field is significant and is supported by the energy strategy of Russia to 2035 [8].

2 Experimental section

For experimental studies 3B-grade coal of the Balakhtinsky deposit and waste products of timber produced by OOO Dzerzhinsky LPK, Tomsk, were used. Preparation of samples was carried out according to GOST 10742-71 with subsequent sieving according to GOST 3306-88. The particle size of the investigated coal samples was less than 80 μm, particle size of wood and flour were less than 200 μm [9].

Three different compositions of fuel were prepared with different mass concentrations of components (Table 1). A mixture of coal and wood particles was placed into a galvanized drum with 5 mm diameter spherical grinding bodies of Pulverisette 6 mill at 1:1 weight ratio. The mixing process was carried out at 500 rpm rotation frequency for 7 minutes [9].

Table 1. Composition of studied sample.

Sample	Volumetric fraction, %			
	a	b	c	d
Brown coal 3B	47.5	45.0	42.5	50.0
Fine wood waste	47.5	45.0	42.5	50.0
Flour waste	5	10	15	0

To form a composite fuel, a manual mechanical press was used. The obtained fuel samples are shown in fig. 1.

Fig. 1. Composite fuel appearance. Brown coal content to timber and flour waste ration:
a) 47.5/47.5/5.0; b) 45.0/45.0/10.0; c) 42.5/42.5/15.0.

Technical analysis of the obtained samples of solid mixed fuels was carried out. [10, 12]

Thermogravimetric analysis was carried out to obtain experimental data on the degree of fuel burnup and the energy potential of solid mixed fuels [11, 12] using a combined TG-DSC analyzer Netzsch STA 449 F3 Jupiter (Netzsch, Germany). Samples with 20 mg mass were used for experiments. The heating rate was 10 K/min and the maximal temperature was 1373 K [13, 14].

DOI: 10.1051/matecconf/201714101027

3 Results and discussion

A series of experiments was carried out to obtain experimental curves using differential thermal gravimetry. Obtained curves of DTG and DSC are presented in fig. 2 and 3.

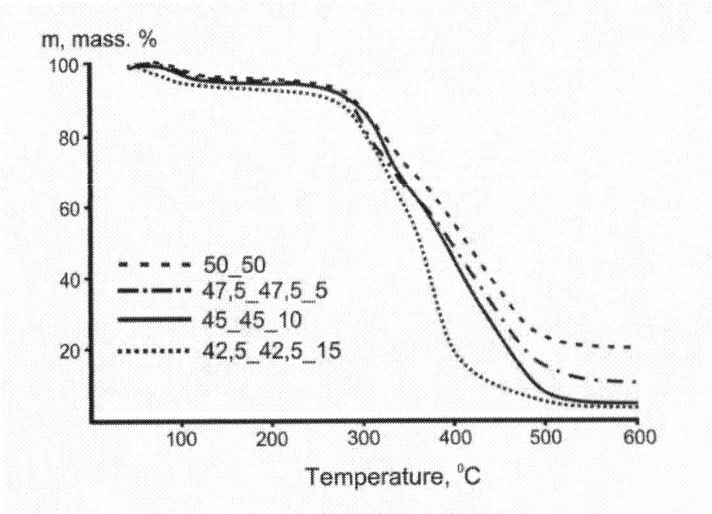

Fig. 2. TG curves of composite fuels with the different ratio of the main components (coal / wood / flour waste).

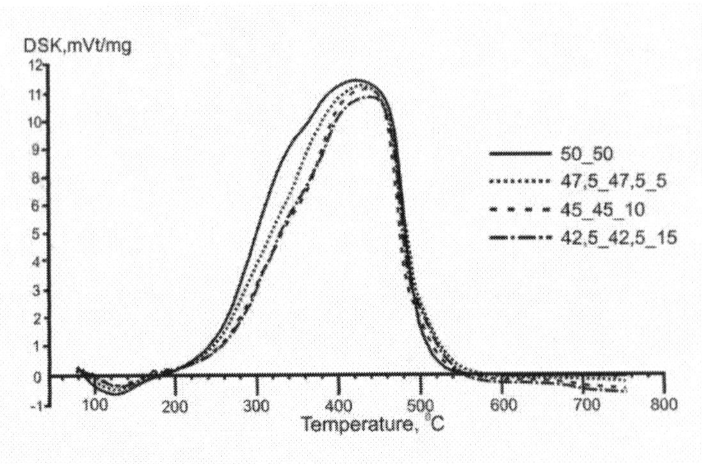

Fig. 3. DSC curves of composite fuels with the different ratio of main components (coal / wood / flour waste).

Analysis of fig. 2 allows to determine that with the increase of the third component (flour waste) from 0% to 15%, the mixture burn up degree increases from 80.3% to 97.3%. This effect appears due to decrease in the carbon content and an increase in the yield of volatiles in the composite fuel. The characteristic temperature of 10 % sample mass loss for samples without flour content is higher than the corresponding characteristic temperature for the sample with 15% of flour waste by 24 °C and is 292 ° C.

DSC-curves in fig. 3 allow estimating energy release of the composite fuels. With an increase in the flour waste content to 15% in composite fuels the energy release is reduced by less than 6%. The temperature of the thermal decomposition beginning remain at the level of the initial fuel based on coal and wood (without flour waste) as well as temperature of complete thermal decomposition is stays in same temperature range.

4 Conclusion

Composite fuels based on low-grade 3B brown coal of Balakhtinsky deposit with addition of fine-dispersed wood and flour industry waste were investigated. It was found that the addition of 15 wt.% of flour waste to initial fuel based on coal and wood improves strength properties of fuels and increases the degree of mixture burnup up to 97,3 %. However, such increase in the third component leads to a decrease in heat release by 6%. It also moves temperature range of fuel thermal decomposition process into higher temperature area. The investigated fuel mixtures appear to have improved strength characteristics with lower energy spends for their preparation. They are also more environmentally friendly compared to coal. The results of the analysis make it possible to conclude that the using of composite fuels based on coal and wood with the addition of flour waste for combustion in the layer furnaces of boiler installations is promising.

The work was carried out with financial support of Minobrnauki RF in framework of project №13.7644.2017/BCH.

References

1. Dr. Fatih Birol *Key World energy statistics* (IEA, 2016)
2. E. Pavelyev, S. Lavrinenko, MATEC Web of Conf., **72**, 01085 (2016)
3. S. Lavrinenko, P. Polikarpov, A. Matveeva, V. Martyshev, MATEC Web of Conf., **110**, 01049 (2017)
4. A.I. Gabdullina, N.N. Galashov, S.A. Tsibulskiy, I.A. Asanov, A.S. Kiselev, MATEC Web Conf., **91**, 01004 (2016)
5. N.V. Baranovskiy, A.E. Nee, EPJ Web Conf., **82**, 01004 (2014)
6. N.V. Baranovskii, A.E. Ni, J.Eng.Thermophys., **87**, 1438 (2014)
7. A.E. Nurpeiis, E. Orlova, K. Ponomarev, MATEC Web of Conf., **110**, 01062 (2017).
8. A.N. Kozlov, D.A. Svishchev, G.I. Khudiakova, A.F. Ryzhkov, Solid Fuel Chem., **51**, 205 (2017)
9. S. Jankovskiy, A. Tolokolnikov, V. Zaytsev, MATEC Web Conf., **110**, 01080 (2017)
10. I.A. Dolgushin, Research and modernization of TPP scheme with CFB boiler to improve its efficiency and ecological characteristics (Moscow, 2014)
11. S.A. Jankovskiy, G.V. Kuznetsov, SPbPU scientific and technical bulletin. Fundamental and engineering science, **23**, 29 (2017)
12. K. Larionov, I. Mishakov, A. Gromov, A. Zenkov, V. Glaktionov, MATEC Web Conf., **110**, 01048 (2017)
13. K.B. Larionov, I.V. Mishakov, A.A. Gromov, A.V. Zenkov, MATEC Web Conf., **91**, 01007 (2016)
14. A.A. Dekterev, P.V. Osipov, M.Y. Chernetskiy, A.F. Ryzhkov, Solid Fuel Chem., **51**, 17 (2017)
15. M.Y. Chernetskiy, V.A. Kuznetsov, A.A. Dekterev, N.A. Abaimov, A.F. Ryzhkov, Thermophys. Aeromech., **23**, 591 (2016)

MATEC Web of Conferences **141**, 01028 (2017) DOI: 10.1051/matecconf/201714101028
Smart Grids 2017

Research of impact of kind resuperheat and structure of system regenerative feed water to thermodynamic efficiency of cycle with steam-coolant reactor

Svetlana Maykova[1,*], *Vassily* Borisov[1], *Aleksandra* Antonova[1], *Aleksandr* Vorobiev[1], and *Alexey* Abramovskikh[1]

[1]National Research Tomsk Polytechnic University, 634050 Tomsk, Russia

Abstract. The first key problems of modern nuclear reactors are inability of closed nuclear cycle, problems with spent nuclear fuel, poor effectiveness of nuclear fuel and heat-exchange equipment usage. Dealing with problems consists in usage of fast-neutron reactors with steam coolant. Scientific men analyzed neutron-physical processes in steam-cooled fast reactor and consulted that creation of the reactor is viable. In consequence of low steam activation a single-loop steam cycle may be create. The cycle is easy and fool-proof. Core thermomechanical equipment has mastered and has relatively low metal content. Results of calculation are showing that nuclear unit with steam-coolant fast neutron reactor is more efficient than widely used unit with reactor VVER. Usage of simple scheme with four regenerative feedwater heaters the absolute efficiency ratio is more than 43%.

1 Introduction

Currently, work is underway to improve the efficiency of power plants in General and its separate elements [1-4]. For analyze of thermodynamic efficiency of the single-loop nuclear power plant unit with steam-cooled fast reactor we must to calculate of plant cycle arrangement with variable types of resuperheat and contrasting characteristics and structure of regenerative feedwater heating.

Active zone of the reactor is cooled by dint of superheated steam. Feed water enters in ejectors with the aid of feedpumps. Ejectors are necessary for dispersal of feed-water and mixed it with steam from active zone [1]. Slightly overheated steam enters in reactor active zone where the steam is heated up to high temperature. At reactor outlet the overheat steam is divided into two parts: the first part is going to ejectors as coolant and the second part is sent to a turbine. There are different kinds of steam resuperheat.

1.1 Bled steam reheating

* Corresponding author: anton@tpu.ru

Steam from reactor comes in high-pressure casing, depresses to 10 MPa and then it is divided to two streams: the first stream comes in reheated like as coolant, condense and is sent to feedwater system, the second stream depresses to output pressure of the high-pressure cylinder, comes in reheated like as heated medium and then comes in low-pressure cylinder. After reheater the main stream has temperature less then temperature at saturation point by superheater approach. However, extra-work in turbine can be obtained into the reheating scheme. Thermal efficiency grows at 1-2%.

1.2 Superheated steam reheating

Steam from reactor is divided to two streams: the first stream comes in reheated like as coolant, condense and is sent to feedwater system, the second stream comes in high-pressure casing, depresses to output pressure of the high-pressure cylinder, comes in reheated like as heated medium and then comes in low-pressure cylinder. After reheater the main stream has temperature less then temperature at saturation point by superheater approach. Nevertheless, heat efficiency is bigger (2-3%) then in (a) because the coolant has a higher pressure.

1.3 "Gas" reheating

Steam from reactor is comes in high-pressure casing, depresses to output pressure of the high-pressure cylinder and then the steam again comes in reactor, where temperature is rising to inlet temperature. After that the main stream comes in low-pressure cylinder. The kind of reheating has main advantage- high heat efficiency, but the scheme has disadvantages: very long pipelines, energy dissipation in the pipelines, complex reactor design.

2 Method

Analysis is based on method of relative discharges [3], where steam discharge for turbine is equal 1.

Method of the relative flow rate: flow rates of steam, condensate water and water are found like a part of steam flow rate to turbine which is found from energy equation for turbine.

Allowances:

- hydraulic resistance of pipelines is absent;
- regenerative feedwater heaters are open heaters;

Effect of final moisture content is made provision.

Table 1. Design conditions.

Conditions	Magnitude
Initial pressure P_0, MPa	13
Initial temperature t_0, °C	555
Condenser backpressure p_k, MPa	0.0045
Steam temperature after reheater, t_{rh}, °C:	
Bled steam reheating	296
Superheated steam reheating	316
"Gas" reheating	555
Efficiency ratio of turbine for superheated steam area, η_{oi}	0.87
Maximum allowed final moisture content, %	10

 DOI: 10.1051/matecconf/201714101028

3 Results

At the result of calculation characteristic curves of thermodynamic efficiency have plotted:

1. Characteristic curves of absolute efficiency ratio of cycle from output pressure of the high-pressure cylinder.
2. Characteristic curves of final moisture content from output pressure of the high-pressure cylinder.
3. Characteristic curves of absolute efficiency ratio of cycle from number of regenerative feedwater heaters.

3.1 Characteristic curves of absolute efficiency ratio of cycle from output pressure of the high-pressure cylinder

The calculation was made for pressure range from 0.3 MPa to 1.35 MPa with step 0.15 MPa (fig. 1).

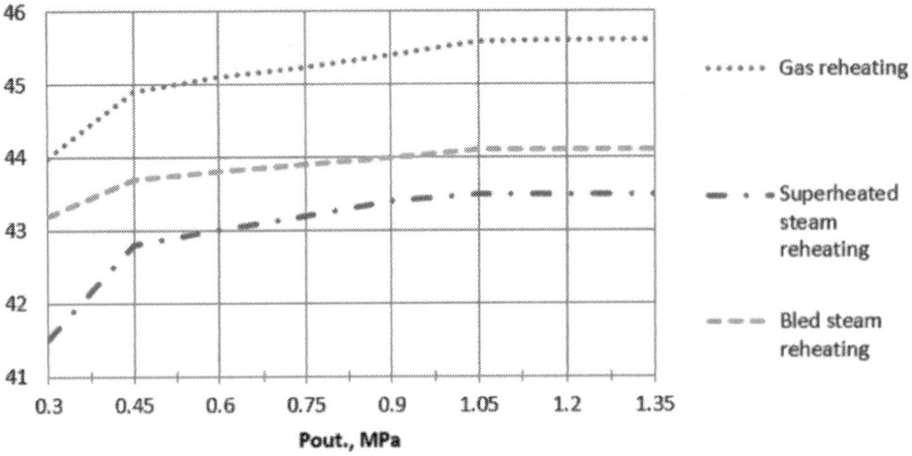

Fig. 1. Characteristic curves of absolute efficiency ratio of cycle from output pressure of the high-pressure cylinder.

The characteristic curves have nonlinear dependence. The most intensive increase of absolute efficiency ratio is in range from 0.3 MPa to 0.75 MPa. This is because with increasing of output pressure of the high-pressure cylinder heat amount is decreased (final temperature of resuperheat is constant). Post increase of the pressure is ineffective.

DOI: 10.1051/matecconf/201714101028

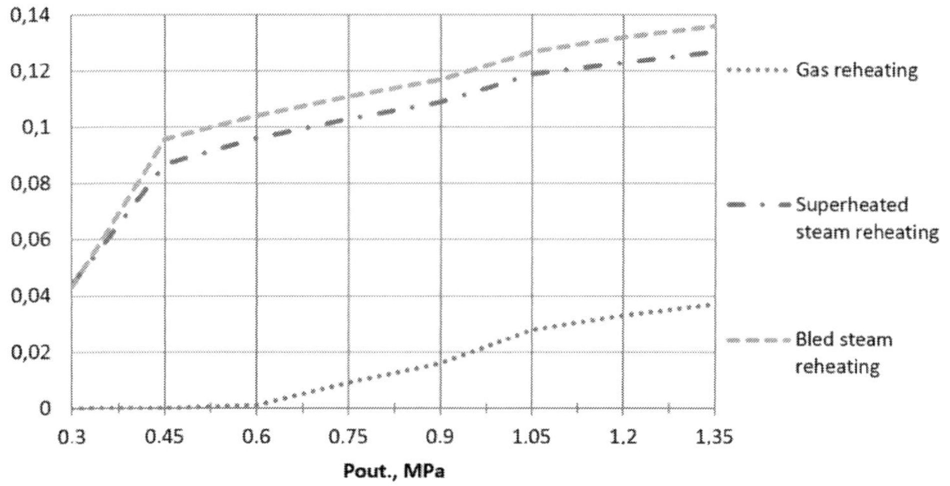

Fig. 2. Characteristic curves of final moisture content from output pressure of the high-pressure cylinder.

The characteristic curves have nonlinear dependence. With increasing of output pressure of the high-pressure cylinder, heat efficient of cycle is growing but final moisture content is increased too. The cycles have maximum efficiency at the pressure 1.35 MPa, but final moisture content is more than maximum allowed final moisture content.

3.2 Characteristic curves of absolute efficiency ratio of cycle from number of regenerative feedwater heaters

For impact assessment of the regenerative feedwater heating system structure the calculation was made for different number of regenerative feedwater heaters (in range from 4 to 10) for the output pressure of the high-pressure cylinder 0.75 MPa. The pressure was taken from condition of allowed final moisture content for different types of resuperheat. Thermodynamic optimum temperature of feed water was calculated with help the equation (10) and magnitude of the temperature varied in range from 207 to 303 °C.

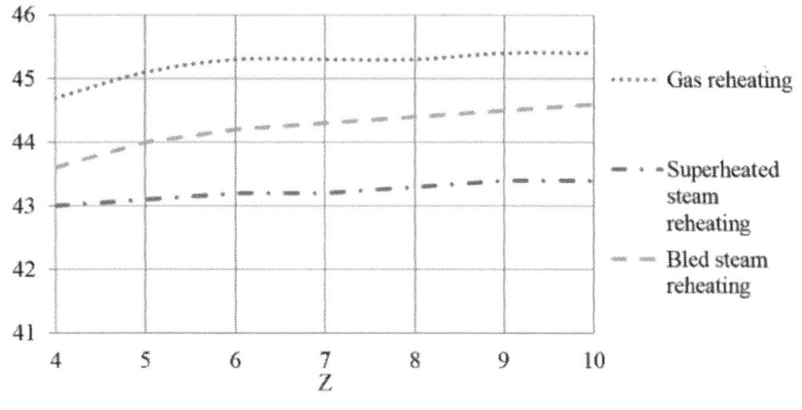

Fig. 3. Characteristic curves of absolute efficiency ratio of cycle from number of regenerative feedwater heaters.

The characteristic curves have nonlinear dependence. For "Gas" reheating and bled steam reheating the most intensive increase of absolute efficiency ratio (0.6% abs.) is in range from 4 to 6 regenerative feedwater heaters. Every next heater gives a very small increase of efficiency.

For superheated steam reheating increasing of heater number from 4 to 10 give a small increase of absolute efficiency ratio (0.4% abs.).

4 Conclusion

Results of calculation are showing that nuclear unit with steam-coolant fast neutron reactor is more efficient than widely used unit with reactor VVER. Usage of simple scheme with four regenerative feedwater heaters the absolute efficiency ratio is more than 43%.

However, question about range of turbine power is open. The large value of absolute flow rates at the turbine create necessaire to calculate of admissible value range of power [2].

Goals of next experiment are size range of equipment, calculation of ejector and impact assessment of deaerator plant to thermodynamic efficient of cycle. "

References

1. D. Gvozdyakov, V. Gubin, S. Shvab, A. Tanishev, MATEC Web of Conf., **72**, 01037 (2016), DOI: 10.1051/matecconf/20167201037
2. S. Lavrinenko, P. Polikarpov, A. Matveeva, V. Martyshev, MATEC Web of Conf., **110**, 01049 (2017), DOI: 10.1051/matecconf/201711001049
3. E. Pavelyev, S. Lavrinenko, MATEC Web of Conf., **72**, 01085 (2016), DOI: 10.1051/matecconf/20167201085
4. A.V. Vorobiev, E.M. Dutova, A.S. Latyshev, MATEC Web of Conf., **72**, 01125 (2016), DOI: 10.1051/matecconf/20167201125
5. A.A. Prosvirnov, M.M. Naftal. The concept of steam-coolant fast reactor of low power. OJSC "RSRIONPP", Moscow
6. E.V. Gukov, Power Engineering: efficiency, reliability, safety: collection of reports, 241 (2013)
7. A.M. Antonova, A.V. Vorobev, *Nuclear Power Plants: study guide for discipline "Nuclear Power Plants"* (TPU, Tomsk, 2009)

MATEC Web of Conferences **141**, 01029 (2017) DOI: 10.1051/matecconf/201714101029
Smart Grids 2017

Selection of optimal reheat temperature in heat recovery boiler at combined heat and power plant with GT-topping

Olga Romashova[1], *Alexander* Tubolev[1,*], *Anastasia* Matveeva[1], *Andrey* Zenkov[1], and *Valentin* Asmolovsky[1]

[1]National Research Tomsk Polytechnic University, 634050 Tomsk, Russia

Abstract. Calculations to determine optimum value of steam reheat temperature in relation to the cycle arrangement of steam turbine power plant with gas turbine topping are conducted in this paper. Operation of T-250-240 turbine unit in cogeneration mode has been considered during steam reheat displacement from steam boiler to heat recovery boiler.

1 Introduction

There are various possibilities for reconstruction and extension of steam power stations durability, some of which are discussed in [1-6]. One of them involves gas turbine (GT) topping to increase efficiency and capacity. Such cycle arrangements are used today at coal-fired power plants, where natural gas is used as backup fuel. The most common cycle arrangement is heating of feed water and condensate in heat recovery boiler (HRB), which is called GT-topping with regeneration displacement.

Analysis of CCGT operation with K-300, T-250 steam turbine units shows that displacement of regeneration has significant disadvantages [1, 2]:

1) efficiency of the steam cycle is reduced as a result of heat loss increase in condenser;

2) capacity of steam turbine decreases due to limitation of the maximum steam flow through the last stage of the turbine.

An alternative to regeneration displacement at unit thermal power plants is displacement of steam reheat (RH) from steam boiler (SB) in HRB using the residual heat of gases to heat the main condensate in a group of low-pressure preheaters (LPP).

Advantages in the economics of such a solution for condensing units in comparison with displacement of regeneration are shown in [1]. At the same time, maximum efficiency of CCGT is reached at RH temperature equal to the nominal value.

Relevance of reheat and selection of its parameters is not obvious for cogeneration steam turbine plants (CSTP), which operate most time of heating season with maximum heat output to external consumer, because the effect from annual power generation increase due to RH does not always exceed the loss of profitability. This is explained by the

*Corresponding author: tubolev@tpu.ru

influence of RH on steam temperature at the inlet in low-pressure section (LPS) and an increase of minimal vent pass to cool the blades of turbine's last stages.

For this reason, extraction turbines operating at initial pressure of 13 MPa, such as T-110-130, T-175-130, etc., are constructed without RH in contrast to condensing units operating at the same pressure (K-210-130, K-160-130).

2 Experimental setup and limits

Considering the fact that temperature of RH has an ambiguous effect on the economics of the units in cogeneration mode, calculations in this paper are conducted to determine its optimum value in relation to the cycle arrangement of T-250-240 steam turbine power plant with GT-topping with RH displacement from steam boiler to HRB. The cycle arrangement with heat transfer on the surfaces of HRB is shown in Figure 1.

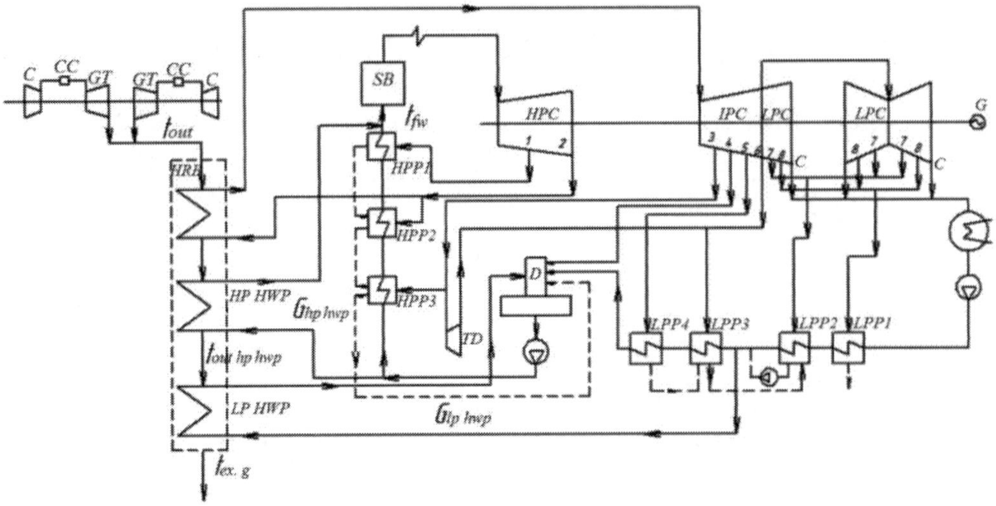

Fig. 1. Developed T-250/300-240 cycle arrangement with GT-topping.

Siemens V64-3A was considered as a gas turbine (GT) in this paper. Gas flow rate of GT is 154 kg/s and remains constant in variants calculations. Decrease in RH temperature at constant gas parameters after GT leads to decrease in RH thermal capacity and to increase of gases temperature after it. Residual heat is used to heat the main condensate and feed water.

Dependences of change in steam ventilation pass into condenser from steam temperature at LPS input for T-250-240 turbine are considered on the basis of this turbine operation analysis in cogeneration mode with a shut-off valve at LPS input and its supply from external source after humidification and steam separation [7], and also taking into account experimental data of the manufacturer.

3 Equations and mathematics

Figures 2 and 3 show the main calculations results of considered regimes for T-250-240 cogeneration unit. Minimal fuel consumption of CCGT during operation in cogeneration mode corresponds to the temperature of 480-490°C, which is significantly lower than the nominal value of this parameter in traditional mode of T-250 turbine (t_{rh} = 540°C - in all modes).

 DOI: 10.1051/matecconf/201714101029

Availability of optimal RH temperature and value close to 480°C has also been confirmed by similar calculations for K-210-130 turbine unit of Surgut GRES-1, switched to operation mode with maximal heat release from controlled extraction. Figure 4 presents dependence of steam flow rate for LPS cooling from steam enthalpy after the last stage of IPS, obtained during testing of K-210-130 turbine. Calculations for this turbine are based on this dependence.

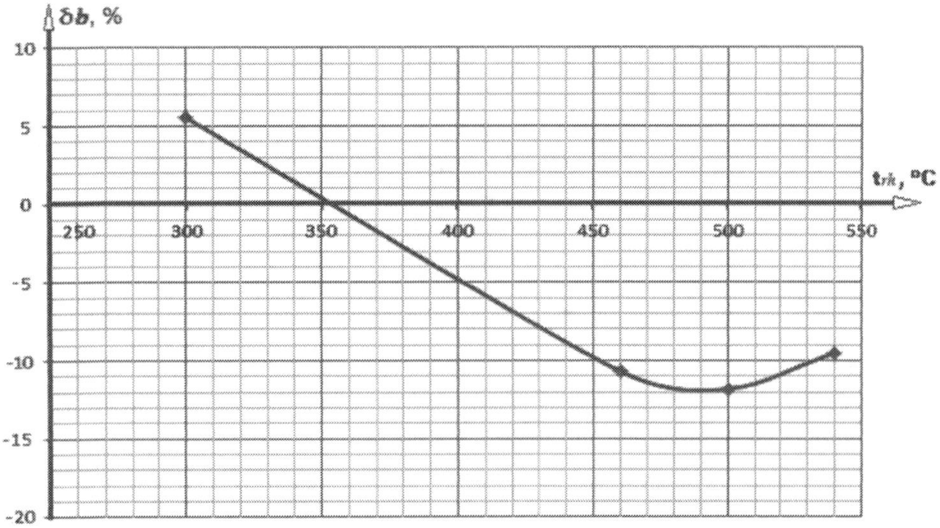

Fig. 2. Change of specific reference fuel consumption from reheat temperature in HRB for T-250/300-240 turbine with GT-topping at constant gas flow rate (operation in cogeneration mode).

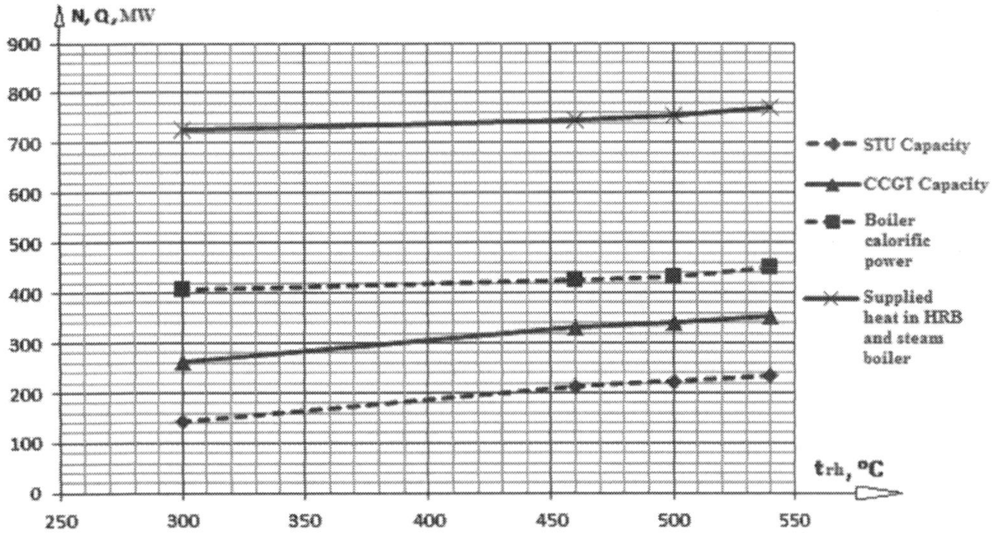

Fig. 3. Change of CCGT performance with T-250 units from reheat temperature.

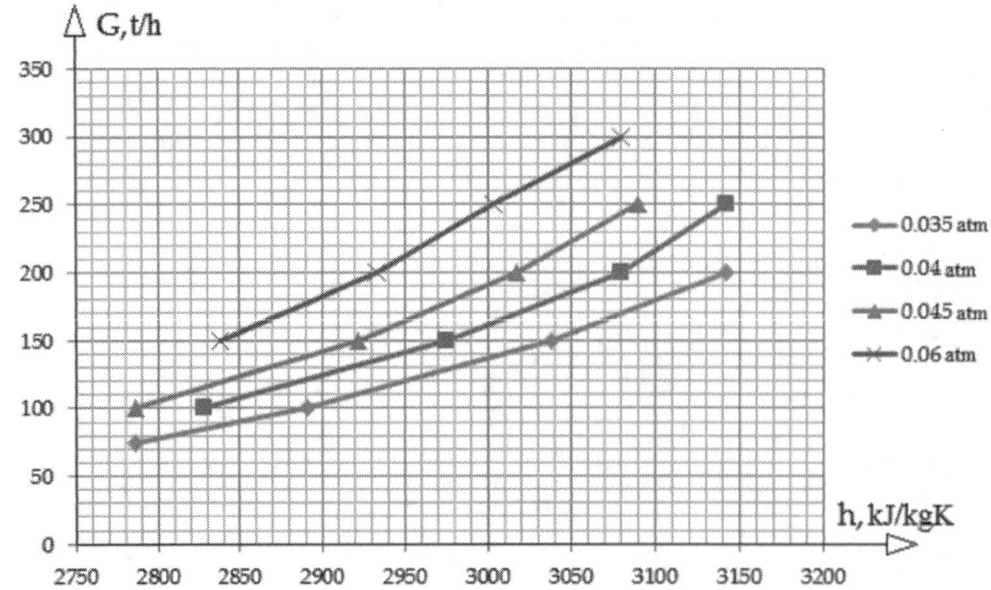

Fig. 4. Dependence of minimal steam flow in LPC of K-210-130 turbine from enthalpy before LPC at different pressures in condenser.

3 Conclusion

1. Optimal temperature of reheat in heat recovery boiler was determined by variants calculations for cogeneration power units with reheat at thermal power plant with GT-topping during operation in cogeneration mode that is well below nominal value (540°C) and lies in the range of 480-490°C.
2. Specific consumption reduction of reference fuel at optimal RH temperature is about 4% in comparison with nominal reheat temperature in traditional cycle arrangement.

References

1. O. Romashova, A. Minor, V. Martyshev, A. Tubolev, A. Katalevskaya, MATEC Web of Conference, **110**, 01054 (2017)
2. O. Romashova, A. Tubolev, L. Belyaev, E. Skrebatun, *Smart grids: IV International youth forum*, Tomsk, **1**, 260 (2016)
3. K. Larionov, D. Gvozdjakov, A. Zenkov, V. Zaytsev, EPJ Web of Conference, **110**, 01034 (2016)
4. K. Larionov, I. Mishakov, A. Gromov, A. Zenkov, V. Glaktionov, MATEC Web of Conference, **110**, 01048 (2017)
5. N. Galashov, S. Tsibulskiy, T. Serova, EPJ Web of Conference, **110**, 01068, (2016)
6. A. Gabdullina, N. Galashov, S. Tsibulskiy, I. Asanov, A. Kiselev, MATEC Web of Conference, **91**, 01004 (2016)
7. A.D. Trukhniy, *Extraction steam turbines and turbine units*, 540 (2002)

Effective fuel temperature of WWER-1000

Alexander Kudrov[1*], *Arian* Kuz'min[1], and *Yuri* Rakov[1]

[1]National Research Tomsk Polytechnic University, 634050 Tomsk, Russia

Abstract. The main temperature characteristics of a pressurized water reactor are distinguished, supporting its safety and reliable operation. The special role of the uranium fuel effective temperature is emphasized and the accuracy of the analytical determination of the power effect is increased. The calculation of the temperature distribution along the radius of the fuel rod was carried out taking into account the temperature dependence of the thermal conductivity UO_2. The design procedure was corrected for using the Finca-Ronchi dependence for the thermal conductivity of 95% density of the theoretical one.

1 Introduction

Safety operation of nuclear power plants is the reliability and efficiency of the individual components: staff, heat transfer equipment [1-6], characteristics of nuclear fuel [7] and many others.

They deserve special consideration of characteristics affecting the internal safety of a nuclear reactor: power and temperature reactivity effects [8, 9].

The temperature effect of reactivity (TER), which is defined by the difference of reactivity, caused by the same temperature change from the cold to the hot state of all materials in reactor core (RC). The initial temperature of the cold reactor is set to 20-40 °C. Hot reactor temperature varies with a minimum controlled power level due to external heat sources. Fuel reactivity effect is responsible for forming the neutron spectrum and its leakage.

The power reactivity effect (PRE) ρ_N is determined by the mean or effective temperature of uranium fuel T_{eff} and the actual presence an isotope ^{238}U in it. Due to the Doppler effect, there are resonance levels of the uranium isotope broadening with temperature rise, which increases the neutron absorption probability, thereby reducing reactivity. The higher the fuel temperature and the lower its concentration, the greater the effect. Also, magnitude ρ_N is important for to assess the mode extension fuel campaign using the power effect reactivity.

2 Differential characteristics

Isothermal $\alpha_T = d\rho/dT$ and power $\alpha_N = d\rho_N/dN$ reactivity coefficients determine self-regulation, self-protection, reliability and safety of the nuclear reactor. Both coefficients to

* Corresponding author: alex12kudrow@gmail.com

provide these nuclear reactor properties must be negative: the first is near the operative point, the second is on the entire range of power transient.

The fuel effective absolute temperature is determined by the exact integral formula

$$T_{eff} = \frac{\int_0^{r_T} T_T(r) 2\pi r dr}{\int_0^{r_T} 2\pi r dr} \tag{1}$$

Or approximate expression [1]

$$T_{ef} = T_c + 0.4(T_{max} - T_c) \tag{2}$$

where

$T_T(r)$ – temperature distributions over the fuel block;

T_{max} – temperatures on the axis of the fuel block;

T_c – temperatures on the surface of the fuel block.

2 Characteristic comparison on the coefficient of thermal conductivity

The method of direct explicit calculation $T_T(r)$ using the $\lambda_{UO_2}(T)$ is presented in handbooks [10, 11]. Research activity was based on the results [12], where the theoretical dependence of the thermal conductivity UO_2 was used as the equation:

$$\lambda_{UO_2} = \frac{10^2}{3.77 + 0.0258 \cdot T} + 1.1 \cdot 10^{-4} \cdot T + 1.01 \cdot 10^{-11} \cdot T^3 \cdot \exp(7.2 \cdot 10^{-4} \cdot T) \tag{3}$$

There is also the empirical formula:

$$\lambda_{UO_2} = 11.5 - 1.14 \cdot 10^{-2} \cdot T + 4.4 \cdot 10^{-6} T^2 - 5 \cdot 10^{-10} \cdot T^3 \tag{4}$$

which could be approximated by

$$\lambda_{UO_2} = \lambda_a \mathrm{ch}\left[(T - a)/b\right] \tag{5}$$

choosing the constants: $\lambda_a = 2{,}3$ W/(m·K); $a = 2000$ K ; $b = 870$ K .

More recent works are recommended to use the Fink-Ronchy formula to determine the thermal conductivity with density of 95%:

$$\lambda_{UO_2} = \frac{100}{7.5408 + 17.692 \cdot \tau + 3.6142 \cdot \tau^2} + \frac{6400}{\tau^{5/2}} \exp\left(-\frac{16.35}{\tau}\right) \tag{6}$$

where

$\tau = (T/1000)$K.

The equations (3) - (6) are presented in Fig. 1 and show the necessity of the coefficients adjustment in dependence (5).

3 Updating of the design formula for the fuel block

We take the Fink-Ronchy formula (6) as the main dependence for the coefficient of the thermal conductivity UO_2 and approximate it by the trigonometric function (5)

$$\lambda_{UO_2} = \lambda^*_a \text{ch}\left[\left(T - a^*\right)/b^*\right] \tag{7}$$

which has the following constants: $\lambda^*_a = 2{,}1$ W/(m·K); $a^* = 2070$ K; $b^* = 980$ K.

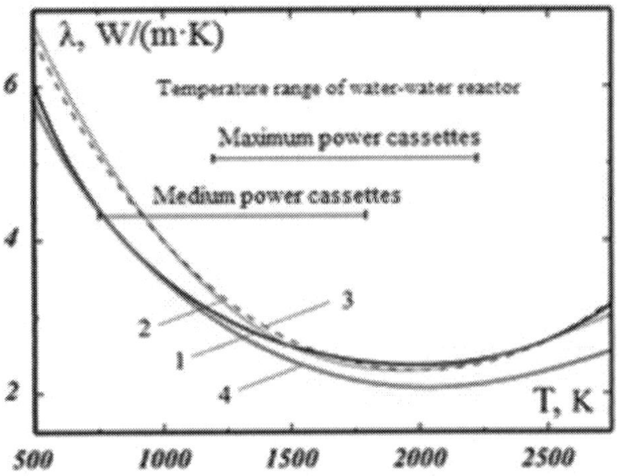

Fig. 1. Comparison of approximations using the thermal conductivity UO_2:1 – theoretical dependence (3) of work [12]; 2 – empirical formula (4), [12]; 3 – formula (5); 4 – Fink-Ronchi ratio (6).

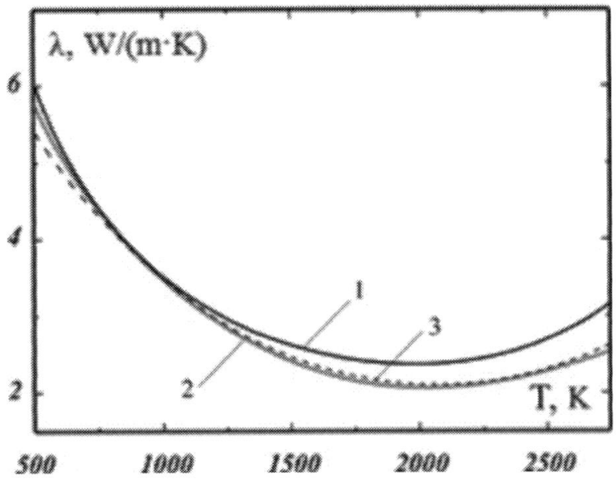

Fig. 2. Updating of approximation using the thermal conductivity UO_2: 1 – theoretical dependence (3) of work [11]; 2 – Fink-Ronchi ratio (6); 3 – formula (7).

4 Analysis of results

Carrying out estimations on the proposed dependencies allows us to draw the following conclusions:

- the approximation formula for determining the effective fuel temperature (2) is in error slightly more than 3%;
- the acceptable trigonometric approximation (7) of the Fink-Ronchi formula has been received; it will allow calculating the radial temperature distribution using the temperature dependence of the uranium dioxide thermal conductivity (8);
- comparison with the known results of thermal design for the reactor fuel rods (WWER–1000 [13]) showed slightly lower values of the temperature field; this indicates, first of all, the necessity of using the effect of density changes and deviations from stoichiometry, which requires further analysis.

References

1. O.S. Dmitrieva, I.N. Madyshev, A.V. Dmitriev, J. Eng. Phys. Thermophys., **90**, 651 (2017), DOI: 10.1007/s10891-017-1612-z
2. A.V. Dmitriev, O.S. Dmitrieva, I.N. Madyshev, Thermal Engineering, **63**, 674 (2016), DOI: 10.1134/S0040601516080036
3. A.M. Antonova, A.V. Vorobiev, A.S. Orlov, MATEC Web of Conf., **23**, 01069 (2015), DOI: 10.1051/matecconf/20152301069
4. A. Uvarov, A. Antonova, A. Vorobjev, MATEC Web of Conf., **37**, 01062 (2015), DOI: 10.1051/matecconf/20153701062
5. S. Lavrinenko, P. Polikarpov, A. Matveeva, V. Martyshev, MATEC Web of Conf., **110**, 01049 (2017), DOI: 10.1051/matecconf/201711001049
6. E. Pavelyev, S. Lavrinenko, MATEC Web of Conf., **72**, 01085 (2016), DOI: 10.1051/matecconf/20167201085
7. V.A. Ruppel, Yu.S. Tretyakova, S.V. Lavrinenko, A.A. Matveeva, V.N. Martyshev, MATEC Web of Conf., **37**, 01059 (2015), DOI: 10.1051/matecconf/20153701059
8. G.G. Bartholomew, G.A. Bat', V.D. Bajbakov, M.S. Alhutov, *Fundamentals of the Theory and Methods for Calculating Nuclear Power Reactors* (Energoizdat, Moscow, 1982) [in Russian]
9. V.I. Vladimirov, *Physics of Nuclear Reactors: Practical Tasks for Their Operations* (Knizhnyy dom "LIBROKOM", Moscow, 2009) [in Russian]
10. P.L. Kirillov, Yu.S. Yur'ev, V.P. Bobkov, *Handbook of Thermohydraulic Calculations (Nuclear Reactors, Heat Exchangers, Steam Generators)* (Energoizdat, Moscow, 1990) [in Russian]
11. P.L. Kirillov, A.V. Zhukov, N.I. Loginov, V.M. Makhin, I.L. Pioro, Yu.S Yur'ev, *Handbook of Thermohydraulic Calculations in Nuclear Power. Volume 2. Nuclear Reactors, Heat Exchangers, Steam Generators* (IzdAt, Moscow, 2013) [in Russian]
12. V.I. Kolyadin, E.P. Il'in, A.G. Kharlamov, V.V. Yakovlev, Atomic energy, **36 (1)**, 59 (1974)
13. A.M. Afrov, S.A. Andrushechko, V.F. Ukraintsev, Nuclear Fuel, **7**, 178 (2006)

Influence of insoluble aerosols on the heat-exchanging characteristics of the SPHR CV in the case of severe accidents

Aleksandr Nigay[1,*], and *Natalya* Ivanova[1]

[1]National Research Tomsk Polytechnic University, 634050 Tomsk, Russia

Abstract. Analysis of the processes in severe accidents at nuclear power plants has allowed distinguishing four characteristic stages of the accident, which differ in condensation rates, concentration and composition of aerosols (table 1) [1]. The released aerosol may condense on the heat transfer surfaces and reduce the effectiveness of the work of PCHRS under severe accident conditions at the deposition of insoluble aerosols on its heat transfer surfaces.

1 Introduction

Analyzing the consequences of the accident at Fukushima-1 occurred on 11 March 2011 in Japan, in [2] assess the environmental, social and economic damage to the population and the state. Received IPSDAE of RAS result of processing of the accumulated by the middle of 2012 information helped to identify the main causes of the accident: deprecated NPP project, the lack of passive safety systems [2].

Table 1. Stages of the accident at the NPP.

Mode	Stages of the accident			
	I	II	III	IV
Explanation	Main output of aerosol fission products	Output of aerosols from melt at bottom of reactor vessel	Output of aerosols from melt to site of localization of melt	Deposition of aerosols after their release in containment
The duration of deposition of aerosols, sec	2000	6000	7000	35000
Fission products	Cs_2MoO_4 (100 %)	Fe (20.58 %); Cr (15.81 %); Ni (2.21 %); Nd_2O_3 (1.23 %); Zr (0.19 %); Mn (59.98 %)	Fe (17.7 %); ZrO_2 (8.5 %); MnO (51.2 %)	–

* Corresponding author: nigayalexandr@mail.ru

The aim of the present work is to prove efficiency of PCHRS under severe accident conditions at the deposition of insoluble aerosols in the heat exchanging structures.

2 The methodology of experiment

For the research made experimental unit, shown in a simplified technological scheme (figure 1).

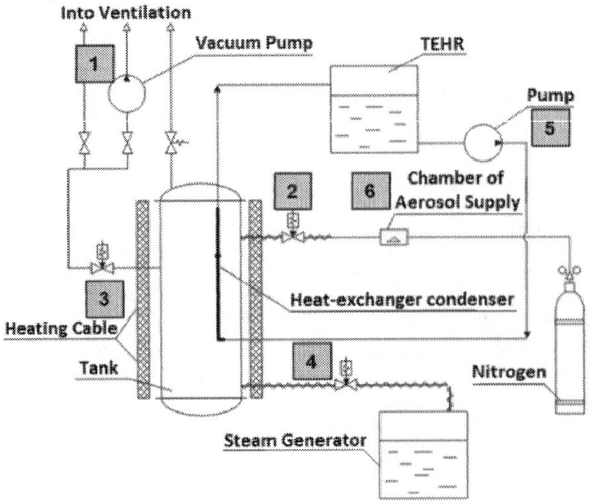

Fig. 1. Simplified technological scheme of experimental unit for research.

The methodology of the experiment:

Heat-exchanger condenser is installed into the tank.

1) Firstly, air is evacuated from the tank;

2) Then nitrogen is supplied;

3) Then tank heats to a predetermined temperature;

4) Next, in the heated tank supply steam. When temperature and pressure reached the predetermined value, the tank of aged 10-20 minutes for stabilize all parameters.

5) Further to the heat-exchanger condenser water is supplied.

6) Upon reaching stationary conditions (water temperature, temperature and pressure of the vapor environment) in the tank is start model of mixture of powders in the carrier gas is nitrogen. Turn on the countdown.

Simulators of fission products for the produce of aerosol take from industrial micron and nanosized powders. In the tank under the pressure of gas-vapor mixture on the heat-exchanger condenser of the condensation of vapor and aerosol. In the transition between stages, the temperature inside the tank is reduced by reducing the temperature of the container wall operated from a PC. Release of excessive pressure from the tank is carried on in the ventilation system through vent line. All operation in automatic mode [3-4].

3 The results of the research

Analysis of aerosols deposited on the heat exchange surface.

Heat exchanger is made of stainless steel 08X18H10T, its chemical composition is shown in Table 2 in mass %.

Table 2. Chemical composition of the steel 08X18H10T as per GOST Standard 5632 – 72.

DOI: 10.1051/matecconf/201714101031

Chemical element	C	Si	Mn*	Ni*	S	P	Cr*	Cu	C	Ti	Fe*
Element ratio	Up to 0.08	Up to 0.8	Up to 2	9 -11	Up to 0.02	Up to 0.035	17 - 19	Up to 0.3	5	0.7	The rest

*- matching components included in the composition of steel and aerosol

The steel contains some elements as basis and alloying, which are also included in the aerosol components, and it is not possible to discover their distribution in the aerosol deposited on the heat exchanger surface. In this regard, replicas were prepared from the tube surface on the conductive carbon tape in order to determine the elemental composition of the aerosol deposited on the heat exchanger. Then, the replicas were analyzed on electron microscope JEOL JSM 6000 Neo Scop with the attachment EDS (elemental analysis). The results of the elemental composition and morphology of aerosol deposited on the heat exchanger-condenser are presented in the figures below.

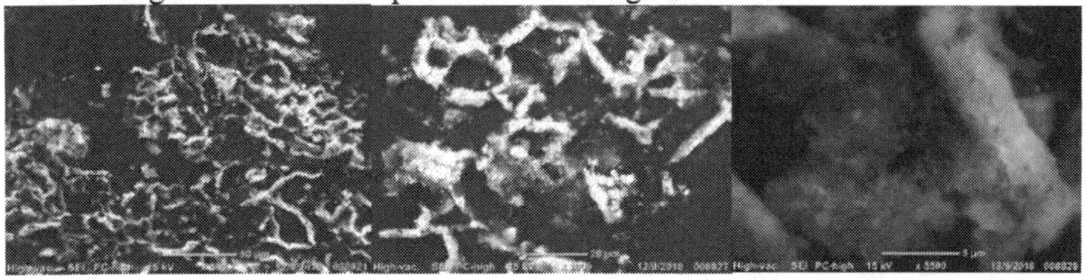

Fig. 2. Micrograph of the aerosols deposited on the heat exchanger-condenser with the low part.

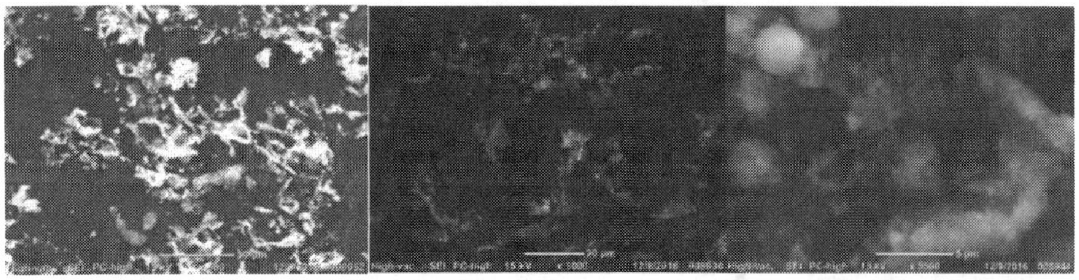

Fig.3. Micrograph of the aerosols deposited on the heat exchanger-condenser with the middle part.

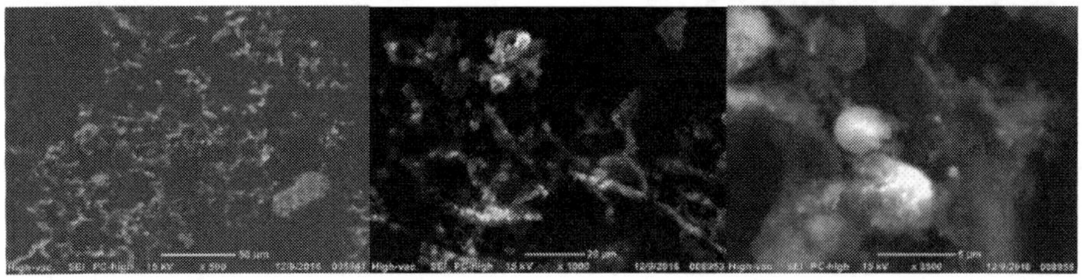

Fig.4. Micrograph of the aerosols deposited on the heat exchanger-condenser with the high part.

Aerosols from the vapor-gas space deposit on the surface of the heat exchanger-condenser, the nature of deposition and composition of the deposited aerosols along the length of the heat exchanger depends on the experimental parameters and the location of sampling collection on the heat exchanger.

At the top of the heat exchanger, the aerosol does not create a continuous coating, but precipitates in an isle structure. These isles present elongated formations, with the length (20-30 μm) several times greater than the width (3-5 μm). These formations have a loose structure, single particles of 0.3-0.7 μm and 5-7 μm in size are seen on the surface.

DOI: 10.1051/matecconf/201714101031

Elemental composition of the aerosol is represented by Cs and Mo as basic elements and Si and Fe as impurities from the initial powder Cs_2MoO_4.

In the middle part of the heat exchanger, the area covered by the aerosol becomes smaller, the length and thickness of the formations decreases. With a larger magnification, a loose surface layer is seen, where individual particles of both micron and nanoscale are distinguishable. Elemental composition of the aerosol is similar to the composition from the top of the heat exchanger, but the intensity of the main elements is lower.

In the lower part of the heat exchanger, the aerosol is deposited by even smaller fragmented "isles". The ratio of length and width ((10-20) μm to (1-3) μm) of individual elements decreases significantly. Single aerosol particles with sizes ranging from fractions of microns to several microns can also be observed on the surface loose layer. The elemental composition of the aerosol is similar to that of the upper and middle parts of the heat exchanger.

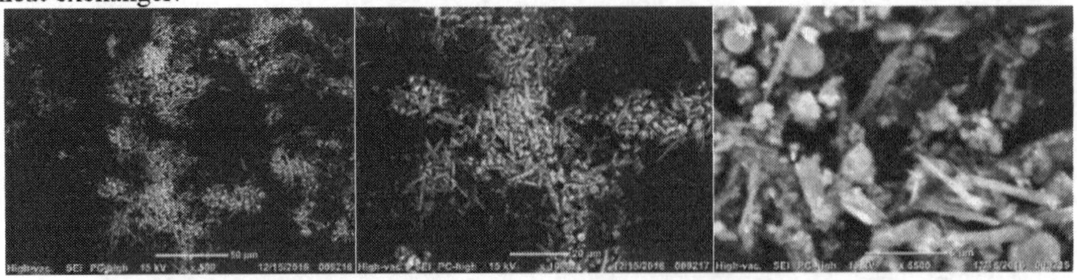

Fig. 5. Micrograph of the aerosols deposited on the heat exchanger-condenser with the low part.

Fig. 6. Micrograph of the aerosols deposited on the heat exchanger-condenser with the middle part.

Fig. 7. Micrograph of the aerosols deposited on the heat exchanger-condenser with the high part.

For experiment No. 4, aerosol deposition was carried out in four stages. At the end of the experiment, samples were collected and replicas were prepared.

According to electron microscopy, local deposition of the aerosol with the sizes of up to 40-50 microns occurs in the upper part of the heat exchanger. These structures consist of individual particles of an elongated shape with well-defined boundaries and with different sizes from 5 to 20 μm. Nano sized particles are located on the surface of elongated particles. Elemental analysis of the aerosol of the heat exchanger deposited on the surface showed the presence of the following elements in the deposited aerosol: Mn - basic, Cr - basic, Fe - basic, Zr - basic. The rest of the aerosol components were not detected.

In the middle part of the heat exchanger, the deposited aerosol covers a much larger area. The aerosol is disposed on the heat exchanger in the form of elongated irregularly shaped particles, and in the form of initial particles of an irregular detrital form. The elemental composition of the aerosol shows the presence of the following elements introduced during the experiment: Mn - basic, Cr - basic, Fe - basic, Zr - basic. The rest of the aerosol components were not detected.

In the lower part of the heat exchanger, the aerosol is deposited as a cluster of separate particles and in the form of agglomerates. Elemental composition of the aerosol is similar to the composition of the aerosol from the middle part of the heat exchanger: Mn - basic, Cr - basic, Fe - basic, Zr - basic. The intensity of reflexes is significantly lower than on the middle and upper parts of the heat exchanger.

Compiled tables of thermophysical characteristics [5-6].

4 Conclusion

1. The study of the structure of aerosol deposited on the heat exchanger tube at different stages of the heat exchange process simulating the development of severe accidents showed:

a) The maximal deposition of the aerosol is fixed in the middle part of the heat exchanger tube;

b) The deposited aerosol does not form a continuous coating on the heat exchanger tube, the thickness of the local sections of the deposited aerosol is up to 10 μm;

c) Element analysis of the deposited aerosol showed the presence of all elements introduced in the composition of the FPI.

2. The experimental determination of the coefficient of thermal conductivity of the witness samples, for which the most conservative conditions for aerosol deposition were realized, did not reveal significant changes in the thermal conductivity coefficient.

3. The study of the structure of the aerosol deposited on the tube and the measured coefficients of thermal conductivity of the witness samples with aerosol makes it possible to state that the deposited aerosol does not significantly influence the heat exchange process between the vapor-gas mixture and the water flowing in the heat exchanger.

4. The introduction a boric acid aerosol into the vapor-gas mixture with simultaneous presence of an FPI aerosol slightly increases the heat transfer coefficient from vapor to water.

References

1. I.L. Kozlov, Works of the Odessa Polytechnic University, **1** (45), 65 (2015)
2. D.V. Aron, M.V. Pavlova, S.V. Panchenko, Bulletin of the Russian Academy of Sciences, Energy, **5**, 65 (2013)
3. E.V. Kravchenko, A.G. Nigay, MATEC Web Conf., **37**, 01037 (2015), DOI: 10.1051/matecconf/20153701037
4. A.G. Nigay, J.Z. Vasileva, Smart energy systems: proceedings of the IV International youth forum, **1**, 119 (2016)
5. V.V. Medvedev, N. Galchenko, A.G. Nigay, D. Sivkov, MATEC Web Conf., **72**, 01032 (2016), DOI: 10.1051/matecconf/20167201032
6. N.A. Ivanova, E.E. Bulba, MATEC Web Conf., **110**, 01020 (2017), DOI: 10.1051/matecconf/201711001020

A method for measuring the amount of hoar frost formation in the recuperation channels of ventilation systems using the adjustable mathematical model of this process

Aleksandr Shilin[1,*], *Sergey* Prokhorov[1], *Andrei* Shabo[1], and *Viktor* Bukreev[1]

[1]National Research Tomsk Polytechnic University, 634050 Tomsk, Russia

Abstract. The paper presents a method for measuring the amount of hoar frost formation in the recuperation channels of ventilation systems using the adjustable mathematical model of the hoar frost process. The principle is based on the fact that the contour of the adjustment of the hoar frost model is included in the measurement in accordance with the measured pressure drop, which is proportional to the amount of hoar frost. Unlike the known measurement methods, it is proposed to use the state variables of the mathematical model as the measured value. These state variables are not subject to non-deterministic interferences and random influences. The paper presents simulation results confirming the adequacy of the dynamic model. In conclusion, an example of the use of a recuperation channel in the defrost management system is given.

1 Introduction

The problem of frost formation in recuperation systems is known in many countries such as Russia, Canada and other northern countries, where the temperature of the environment can be below −25 ◦ C. The work [1] include the detail description of the hoar frost formation, dynamic processes, experimental data and many other features of such systems and other [2].

A several methods for solving this problem can be found in the works [3, 4], where it is proposed to add an additional heating element to the system, which defrosts the channel. The main problem of this method is providing the well-timed switching of modes to avoid the complete blockage of the channels. Switching time depend on the degree of hoar frost of the channel. The problem of switching time estimation are described in the works [5–7], where the problem of measuring the degree of hoar frost formation and its derivative is actual.

In this paper, a method is proposed for measuring this degree using a tunable mathematical model, where the variable states qualitatively reflect the process of dynamics.

* Corresponding author: shilin@tpu.ru

 DOI: 10.1051/matecconf/201714101032

2 The dynamic properties of process

For the formation of state variables, a qualitative mathematical model of the process is required. To synthesize the model, we used the results of [8], which shows the results of mathematical modelling of the recuperation system and the results of experimental studies.

The mathematical model is created in the Xcos environment of the mathematical package SciLab on the functional blocks and is presented in Fig. 1 The model has the following blocks:

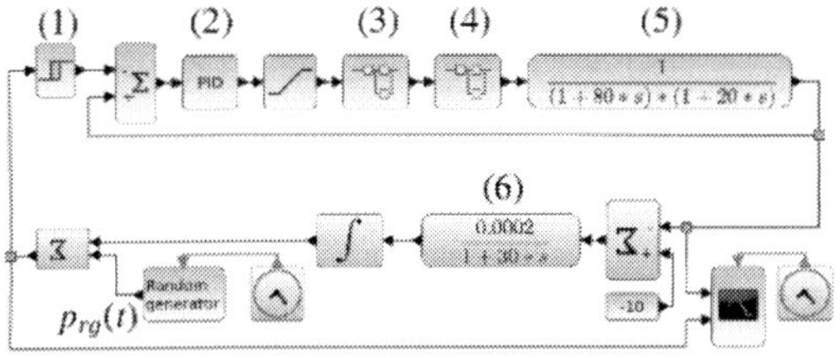

Fig. 1. Mathematical model of defrosting control in a recuperation system.

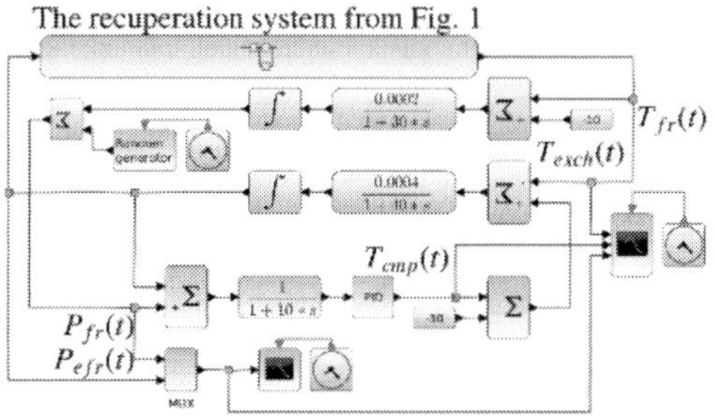

Fig. 2. Structural diagram with the observer of state variables.

Block 1: The Hysteresis element, designed to define the boundaries corresponding to the values of the layers at which the system will enter the defrost and frost operation mode, to provide the maximum efficiency mode. Block 2: PID-regulator, to maintain the modes. Block 3: The block of actuators "Frequency converter - Asynchronous motor". It is represented by a simplified differential equation in the form of a transfer function

$$W_{am}(s) = \frac{1}{(1+sT_{ft})(1+sT_{am1}+s^2T_{am2}^2)} \tag{1}$$

where time of the frequency transformer T_{ft}=0.2 s, times of Asynchronous motor T_{am1}=0.15 s and T^2_{am2}=0.03 s. The executive element of the system is a pump that ensures the circulation of the heat transfer medium between heat exchangers. This pump is made on the basis of an asynchronous motor. Since the transition process of hoar frost formation and defrosting is much slower than the processes in the engine itself. It is perfectly permissible to use aperiodic links of the first order as a mathematical dynamic model of the engine.

Block 4: The unit realizes the static connection from the control signal from the PID controller output in the "Frequency converter - Asynchronous motor", to the temperature at the output of the heat exchanger. In the block, the function proposed in [9] is implemented by expression:

$$T_{exch} = T_{room} + (T_{out} - T_{room})K_{ef}0.1U_{pid} \tag{2}$$

where the room temperature T_{room} = 20°C, Ambient air temperature T_{out}=−30°C; Efficiency coefficient of recuperator K_{ef}, U_{pid}∈[0, 10.0] – output value of PID-regulator.

Block 5: The link characterizing the process of increasing frost is represented by two successively connected inertial links of the first order, realizing the transfer function

$$W_{exch}(s) = \frac{1}{(1+sT_{o1})(1+sT_{o2})} \tag{3}$$

with time constants (T_{o1}=80 s) and (T_{o2}=20 s).

Block 6: The defrost process describes an aperiodic link of the first order with a time constant (T_o = 30 s) and a coefficient (K_p = 0.0002) proportional to the rate of increase in frost.

The results of the simulation are presented in the form of transition process Fig. 3 ($p_{rg}(t)=0$). From the processes it is clear that the model reflects the processes of growth and thawing of frost. The operation of the PID controller ensures that the temperature at the heat exchanger outlet is maintained in both modes. The form of the obtained transient processes was also confirmed by specialists servicing the system under investigation in Yakutia (Russia). However, in real conditions, due to the influence of many external factors, it was impossible to qualitatively evaluate the degree of hoar frost formation directly on the basis of a pressure differential sensor. The measured value of the differential pressure in the actual system includes significant interference from the mechanical part of the recoperation system. This problem caused random switching between modes.

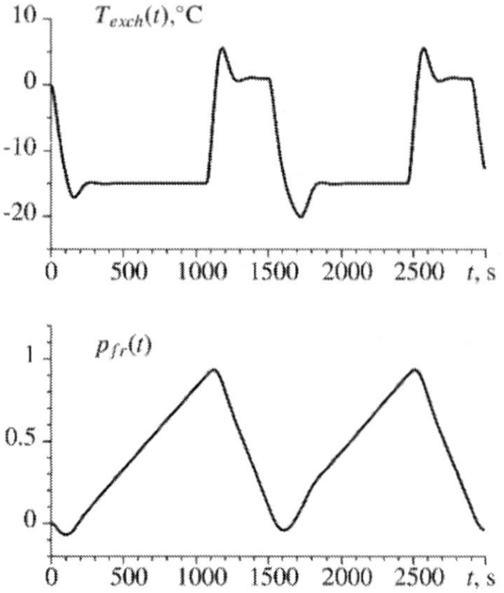

Fig. 3. Results of simulation of a deterministic system ($p_{rg}(t)=0$).

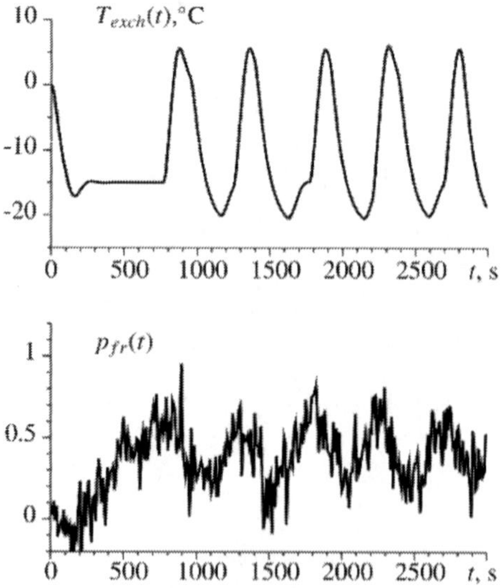

Fig. 4. Results of simulation under interference conditions ($p_{rg}(t)=10\%$).

Transition process resulting from the simulation of a system with interference are shown in Fig. 4 ($p_{rg}(t)=10\%$). The results of the simulation reflect the reason for the unstable operation of the system as a whole, where untimely operation of the hysteresis element does not allow to ensure the efficiency of the recuperator. To solve this problem, the sensor can be represented in the form of differential equations or a dynamic model. Such a method of evaluation makes it possible to observe the tendency of the hoar frost formation process, due to the use of state variables. This solution will be efficient for the control systems with thawing, since the mathematical model of the process is not subject to interference from the mechanical part of the system.

3 Observer of state variables of the hoar frost process

The main factor affecting the process of hoar frost formation is the temperature of the environment. Thus, for a certain value of this parameter, the process of frost formation begins. In addition to the temperature of the outside air, the humidity of the air in the installation has a direct influence on the process of frost formation. In very dry air, the temperature of the frost can drop to -30 C. And if the humidity is too moist, the temperature of the frosting hardly drops below the dew point temperature. This relationship between temperature and humidity was noticed during the operation of the automatic recuperation system in Yakutia (Russia). The main task of the presented model is to find the moment of

 DOI: 10.1051/matecconf/201714101032

switching from one mode to another, based on the value of the formed frost. In general, other values (humidity, pressure, e.t.c.) that affect the model have not been investigated, since they do not have a significant effect on switching moment calculation.

The main indicator influencing the hoar frost formation process is the temperature (T_{fr}) of the environment at which this process actually begins. For different objects this temperature can vary considerably. During the operation of the automatic recuperation system in Yakutia (Russia), a link was observed between the temperature (T_{out}) and the humidity of the exhaust air. In particular, for dry outgoing air, the temperature (T_{fr}) can drop to -30 C, and also the rate of increase of hoar frost formation (P_{fr}) falls off sensitively. And with moist air, the rate of frost rises with the temperature ($|T_{out}|$). Fig. 2 presents structural diagram of the system with an observer of the state variables of the process of hoar frost. The upper part of the structural diagram is taken from Fig. 1. A reference model, is presented where the dynamic characteristics of the reference model differ from those adopted in the system. In particular, the rate of hoar frost formation is overestimated by a factor of two, and the time of inertia is increased by one third. In the system, the frost-formation temperature adjustment circuit (p_{efr}) is added, which includes a damping unit and a PID controller. The goal of a closed loop is to adjust a (T_{cmp}) at which the damped mismatch tends to zero.

Stability and speed of the tuning loop can provide the optimal parameters of the PID controller. It is pssible by any available method adopted in the theory of automatic control. The transfer function of the open system was assumed in the following form

$$W_{cl}(s) = (k_p + \frac{k_i}{s} + \frac{T_d s}{1+0.1T_d s})\frac{1}{s(1+s10)(1+s40)} \quad (4)$$

where k_p=50, k_i=0.1s^{-1}, T_d=10 s.

In Fig. 5, the simulation results are presented in the corresponding structural scheme of Fig. 2.

As can be seen from the figure, the compensating value (T_{cmp}) at the output of the PID controller should be T_{cmp} = 20°C. But due to the fact that the parameters of the model are significantly different from the parameters of the object describing the process of a clogging an absolute coincidence is not observed. Nevertheless, the main purpose of solving the problems of false triggering of the hysteresis element is solved and it is possible to observe the restoration of the operation of the thawing circuit. Indeed, the processes of transition from regime to regime practically correspond to Fig. 3. and there are no false positives in them, which are observed in Fig. 4.

As a result, the structural diagram of the system with the observer of state variables of the degree of hoar frost formation provides a qualitative estimate of the actual degree of contamination over a sufficiently noisy value of the pressure drop across the heat exchanger. In Fig. 6 shows the correspondence of the pressure drop ($p_{fr}(t)$ – dotted line) and the evaluation of the degree of hoar frost formation ($p_{efr}(t)$ – solid line), which has a smooth character of change.

4 Conclusion

The article describes a solution for eliminating random switching in the system. This result is achieved by observing the trend of the incoming signal through the introduction of a damper. During operation, the control system has been able to cope with the tasks correctly. In highly specialized systems, problems may occur where measurement of non-trivial physical quantities is required to organize a process or an automatic system. These physical quantities can have quite understandable relationships with quantities that can be measured

DOI: 10.1051/matecconf/201714101032

by standard sensors under external conditions and described in the form of differential equations.

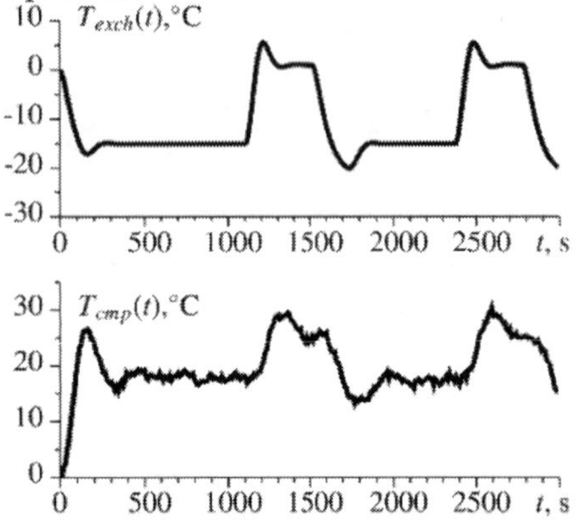

Fig. 5. The results of modeling using the state variable as the degree of hoar frost formation.

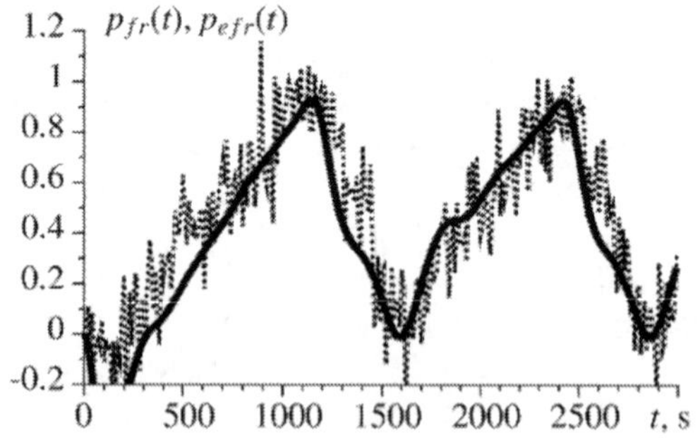

Fig. 6. Differential pressure and evaluation of the degree of hoar frost formation.

In such cases, a complex technique is required to estimate the value of a nontrivial physical quantity, taking into account both static and dynamic relationships. In these conditions it is quite difficult to find a suitable prototype of the solution in the scientific and technical literature. In this paper, a methodology for performing similar tasks is presented, which can be applied in other technological objects, where the measurement of non trivial physical quantities is required. Also, the material of the article can be useful to specialists in the field of synthesis of automatic recuperation systems.

References

1. M. Rafati Nasr, F. Fathieh, D. Kadylak, R. Huizing, R. Besant, C. Simonson, Exp. Therm. Fluid Sci. **77**, 100 (2016)
2. K. Smith, S. Svendsen, Energy and Buildings, **107**, 1 (2015)
3. A. Zandeckis, K. Klavina, M. Dzikevics, V. Kirsanovs, G. Zogla, J. Eng. Sci. Technol. Review, **8**, 98 (2015)
4. B. Ouazia, R. Glazer, F. Szadkowski, C. Hoyme, Int. J. Ventilation, **14**, 219 (2015)
5. S. Koziol, S. Zacharski, Solid State Phenom. **237**, 148 (2015)
6. K. Allaerts, J. Al Koussa, J. Desmedt, R. Salenbien, Energy and Buildings, **138**, 559 (2017)
7. A. Krofova, P. Kic, Eng. Rural Dev. 56 (2016)
8. F. Savrasov, S. Prokhorov, A. Shilin, J. Phys.: Conf. Series, **803** (2017)
9. A. Shilin, V. Bukreev, Commun. Comput. Inf. Sci. **487**, 387 (2014)

Survival analysis of A.C. machines in the diamond industry using the Kaplan-Meier estimator

Vladislav A. Shevchuk[1], *Oleg* P. Muravlev[1], and *Olga* O. Stolyarova[1,*]

[1]National Research Tomsk Polytechnic University, 634050 Tomsk, Russia

Abstract. The article deals with the survival analysis of a.c. machines operating under the conditions of the underground mine "Mir" (Yakutia). The estimation of their survival is carried out using the Kaplan-Meier estimator. The main advantage of this method is in the opportunity to process censored (incomplete) data, which in turn leads to the accuracy of results. The statistical data used in this article were obtained from maintenance books and reports made by a repair team working in the considered mine. This analysis allows not only identifying faulty electric machines and their fault-free operation but also the faulty components of a certain electric motor.

1 Introduction

Significant advances have been recently achieved in the field of diagnostics and maintenance of electric machines. It is associated with the improvement of the system of planned preventive maintenance and the introduction of modern high-technology diagnostic techniques and information technologies. However, electric motors operating in the diamond industry can be occasionally exposed to some sudden breakdowns and emergencies. These problems lead to economic losses during the period of the machine down time.

The goal of this article is to carry out the survival analysis of a.c. machines operating in the underground mine "Mir" (Yakutia) using the Kaplan-Meier estimator.

Initially, the methods of survival analysis were widely used in medicine and biology, as well as in economics and social science. Then, they gradually gained in popularity and found their application in mechanical engineering in the form of reliability analysis. However, the methods of survival analysis are still not widespread in Russia [1, 2].

2 Experimental procedure

The key feature of survival analysis is the possibility to process censored (incomplete) data, which enables to enhance the data selection and leads to the accuracy increase of results.

* Corresponding author: muravlevaoo@tpu.ru

Censored data are observations which do not contain any data about the development of the corresponding event. The survival function for data selection with censored data can be presented as a formula of the Kaplan-Meier estimator [3 - 6]:

$$S(t) = \prod_{j=1}^{t} \{[n - j] / [n - j + 1]\}^{\sigma(j)} \quad (1)$$

where n is the number of events (failures), j is the order number of a certain event; $\sigma(j)=1$, if the observation is complete and 0 if the observation is censored.

S(t) from (1) plot of function is called a survival curve. Sloping plot indicates high survival; abrupt plot shows low survival. The survival curve construction allows comparing the distribution for two or more groups [4].

The underground mine "Mir" was chosen as the object for analysis, since a major share of electric machines failures has occurred within the period from February 2012 till August 2017. Electric machines ranging from pumps to industrial combines that operate in different mining equipment were chosen for data selection. Figure 1 illustrates the comparison of survival of two most widely used electric machines.

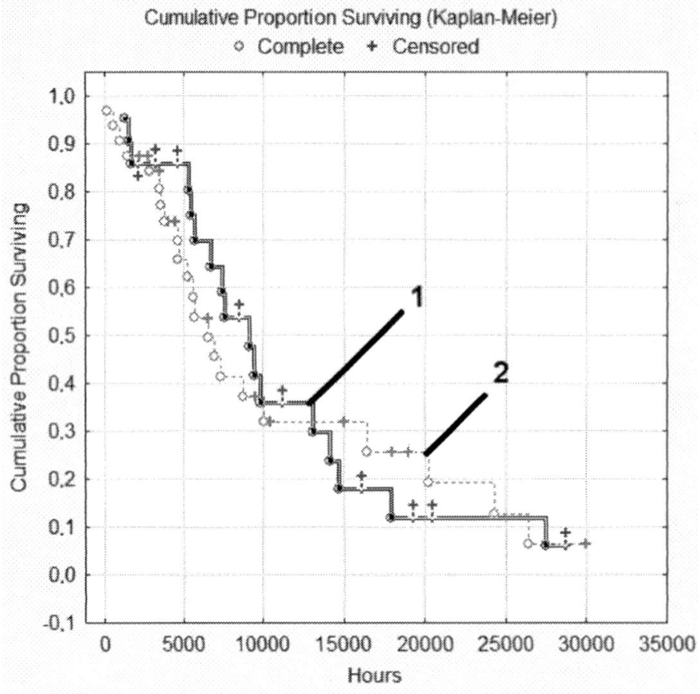

Fig. 1. Comparison of survival of electric machines in the underground mine "Mir".

The diagram shows that for electric machines of d2PRW X250/31-KL model (presented by curve 1) the first 25 000 hours of operation are critical. The survival probability abruptly drops to 85 %, and the survival probability until 10000 is equal to 35%. Despite the fact that curve 2 for d2PRWX280 model is located more left-sided than curve 1, it is longer than curve 1. This indicates that curve 2 has higher survival index. Thus, it can be assumed, that electric model of d2PRW X250/31-KL is more defective.

 DOI: 10.1051/matecconf/201714101033

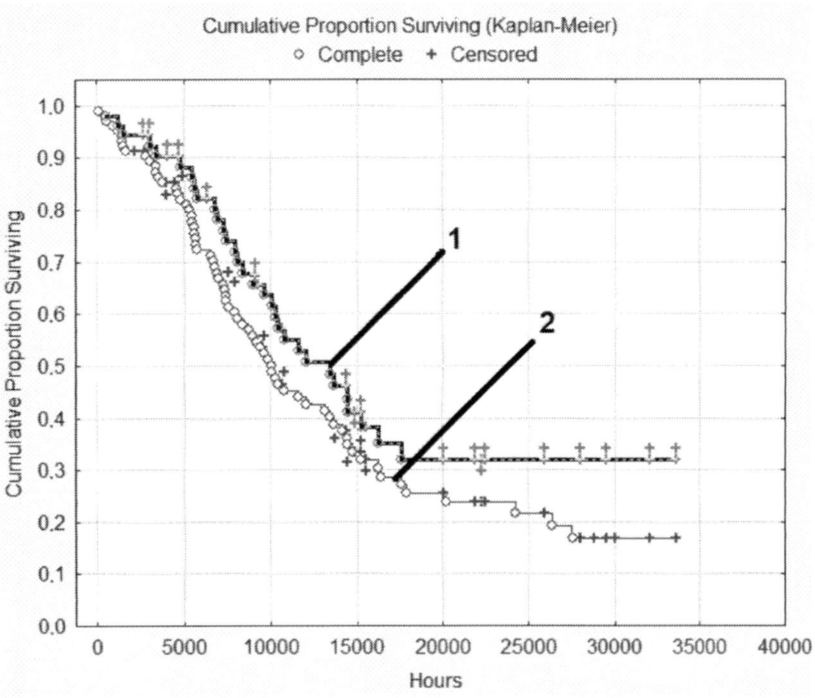

Fig. 2. Comparison of electric machine survival in the underground mine "Mir" with and without an electric motor of d2PRW X250/31-KL model.

Fig. 2 presents the survival plot of an electric machine operating at the mine "Mir" without an electric motor (curve 1) and with an electric motor of d2PRW X250/31-KL model. The rate of survival has considerably increased since the survival probability till 34000 hours is equal to 31% compared to 18%. Thus, it can be assumed, that in order to increase the survival rate some extra diagnostic and preventive measures are required.

The faultiest component as well as the critical time for each component can be detected using the Kaplan-Meier estimator. To do this we should know some supplementary data about faults. These measures in turn help to plan scheduled preventive maintenance for a particular electric motor [1].

The faultiest components (Fig.3) in our case with the lowest survival rate are the elements of bearing assembly (curve 1). It was defined that by achieving some certain operating time there are some sharp drops of survival curve during the motor operation. The survival probability is 80 % if the operating time equals 5000 hours and 30 % relatively if the operating time is 9000 hours.

There is also necessary to pay attention to the survival curve of shaft components (curve 2). It is obvious from this curve that the main number of failures occurs within the first 5000 hours. The survival probability till 5000 hours is equal to 41 %. In accordance to the survival curve of stator components (curve 3) the abrupt drop led to 28 % of survival probability till 10 000 hours.

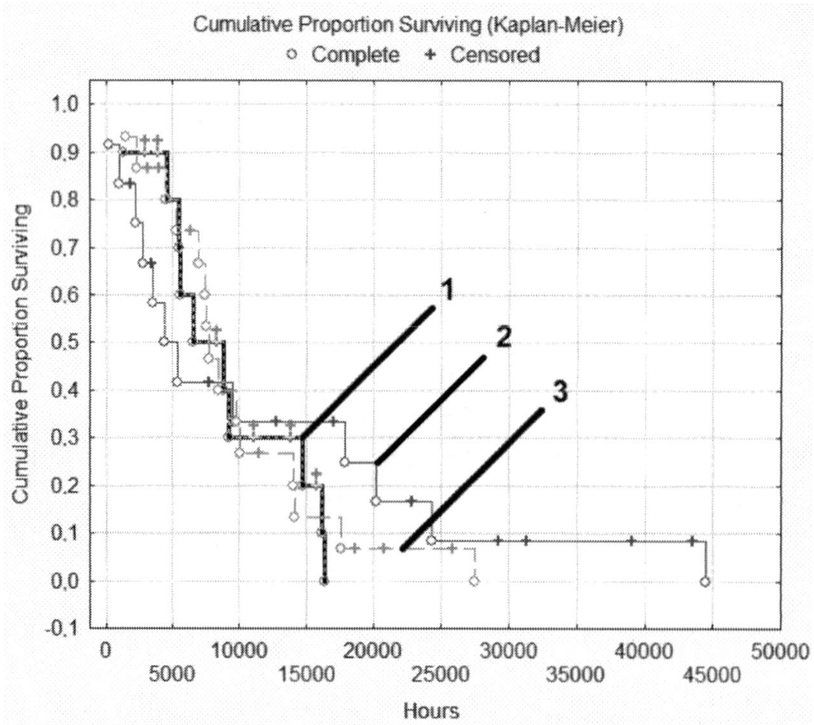

Fig. 3. Comparison of survival for electric motor components of d2PRW X250/31-KL model.

3 Conclusion

To sum up, using the Kaplan-Meier estimator it is possible to detect the faultiest model of electric motors in a system as well as the faultiest components in a certain electric motor. Since the survival estimation is calculated with reference to time, some certain faulty components can be defined in advance. The Kaplan-Meier method was recommended for application and adopted by the department of technical diagnostics of Specialized Repair and Construction Administration (Mirny, Yakutia).

References

1. J. Huang, M. Pflueger, "Electrical Submersible Pump Survival Analysis" (2011)
2. B. Tsybikov, E. Beyerleyn, P. Tyuteva, MATEC Web Conf. **91**, 01034 (2017), DOI: 10.1051/matecconf/20179101034
3. J. P. Klein, H. C. van Houwelingen, J. G. Ibrahim, T. H. Scheike, *Handbook of survival analysis* (CRC Press, 2014), DOI: 10.1002/bimj.201400252
4. A. van den Hout, *Multi-state survival models for interval-censored data* (CRC Press, 2017)
5. V. Borovikov, *STATISTICA. The art of data analysis through a computer* (Piter, 2003)
6. V. A. Shevchuk, O. P. Muravlev, O. O. Stolyarova, V. P. Shevchuk, MATEC Web Conf. **91**, 01033 (2017), DOI: 10.1051/matecconf/20179101033

MATEC Web of Conferences **141**, 01034 (2017) DOI: 10.1051/matecconf/201714101034
Smart Grids 2017

Instrumental evaluation of traction machines sparking

Nikita Kolobashkin[1], *Beligma* Dorzieva[1], *Evgeniy* Beyerleyn[1,*], *Polina* Tyuteva, and *Anatoliy* Cukublin[1]

[1]National Research Tomsk Polytechnic University, 634050 Tomsk, Russia

Abstract. In the article, the authors considered the possibility of the sparking monitoring at a commutator of a traction motor under operating conditions. The authors developed a method for the evaluation of sparking level at the commutator by means of the current flowing through separate, isolated parts of the split brush recording. The relationship between an auxiliary commutation current and the sparkling level is revealed.

1 Introduction

The importance of brushed direct current traction motors (DC TM) on electrified sections of the railway has not decreased in recent years, as it was expected, but on the contrary began to grow. There have appeared new electric locomotives 2ES4, 2ES5, 2ES6 and others, intended for the replacement of outmoded types of machines constructed on the basis of DC machines [1]. At the same time, there are problems related to the commutator unit, monitoring of its quality and magnetic system maintenance at the depot test stations. Tests carried out on the locomotives equipped with DC motor shows that reliability is also affected by design flaws as well as drawbacks associated with control circuits [2].

The failure analysis shows that about 90 % of DC TM damages occur at about 800 thous. km run from the factory repairs, and approximately 30 % of the DC TM fail during the warranty period at the established mileage per one cycle from the beginning of operation (or overall repair) to the next overall repair 1400 thous.km. As experience has shown one of the effective ways to reduce the number of failures is to monitor the DC TM condition after repair and during operation [3]. However, if testing is performed at repair depot there are National standards and rules for machines testing, then for the testing directly on the electric locomotive there are neither normative documents for controlling the DC TM condition during operation, nor technical facilities for its implementation (except armature current measurement). The availability of methods and technical facilities for controlling could prevent the DC TM from being installed on the electric locomotive without an appropriate setting of commutation and magnetic system, which initially guarantees a mileage rate under normal operating conditions. Monitoring of the DC TM condition during operation makes possible to regulate the load and apply the technique of thrust redistribution, and effectively prevent the slippage. The presented work is done by

* Corresponding author: beierleinev@tpu.ru

 DOI: 10.1051/matecconf/201714101034

the specialists of the Electrotechnical Complexes and Materials Department, Tomsk Polytechnic University and devoted to the creation of method and technical facilities for the DC TM condition monitoring at a repair depot and on an electric locomotive.

2 Theoretical study

The main indicator of the DC TM is the commutator sparking. According to the commutation theory, there will be no sparking if the commutation is linear and the additional commutation current is zero. If the machine has overcommutation or undercommutation, this current differs from zero value. It flows in the section closed with a brush, and flows through the brush in the cross direction [4]. Therefore, K.I. Shenfer called it as the cross current of the brush. As a result of the studies, a quantitative relationship was established between the magnitude of the additional commutation current and the sparking energy. The form of the additional commutation current was investigated as on the developed mathematical model, as well as on the real traction motor, that contributed on a measured signal processing technique. The estimation of the armature and excitation currents influence, the armature speed, brush contact area, commutator temperature and other performance parameters of DC traction motor made it possible to create a new method for the sparking evaluation which is based on additional commutation current measuring and correlation with sparking level in accordance with National Standard GOST 183- 74.

The main point of the method involves recording the cross current and identifying the relation with the sparking level of the commutator unit. The application of this method is most convenient when the motor brush is split (for example, TL2K1, NB-514, DPT810U), as for cross current measuring no change in the commutator design is required. Consider the fact that electromagnetic processes in the machine influence the performance of the commutation process, thermal and mechanical, the value and waveform of the additional commutation current should be construed as a universal characteristic of the electric motor condition [5].

To measure the commutation current, a spark monitoring device (SCR) was used, this device consists of a sensor made on the basis of stock-produced split brushes 1 and 2 that are isolated from each other (Fig. 1), a converter device 3 and a signal measuring system consisting of an analog-to-digital converter. The output of signal measuring system is connected to a personal computer. Unlike the well-known devices, for example PKK2, PKK5, this spark monitoring device uses standard composite motor brush in a function of sensitive element, the sparking on this brush-commutator unit is required to be determined. As a primary transformer we use a specially designed current transformer. The brush is connected to this current transformer so that an additional commutation current flows through its primary winding, the frequency of this current depends on the armature speed and the number of commutator segments. The current transformer is mounted on the brush holder. Its installation on brush holder does not cause any difficulty on any direct current traction or auxiliary motor. The secondary winding of the current transformer is connected to an analog-to-digital converter, which in turn is connected to a computing system. As computing system a portable PC with a specialized computing system or a special controller with appropriate memory can be used. The connection diagram is shown on Fig. 1.

A main feature of the spark monitoring device design is (Fig. 2):

1) the galvanic isolation of the measuring circuit and DC TM supply circuits, which makes the operation with the spark monitoring device safe;

2) the absence of an additional power source of the sensor installed in the traction motor, which dramatically simplifies the measurement circuit.

 DOI: 10.1051/matecconf/201714101034

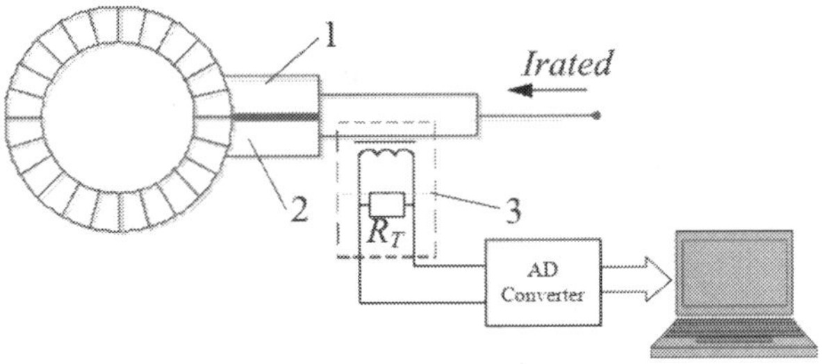

Fig. 1. Connection diagram of spark monitoring device and motor.

Fig. 2. The exterior of the spark monitoring device and its installation on the brush holder.

3 Experimental study

To determine the sparking intensity, the value of the additional commutation current was taken across the load resistor (R_T) of the measuring current transformer mounted on the brush holder. According to the energy theory of commutation [3], the excess of the measured current variable component over a certain acceptable value leads to the sparking appearance. It is known that the sparking intensity or sparking level is determined by the power segregated under the brush per length L_{br} in the time t_k/ϑ_k,(tk is the segment pitch, ϑ_k is the commutator peripheral velocity) [4,6]:

$$\Delta P = \frac{L_c i_k^2}{2} * \frac{\vartheta_k}{t_k L_{br}}$$

According to this equation, if the commutation current i_k is recorded for a particular electrical machine, the power ΔP segregated under the brush will be known; its magnitude can determine the occurrence and sparking level on the commutator. During the tests carried out on experimental model of the traction motor, the cross current curves (Fig. 3) flowing between the individual parts of the split brush were recorded. The sparking intensity was determined visually according to National Standard GOST 183-74.

Next comes the parallel processing of signals, in which two problems need to be solved: the determination of the sparking level and the degree of commutation adjustment. For spark indication, the effective value of the signal is calculated using the RMS method, then the signal is calibrated by recalculating the data on the calibration curve. The result is displayed on the front panel of the devices on the computer screen as shown in Fig. 4. These data are information on the sparking level on the commutator and are graded in accordance with National Standard GOST 183-74. To adjust the commutation, it is necessary to analyze the array of instantaneous values of the spark monitoring device.

DOI: 10.1051/matecconf/201714101034

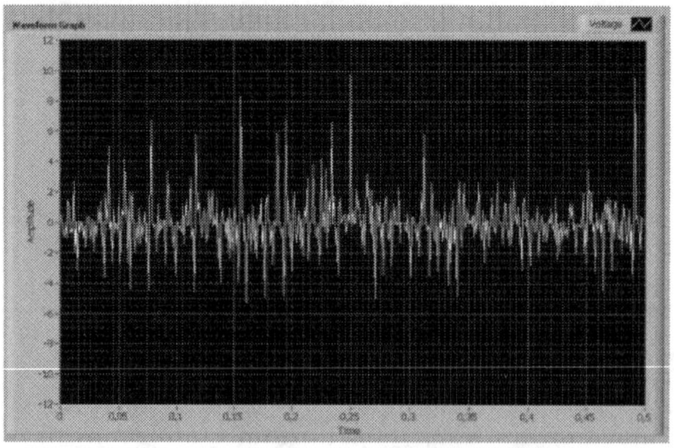

Fig. 3. Cross current curve.

From this array, at each two points, the minimum and maximum values of the signal are calculated, after calculating its absolute value is taken and compared between the minimum and maximum. If the maximum value is greater than the minimum value, the signaling information “Overcommutation” is presented on the front panel, if the condition is not met, Virtual instruments panel signals “Undercommutation” (Fig.4). The difference between the absolute values of the maximum and minimum is displayed on the virtual panel in the form of a pointer device with indication in per unit values. When carrying out a series of tests for each type of traction motor, it is possible to calibrate the indication of this device in millimeters of the diamagnetic gap under the interpoles that significantly simplify the magnetic system tuning. In the suggested study, only a qualitative assessment of the general commutation type is possible. Next, the algorithm implements the test report recording. For this, it is necessary to ensure that only the values specified in National Standard GOST 183-74 are recorded in the test report.

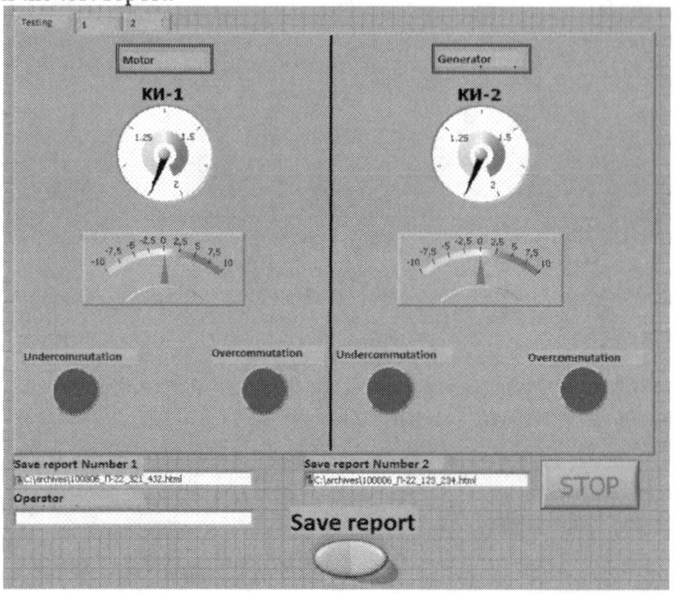

Fig. 4. Virtual instruments panel.

The pilot installation of the spark monitoring device at the test stations of the Ulan-Ude Locomotive and rail-way car repair plant, the Taiga locomotive repair depot of the West Siberian Railroad, and the Ural Locomotives shows that the monitor of sparking level of the DC TM is carried out not only in commutation tests but also in other test modes . It is carried out automatically with normally closed assembling hatches. Sensors installation takes about 10…15 minutes. The recording of the sparking level in grades in the protocol is performed by the operator command.

4 Discussion

The suggested spark monitoring device on direct current traction motors can be used not only at test stations, but also during operation, which makes it possible to use the information obtained for diagnosing and automated DC TM control not only for new locomotives testing at depot but also for monitoring during operation. The additional commutation current change that measured by the spark monitoring device has complex waveform. Studies show that data obtained from spark monitoring device benefit not only the determination of sparking level according to the cross current magnitude, but also provide other important indicators of the DC TM operation, based on the waveform analysis, frequency characteristics, and others [6].

5 Conclusion

As follows from the provided study, the commutation tests and monitoring technology of traction and auxiliary motors of electric locomotives are developed and adjusted.

It should also be noted that the wide use of the spark monitoring system and the diagnosis of traction motors is obstructed by the lack of regulations or rules for instrumental sparking detecting. National Standard GOST 2582-81 in the annex of 1991 permits instrumental indication, but this was not reflected in industry Standards. The most important unit of an electric locomotive is a traction electric motor, nowadays it does not have a regulatory framework for the sparking control during operation and diagnosis organization.

References

1. A.Kotelnikov, A.Nestrahov, *Bulletin of VNIIGT*, **5**, 15 (2000) [in Russian]
2. S.Obuhov, I.Plotnikov, M.Surkov, L.Sumarokova, K.Savkin, MATEC Web Conf. **91**, 01036 (2017)
3. V.Shevchuk, O.Muravlev, O.Stolyarova, V.Shevchuk, MATEC Web Conf. **91**, 01033 (2017)
4. K.Shenfer, *Dymanomachines and DC motors* (Enorgoizdat, Moscow, 1927) [in Russian]
5. A. Girnik, A.Fedyanin, A.Kiselev, E.Dorokhina, J. Mash. manuf. and reliab., **46**, 3 (2017)
6. D.Shibaev, A.Shibaev, O.Rapoport, A.Cukublin, Sc. Tech. J. Russian Electromecanics, **6**, 11-13 (2009)

Analysis of the Energy Potential of Renewable Energy Sources Egypt

S. Obukhov[1], and *A.* Ibrahim[1,*]

[1]National Research Tomsk Polytechnic University, 634050 Tomsk, Russia

Abstract. The article reviews the current energy balance of Egypt's resources and sources of consumption, the main factors affecting the structure of the energy balance, the predicted trends of their changes in the short term, the main tasks of the electric power industry of Egypt and the ways of their solution. Based on the analysis of energy potential of renewable energy sources, the proposed types and locations of installations of renewable energy those have the greatest prospects for practical application.

1 Introduction

Egypt is a country in Northeast Africa; From the north, it is washed by the Mediterranean, and from the east by the Red Sea. The area of the country is 1001 thousand km^2. The population of Egypt in 2016 amounted to 90 million people and grows by 1.3% per year. The average population density is about 70 people per km^2, of which 57% live in villages and 43% live in cities [1]. The country is rich in natural resources, among which there are various minerals: precious stones, natural gas, oil, coal and arable land. From the point of view of the energy balance, Egypt has large reserves of fossil energy sources. On the mainland and coastal deposits, oil reserves are 4189 billion barrels, gas reserves are estimated at 77,200 billion m^3.

Currently, about 90% of electricity in Egypt is produced with the use of fossil fuels and mainly of power plants operating on gas. The rapidly growing demand for electricity caused by demographic factors and the widespread use of low-efficiency electrical appliances has exceeded the level of domestic gas production, as well as the capacity of power systems and the generation of electricity, which leads to the regular formation of power shortages and fan outages. This leads to a depletion of fiscal resources of the state, creates a threat to the functioning of the economy of Egypt in the absence of reliable power supply of population and industry of the country. The data in Fig. 1 gives an idea of the annual growth in peak load of the country, which increased on average by 6% per year [2].

* Corresponding author: ibragim@tpu.ru

 DOI: 10.1051/matecconf/201714101035

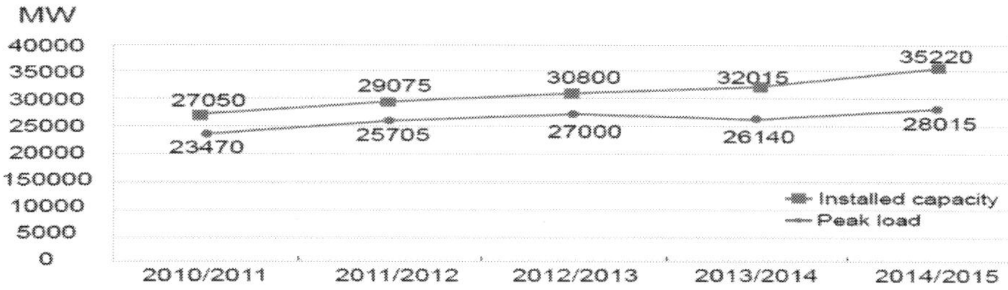

Fig.1. Dynamics of growth of peak load power system in Egypt.

The main objectives of the power sector in Egypt at the present stage are the creation of new generating power capacity and electrification of remote rural areas. In addressing these objectives, it is necessary to take into account, on the one hand, the tightening of requirements for environmental protection, the country's economic possibilities, the difficulties, and due to the limited reserves of fossil fuels, and on the other hand the availability of significant renewable energy sources [2, 3].

2 Research Methods and results

The analysis of energy potential of renewable energy sources shows that Egypt is rich in resources of solar energy, and in the Gulf of Suez includes some of the world's richest resources of wind energy. The development of these abundant renewable energy resources can contribute to increasing power generating capacity and reduce the country's dependence on natural gas imports. Wind and solar Atlas of Egypt (Fig. 2) shows the potential use of renewable energy sources (RES) for generating electricity. The areas along the coast of the Mediterranean Sea in northern Egypt are the most suitable for the use of wind and solar energy. It should be noted that in this place there are new rural areas that are not connected to a centralized network [1].

The increase in the consumption of electricity in Egypt is 6-7% per year [3], and in order to provide the necessary funds, the government is considering plans for the partial privatization of the electricity industry, increasing foreign investment. The total installed capacity of power plants amounted in 2010 to 27.05 GW (Fig.1) by the year 2015 increased to 35.2 GW [1, 2, 3]. The installed capacity of thermal power plants based on natural gas is 52%, and only 1,4% is accounted for by power plants based on wind energy.

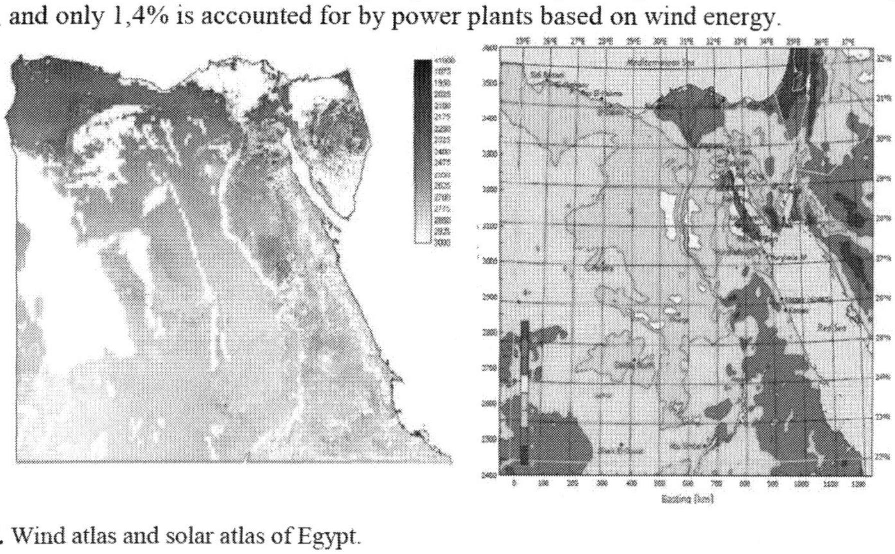

Fig.2. Wind atlas and solar atlas of Egypt.

At present, Egypt provides electricity to 99% of its population. Egypt's need for energy in the period from 1997 to 2008 has grown by an average of 7% per year and will increase in the range of 15-25% per year for the next twenty years. The Egyptian government hopes that by 2020, 20% of the installed capacity in the country will use renewable energy sources [1,4].

Agriculture on new lands in the desert is getting great development. The country sent about 1.26 million hectares for the development of new lands in the desert. Approximately half of this land is located outside the centralized power supply zones. In order to provide power supply to these farms, it is necessary to use autonomous power supply system's [4]. The Main Energy Council in Egypt decided to expand the use of renewable energy sources in order to achieve 20% of the total electricity production in 2020 (hydroelectric power plants 8% and wind power plants 12%) [5].

This can be achieved through the participation of private capital in the creation of wind farms in Egypt. According to annual data from renewable energy sources, the number of hours of sunshine per year in the north of Egypt is 3400 h and 3900 h in the south, and the average annual solar radiation in the north is 1900 kWh / m ^{2}and in the south 2600 kWh / m^2 (Fig.3). Consequently, almost the entire territory of the country the amount of solar radiation is large enough and almost year-round operation of solar power plants. Currently, the total installed solar capacity in Egypt is 145,2 MW, including the capacity of thermal power stations in Kureimat, located 92 km South of Cairo [2], by 2020 it is planned to increase to 2.13 GW with power generation of 7.35 GW·h [7].

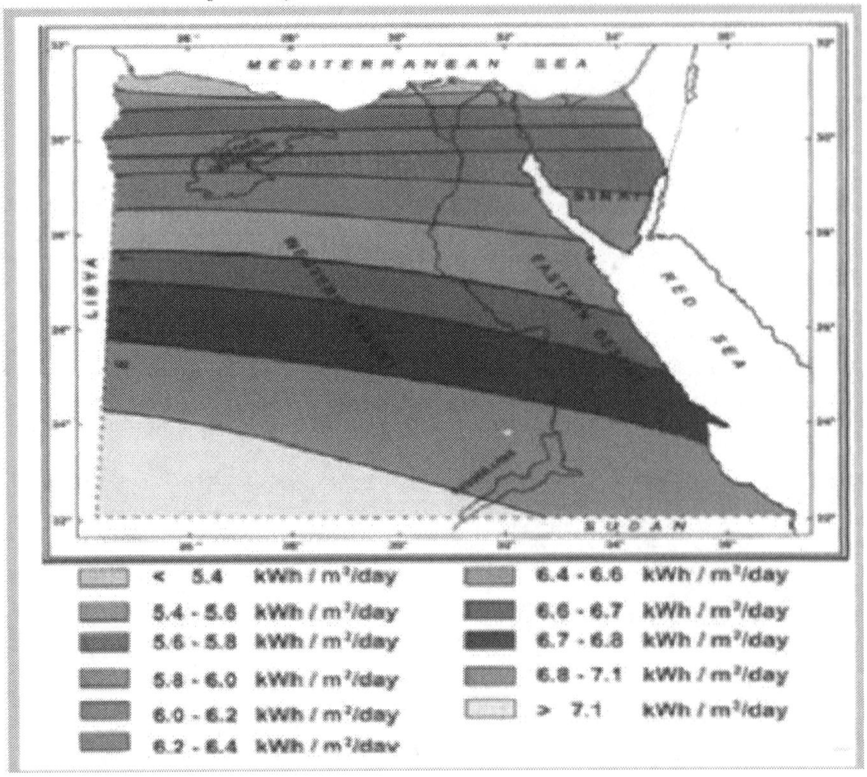

Fig.3. The distribution of solar radiation on the territory of Egypt.

The West of the Gulf of Suez is the most promising place for the construction of large wind power stations due to the wind speed, which varies between 8-10 m / s, and also due to the low population density in this area (fig2). There are other promising places with a wind speed of 7-8 m / s in the east and west of the Nile River near the cities of Beni Suaiwa

DOI: 10.1051/matecconf/201714101035

and Minia. At present, there are two wind farms in Egypt Hurghada and Zafarana wind farms (fig 4) installed wind power plants in Zafaran, their generation of electricity in 2012/2013.

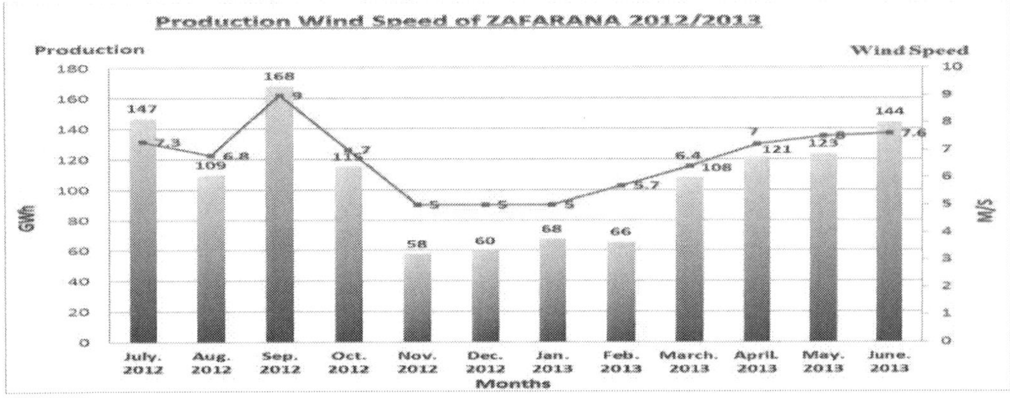

Fig.4. Production of electricity of Zafarana 2012/2013.

3 Conclusion

The increase in electricity generation in Egypt can be ensured by the use of local energy resources, including renewable energy sources. The level of solar radiation in Egypt is high enough that it allows expanding the share of solar power stations in the energy sector of the country. A high average annual wind speed in the east and north of the country, reaching 8-10 m / s, allows using economical wind power plants.

The work was carried out within the framework of the implementation of the Federal Target Program "Research and Development in Priority Areas for the Development of the Russian Science and Technology Complex for "2014-2020", a unique project identifier No. 14.576.21.0098.

References

1. Arab Republic of Egypt, Ministry of Electricity and Energy, Annual Report
2. Arab Republic of Egypt Central Agency for Public Mobilization and Statistics www.msrintranet.capmas.gov.eg.
3. REN21 Renewable Energy Policy Network for the 2E 'Century, (2011)
4. Wind atlas for Egypt. Research and development department, the Egyptian Meteorological Authority, Cairo, Egypt in (2006)
5. Energy in Egypt. Year Book of Egypt. Egypt State Information Service, Cairo (2008)
6. Energy Information Administration (EIA). Egypt Energy Data, Statistics and Analysis Oil, Gas, Electricity, Coal.http://tonto.eia.doe.gov/cfapps/ipdbproject/EEDIndex3.cfm?tid=5&pid=:53&aid.
7. Arab Republic of Egypt. Central Agency for Public Mobilization and Statistics. Annual report (2007/2008)

MATEC Web of Conferences **141**, 01036 (2017)
Smart Grids 2017
DOI: 10.1051/matecconf/201714101036

Obtaining molybdenum-copper composite for effective thermal control in electronic systems

Alexander Sivkov[1], *Yuliya* Shanenkova[1], and *Yuliya* Polovinkina[1,*]

[1]National Research Tomsk Polytechnic University, 634050 Tomsk, Russia

Abstract. In the article, the molybdenum and copper powders, obtained by electric explosion of wires and plasma-dynamic method, respectively, were sintered by spark plasma sintering technology, in order to obtain bulk sample based on them. X-ray diffractometry and scanning electron microscopy were carried out to study the chemical composition and structure of the resulting ceramics. XRD analysis showed that the product consisted of the following main phases such as molybdenum of a cubic crystalline structure (space group Im3m, No. 229) and copper of cubic system (space group Fm-3m, No. 225). The results of scanning electron microscopy indicated that the sintered sample had heterogeneous structure with the presence of small pores. Nonetheless, the mechanical properties such as hardness (Vickers hardness of 292 HV) and thermal conductivity (140.53 W / m · ℃) are good enough that allows considering the proposed synthesizing methods as well as sintering technology as promising for creating the Cu-Mo ceramics with high-mechanical properties.

1 Introduction

Increasing the energy density is one of the biggest problems for thermal regulation in small electronic devices. Indeed, the insufficient cooling negatively affects the reliability and efficiency of semiconductor components [1, 2]. According to statistical data, more than half of the electronic components failure cases are explained with the increased temperature.

Electronic components are usually installed on substrates or plates, which act as the heat removal elements and provide an effective thermal control [3-5]. However, not all materials have suitable properties for reliable heat dissipation. In such substrates, the thermal conductivity and thermal expansion characteristics of the semiconductor must be matched with each other in the optimal way. The better the correlation between the properties, the better the semiconductor module withstands soldering during the production and temperature cycles in operation mode. In this paper, the bulk ceramics sample based on molybdenum and copper powders was sintered using the spark plasma sintering technology. Unlike the existing methods [6, 7], this technology makes it possible to obtain bulk samples based on powders in a short period of time that allows preserving the fine-grained structure of the material and its unique properties. It is assumed that the molybdenum low thermal expansion and the copper excellent thermal conductivity will

* Corresponding author: xyulyashax@mail.ru

provide substrates that can reliably dissipate heat in electrical components and help cool IGBT-modules, RF-modules, LED crystals and other products [8,9]. Also in this paper, a unique method for obtaining ultrafine copper powders based on using coaxial magnetoplasma accelerator is shown [10-12]. In comparison with other methods for obtaining ultrafine powders [13, 14], the discussed technique is more simple (working time less than 1 ms) and doesn't require preliminary preparation of the main precursors.

2 Materials and experimental procedures

In order to obtain the bulk samples, the precursors in the form of ultrafine powders were used. Molybdenum was obtained using the method of electric explosion of wires (commercial powder), as well as copper was synthesized by plasmadynamic method. The unique installation based on the coaxial magnetoplasma accelerator with copper electrodes, which was used to obtain copper powder, is shown in figure 1.

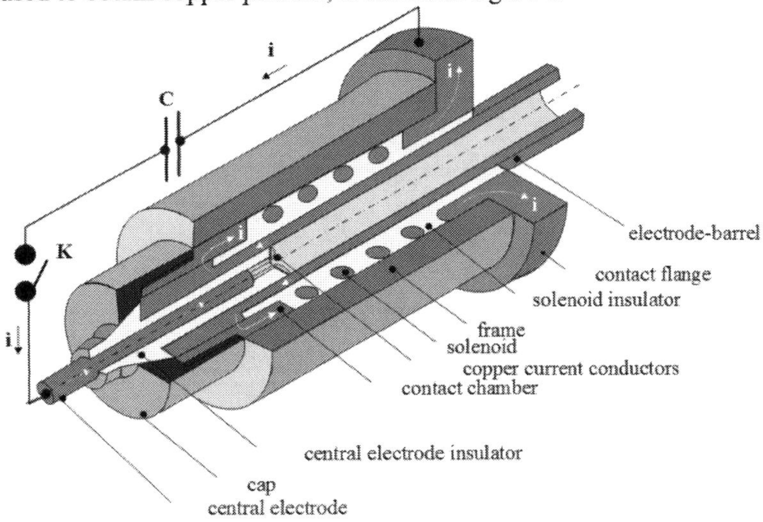

Fig. 1. Sketch-map and elements of coaxial magnetoplasma accelerator.

This device is based on the classical Z-pinch accelerator placed in an external induction system. The power supply of the accelerator is provided by a capacitive energy storage device with the following operating parameters: charge voltage up to U = 5.0 kV and capacitance of capacitor banks up to C = 28.8 mF. The Z-pinch accelerator consists of parts such as a central non-magnetic metal electrode with a copper insert at the end, a copper electrode-barrel, copper current conductors, a central electrode insulator, and power elements, holding this assembly. Z-pinch accelerator is placed in an external induction field, which is necessary to equalize the electroerosion wear along the length of the electrode-barrel.

After closing the power keys (K), the capacitive energy storage device (C) discharges, therefore, the current increases immediately in the circuit; when it reaches to certain critical value, an electric explosion of the copper conductors takes place with the formation of an arc discharge. Due to the magnetic field the formed arc discharge of the solenoid is pulled along the electrode barrel. When the discharge propagates, the plasma jet interacts with the accelerating channel walls, and thus the copper production is performed due to electroerosion wear. Then the plasma escapes from the accelerating channel, getting into the chamber-reactor, where plasma-chemical reaction is implemented. The synthesized product settles down to the wall of the chamber. After one hour, the chamber is opened and

 DOI: 10.1051/matecconf/201714101036

the product is collected. It is worth nothing that the chamber-reactor was preliminary evacuated and filled with inert gas (argon) to prevent the saturation of the resulting material by oxygen (to avoid product oxidation).

Using a mix of such-obtained powders (copper and molybdenum) a bulk sample was obtained using a spark plasma sintering (Advanced Technologies 10-4 SPS). Cu and Mo powders were filled into the graphite molds (maximal pressure P_{max} = 60 MPa, maximal temperature T = 2200 °C) in the following rations of 30 wt % to 70 wt %, respectively. The sintering process parameters were the following: the sintering atmosphere was vacuum, heating rate was 100 °C/min., cooling rate was 500 °C/min, maximum heating temperature was regulated by the current flowing through the sample. Figure 2 shows the curves of temperature, sintering pressure and material density changes.

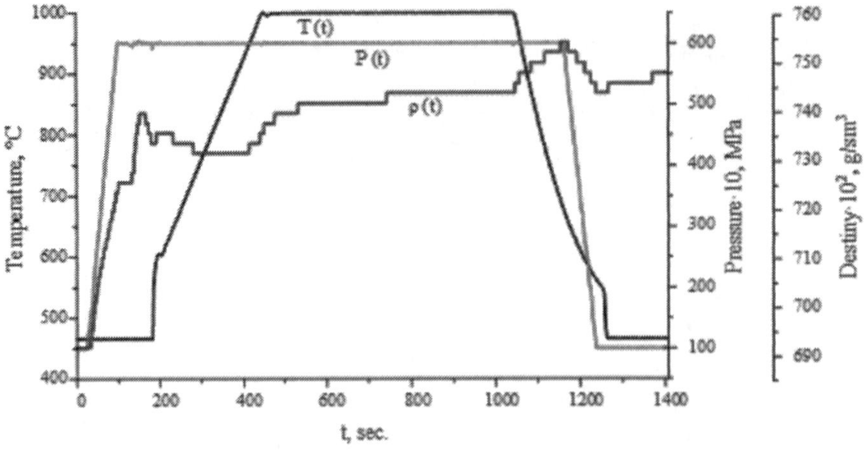

Fig. 2. Graph of temperature, sintering pressure and material density changes.

XRD. X-ray diffractometry analysis was carried out using a Shimadzu XRD-7000S diffractometer with Cu-Kα radiation equipped with the counter monochromator. The phase-structural analysis has been made using the database PDF2 and the program software PowderCell 2.4.

SEM. Scanning electron microscopy of the sintered ceramics was carried out using a Hitachi TM-3000 scanning electron microscope.

Thermal conductivity. Thermal conductivity measurement was studied using a Discover Flash DLF-1200 thermal diffusivity analyzer.

Hardness. The hardness of the bulk sample was determined by a Vickers hardness tester Galileo Durometer Isoscan HV2 with a sample load of 0.5 kgf.

3 Results and discussion

As a result of sintering process, the dark grey ceramics sample was obtained. Its geometrical sizes were found (the diameter of 12.7 mm, the height of 4 mm) in order to calculate the resulting density, which was found to be equal to 7.48 g/sm³. Figure 3 shows the SEM-image of the sintered Cu-Mo ceramics sample. It can be seen that the material is very heterogeneous in structure. There are both nanoparticles and large agglomerations of 20 μm and larger. This unsatisfied structure and a relatively low sample density (ρ = 76%), allows us to conclude that the solid sample obtained under such conditions is not of a sufficient quality because of a significant number of pores, the presence of which can be easily explained by the low sintering temperature. It can be seen that the molybdenum

 DOI: 10.1051/matecconf/201714101036

particles are embedded in the copper matrix. The SEM-image allows also finding the presence of big white inclusions, the nature of which was not understood until the carrying out XRD analysis.

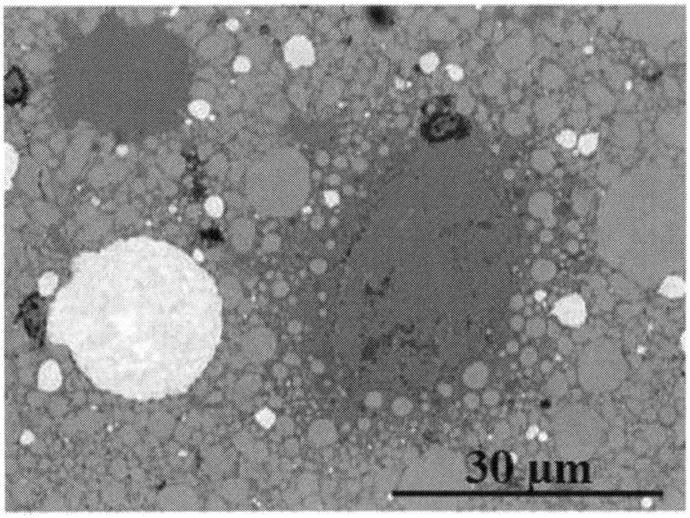

Fig. 3. SEM-image of sample.

Figure 4 shows X-ray diffraction pattern of the synthesized product. This figure shows that the product consists of the following main phases such as molybdenum of cubic system (space group Im-3m, No. 229) and the copper of a cubic crystalline structure (space group Fm-3m, No. 229). There are also a number of low-intensity peaks in the 2θ range from 30 to 40 degrees. The identification of these peaks in the PowderCell 2.4 software allowed attributing these peaks to the phase of tungsten, which can be included in the ceramics sample with the molybdenum powder. The process features electric explosion of wires cannot prevent from obtaining tungsten particles, that is why some of them are found on XRD as well as SEM data. Nonetheless, the dominant phases are Mo and Cu and their content in the resulting ceramics are 70.39% and 29.61%, respectively.

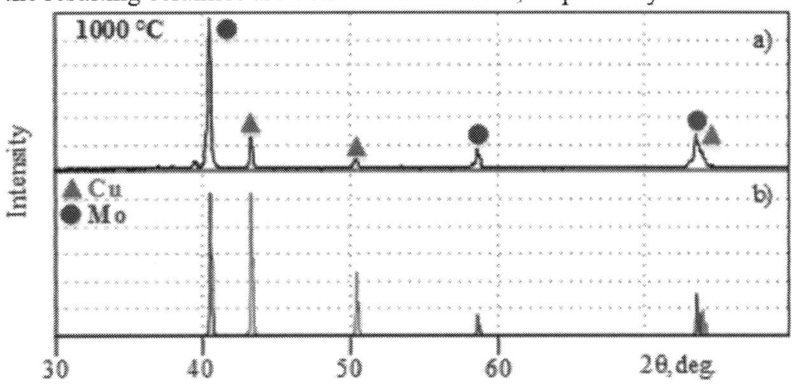

Fig. 4. XRD patterns of: a) sintered product, b) reference phases.

In order to estimate the possibility of using such ceramics in heat-conducting plates, such mechanical properties as thermal conductivity and hardness were studied. The measurements of thermal conductivity, which were carried out at room temperature, showed that the maximum thermal conductivity of the material was 140.53 W / m·℃. This parameter can be considered as good enough in comparison with known data [15], however it can be improved further by increasing the sintering temperature in order to melt all copper, which was found in the structure in the form of large aggregates (fig. 3). The presence of pores between Cu aggregates and Mo particles affects the Vickers hardness value which was equal to 292 HV. This value is much greater than for pure molybdenum (147 HV), nonetheless it can also be improved with increasing the density of ceramics.

4 Conclusion

This paper showed the possibility of obtaining a bulk ceramics material based on ultrafine powders of molybdenum and copper. The methods for obtaining Cu powders and a ceramics sample were considered in detail. The presence of Mo and Cu phases in the final sample was confirmed by the results of X-ray diffraction analysis. The microstructure of the sample was also studied. To sum up, it is necessary to increase the sintering temperature to obtain the sample with the higher density of the material. This fact should have a positive effect on the material heat conductivity, which can increase with decreasing the pore number. Nonetheless, the not too high density does not affect the hardness of ceramics sample, which was found to be equal to 292 HV that is much higher than for pure molybdenum.

References

1. G.V. Kuznetsov, D.V. Feoktistov, E.G. Orlova, K.A. Batishcheva, Colloid J. **78**, 335 (2016)
2. D. Feoktistov, E. Orlova, A. Islamova, EPJ, **110**, 01018 (2016)
3. N.Y.A. Shammas, Microelectron. Reliab. **43**, 519 (2003)
4. A.B. Lostetter, F. Barlow, A. Elshabini, Microelectron. Reliab. **40**, 365 (2000)
5. R. Hocine, A. Boudghene Stambouli, A. Saidane, Microelectron. Reliab. **65**, 293 (2003)
6. A.G. Karengin, A.A. Karengin, I.Y. Novoselov, N.V. Tundeshev, Adv. Mater. Research, **1040**, 433 (2014)
7. A.A. Karengin, A.G. Karengin, V.A. Vlasov, Russian Physics J. **58**, 730 (2015)
8. D. Srinivasan, P.R. Subramanian, Mater. Sci. Eng. **459**, 145 (2007)
9. B. Wang, S.H. Liang, J.T. Zou, X.H. Yang, P. Xiao, Rare Metal Mater. **40**, 1621 (2011)
10. A. Sivkov, A. Ivashutenko, Y. Shanenkova, I. Shanenkov, J. Adv. Powder Techn. **27**, 1506 (2016)
11. A.A. Sivkov, A.Y Pak, D.S. Nikitin, I.A. Rakhmatullin, I.I Shanenkov, Nanotech. in Russia, **8**, 489 (2013)
12. A. Sivkov, A. Saygash, J. Kolganova, I. Shanenkov, IOP Conference Series: Mater. Sci. Engin. **66**, 012048 (2014)
13. E.B. Kasharov, N.N. Nikitenkov, A.N. Sutygina, M.S. Syrtanov, O.V. Vilkhivskaya, etc., J. Surface Technol. **308**, 2 (2016)
14. V.N. Kudiiarov, E.B. Kasharov, M.S. Syrtanov, I.S. Yugova, IOP Confer. Series: Mater. Science Engin. **110**, 012071 (2016)
15. D. Wang, B. Yin, A. Sun, X. Li, C. Qi, B. Duan, J. of Alloys Compounds, **674**, 347 (2016)

The selection of interval boundaries of input and output data method for obtaining complete probabilistic characteristics

Yuly Bay[1], *Anatoliy* Shmoilov[1], *Alexander* Gusev[1], and *Igor* Razzhivin[1,*]

[1]National Research Tomsk Polytechnic University, 634050 Tomsk, Russia

Abstract. The results of development and experimental investigations of the selection of interval boundaries of input and output data method are considered in the paper. The article presents an improved algorithm for obtaining a laws of probability of functional. The experimental part contains examples of finding the full probabilistic characteristics for sums of independent random variables.

1 Introduction

The electric power industry, as well as many others, frequently requires the assessment of risks of overvoltage [3, 9], accidental losses and destruction caused either by real values exceeding standard operating limits or by reduction of service life parameters relative to regulations [4]. This requires the knowledge of total probabilistic characteristics of the parameters of interest in the form of laws of probability.

It is considered rational to address mode parameters (and through them the electrical quantities during failures) through the arguments of active and reactive powers of load buses and active powers and voltages of generator buses. In this case, according to research [2], total probabilistic characteristics in the form of probability distribution functions (PDFs) and probability densities (PDs) fully comply with normal laws of probability (LPs). It is difficult to obtain representative statistics for mode parameters and electrical quantities during breakdowns; however, when the arguments' LPs and mode parameters' functional dependencies (FDs) are known, it is natural to design a method to form the LPs of the mode parameters' functional dependencies according to the LPs of the arguments of the FDs. Such method was designed and called the selection of interval boundaries of input and output data (SIBD). One of its modifications is presented in [5].

2 SIBD Logic

The principle of the SIBD method relies on the logical assumption that if all FD arguments are set as equally probable, the result of FD transformation will have the same probability [8]. This procedure can be implemented through the PDF and the PD of each argument and

* Corresponding author: lionrash@tpu.ru

the result of FD transformation. In this case, the PDF as a non-decreasing function of a random object guarantees the unambiguity of all input and output data, while the PD of this data ensures the correlation of all specified data (arguments and LDs) in the form of the equality of probability of assuming values from the intervals linked respectively with one of its boundaries to the arguments of non-decreasing one-dimensional PDFs. Consequently, second boundaries are also arguments of the specified PDFs that are determined on the basis of the equality of probability of assuming of value by all data (argument as input data and FD as output data), calculated according to the PDs of each of these data. The boundaries of the arguments and the obtained FD transformation result are rigidly fixed by quantiles of orders that are determined by the same values of their PDFs and the same probability of assuming values from these intervals. At the same time, the order of quantiles or argument's PDFs on the other (right) boundary of the intervals is also determined by adding equal probabilities of assuming of values from the intervals to equal values of PDFs on the previous (first) boundary of these intervals; while the result of the FD is determined by adding or subtracting the same probability depending on the increase or the decrease of the FD.

Further, the sequence of calculations to obtain the values of the PD of the mode parameters' functional dependence is given for the proposed fundamental validation of the SIBD method [8].

3 SIBD Algorithm

1. Preparation of types and parameters of one-dimensional PDFs and PDs of random variables $X_1,\dots,X_i,\dots,X_n$, i.e. $\mathrm{f}_1(x_1)$, $\dots,\mathrm{f}_i(x_i),\dots$, $\mathrm{f}_n(x_n)$ and $\mathrm{c}_1(x_1)$, $\dots,\mathrm{c}_i(x_i),\dots$, $\mathrm{c}_n(x_n)$

2. Selection of the quantity and the values of PDF quantile orders 0, p_1, …, p_j, …, 1

3. Forming the RV values as quantiles of the selected orders and values of the RVs of the functional dependence $Y=\varphi(X_1,\dots,X_i,\dots,X_n)$, as well as the RV PD from these quantiles 0, p_1, p_j, : 1. For example quantile 0:

$$x_{10}=\mathrm{f}_1^{-1}(0),\dots,\ x_{i0}=\mathrm{f}_i^{-1}(0),\ \dots,\ x_{n0}=\mathrm{f}_n^{-1}(0),\ y_0=\varphi(x_{10},\dots,x_{i0},\dots,x_{n0}),$$

$$\mathrm{c}_1(x_{10}),\dots,\ \mathrm{c}_i(x_{i0}),\ \dots,\ \mathrm{c}_n(x_{n0});$$

4. The basic components of the joint PDFs and PDs of random FD values or FD quantile orders as probabilities of combining all independent RVs as quantiles of one order. For example quantile 0:

$$0\cdots0\cdots0=\mathrm{f}_{1\dots i\dots n}(x_{10},\dots,x_{i0},\dots,x_{n0})=\mathrm{f}_1(x_{10})\cdots\mathrm{f}_i(x_{i0})\cdots\mathrm{f}_{in}(x_{n0}),$$

$$\mathrm{c}_{1\dots i\dots n}(x_{10},\dots,x_{i0},\dots,x_{n0})=\mathrm{c}_1(x_{10})\cdots\mathrm{c}_i(x_{i0})\cdots\mathrm{c}_{in}(x_{n0});$$

5. Exhaustive search, selection, and summing of joint PDFs and PDs of independent RVs as components of the functional dependence $Y=\varphi(X_1,\dots,X_i,\dots,X_n)$

Indices of the orders of quantiles $(j,\dots,k,\dots,l)=0,\ p_1,\dots,p_{j,\dots,k,\dots,l},\dots,1$

Exhaustive search of quantile-values of RVs and calculation of the FD values compatible with the PDF and PD $y_{j,\dots,k,\dots,l}=\varphi(x_{1j},\dots,x_{ik},\dots,x_{nl})$, so for quantile 0 if $y_{j,...,k,...,l} = y_0$, then $c_{1...i...n}(x_{1j},...,x_{ik},...,x_{nl}) = c_1(x_{1j}) \cdots c_i(x_{ik}) \cdots c_n(x_{nl})$ is calculated and first occupies the cell $c_{1...i...n}(x_{10},...,x_{i0},...,x_{n0})$, and then, during the exhaustive search, is added to the content of this cell, etc.

After the exhaustive search of all joint RV values and the summing, the cells' content will be nothing other than the desired PD values of the functional dependence $Y = \varphi(X_1,\ldots,X_i,\ldots,X_n)$, e.g. $c_y(y_{p_j})$ is the content of the cell

$$c_{1,...,i,...,n}(x_{1p_j},...,x_{ip_j},...,x_{np_j}) + c_{1,...,i,...,n}^{\varphi(x_{1p_j},...,x_{ip_j},...,x_{np_j})}(x_{1p_j},...,x_{ip_k},...,x_{np_l}).$$

Qualitative representations of the beginning and the end of the exhaustive search process provide the possible charts of the PD od the functional dependence.

3 Full probabilistic characteristics for sums of independent random variables

1. Sums of continuous (three, four) uniform random variables

With an increase in the number of RVs of the uniform density involved in the convolution, the resulting form of the LP will tend to form the normal LP [2].

In the example, we define the PD of a quantity $Z = \varphi(X+Y+V)$ and $Z = \varphi(X+Y+V+W)$. Since we still consider the quantities RVs independent, so $f(x,y,v)=f_1(x)*f_2(y)*f_3(v)$ and $f(x,y,v,w)=f_1(x)*f_2(y)*f_3(v)*f_4(w)$ respectively the laws of uniform density are also given on the section (0.1).

2. Convolution of n uniform densities.

Carrying out the experiment and increasing the number of CAs to 4x the same result is observed.

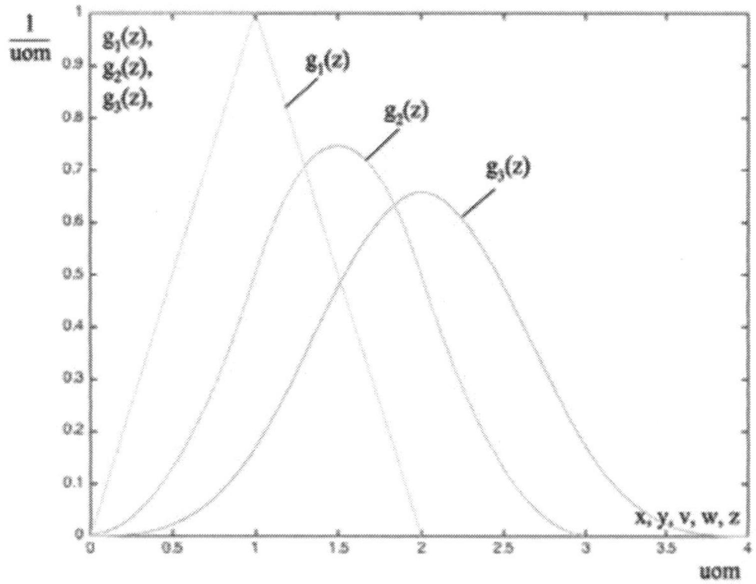

Fig. 1. PDs of the sum of three and four uniform random variables.

DOI: 10.1051/matecconf/201714101037

4 Conclusions

According to the obtained data, comparing with the already available results obtained in classical ways, it can be said with certainty that the PDs found by the SIBD method are correct.

Summation of densities with equal values of the FDs is quite easy and simple method of obtaining full probabilistic characteristics. Unlike standard methods, the method does not require integration, transformation of the original PF. Also, unlike modifications of the Monte Carlo method [1, 7], the SIBD does not require an increase in the number of tests according to the number of random arguments to obtain a statistical representation of the result with the required accuracy [6, 10].

In addition, the SIBD uses not a sample, but a entire assembly (using the method of full enumeration), which allows to capture all areas of the received total probabilistic characteristics without distortion.

References

1. A. Genz, J. Statistics. Computing, **14** (3), 251 (2003)
2. J. Hsu, *Multiple Comparisons* (London: Chapman and Hall, 1996)
3. S. Stavitsky, N. Palukhin, J. Kobenko, E. Riabova, EPJ Web Conf. **110**, 119591 (2015)
4. M. Andreev, Y. Borovikov, N. Ruban, 11th Int. Forum Strategic Technol. 7884223 (2016)
5. Y. Bai, A. Shmoilov, *Analysis of the possibilities of the development of the selection of interval boundaries of input and output data method for obtaining probabilistic characteristics of the functional dependencies of the tasks of the electric power industry* (Proceedings of the VII International Scientific and Technical Conference "Electrotechnical Energy Converters", Tomsk, 2015) [In Russian]
6. A. Suvorov, A. Gusev, A. Sulaymanov, O. Suslova, MATEC Web Conf. **91**, 01052 (2017)
7. A. Gusev, A. Suvorov, A. Sulaymanov, ICIEAM (2016)
8. Y. Bai, A. Shmoilov, *Full probabilistic characteristics of multidimensional Functional dependencies tasks of Electric Power Industry* (Proceedings of the First International Scientific Conference of Young Scientists "Electrical Engineering. Power engineering. Mechanical engineering », Novosibirsk, 2014) [In Russian]
9. M. Andreev, Y. Borovikov, N. Ruban, A. Khlebov, MATEC Web Conf. **91**, 010484 (2016)
10. R. Ufa, A. Vasilev, A. Gusev, A. Suvorov, 11th Int. Forum Strategic Technol. 7884204 (2017)

MATEC Web of Conferences **141**, 01038 (2017)
Smart Grids 2017
DOI: 10.1051/matecconf/201714101038

The concept of usage HRTSIM as a device for setting and certification of AVR

Alisher Askarov[1,*], and *Nikolay* Ruban[1]

[1]National Research Tomsk Polytechnic University, 634050 Tomsk, Russia

Abstract. This paper presents the analysis of application Hybrid Real-Time Power System Simulator (HRTSim), developed at National Research Tomsk Polytechnic University, as a device for setting and certification of automatic voltage regulators (AVR) for synchronous generators. The simulation results demonstrate that the proposed hardware-software complex (HSC) can be used for presented aim.

1 Introduction

The reliability level of the United Power System (UPS) of Russia is determined, among other factors, by reliability and operational characteristics of power facilities, devices and control systems, including the characteristics of the modern microprocessor devices of automated control, regulation, protection and automation.

However, the devices used in practice may have algorithmic or programming errors. Also they are not always adapted to the operating conditions in a real power system. That all leads to increase in the number of technological disturbances in power systems and decrease the reliability of parallel operation of power plants and power systems, as a whole. In particular, disconnection of a generating unit became more frequent due to the loss of excitation (incorrect operation of AVR) in power plants, moreover, due to the fact that interconnecting mains are usually "weak" and heavily loaded, it in turn causes the low-frequency (0,1-0,5 Hz) fluctuations with large amplitude in power systems and leads to a non-effective operation of AVR and false operation of protective relays [1-3].

Therefore, microprocessor devices of AVR used in new or modernized power facilities must be tested in conditions close to the conditions of future operation to ensure system reliability of the UPS of Russia [4], which, moreover, is obligatory, according to the standards of JSC "SO UPS" [5]. In accordance with [5], testing of the AVR settings is carried out – either on a physical model of the power system, either on a mathematical model of the power system using a hardware-software complex of real-time power systems simulation.

But at this moment, the certification of AVR is exclusively performed on the physical model of electric power system.

* Corresponding author: askalibah@gmail.com

2 Research aim

In connection with the foregoing, the aim of this work is the development of the first domestic HSC for setting and certification of AVR of synchronous generators on the basis of Hybrid Real-Time Power System Simulator, developed at National Research Tomsk Polytechnic University, which after adaptation will be meet requirements of JSC "SO UPS" and can be used to solve the set aim [6].

HRTSim is a multiprocessor system based on a modern concept of hybrid simulation of power system and has a number of advantages over widely used digital systems:

1. excluded regimes and processes decomposition;
2. high simulation accuracy, limited only by instrumental error, which depends on the component base;
3. ensures continuous methodically exact solution of adequate three-phase mathematical models of facilities and EPS as a whole in real-time;
4. has no limits on the volume of the simulated power system and the depth of the models.

Properties and possibilities of this HSC allow to use it to solve a wide range of different objectives, which are confirmed by the results of many researches [7-10].

3 Materials and methods

The mathematical model of AVR, shown in Fig. 1 and used in HRTSim, has been used as the basis for the research.

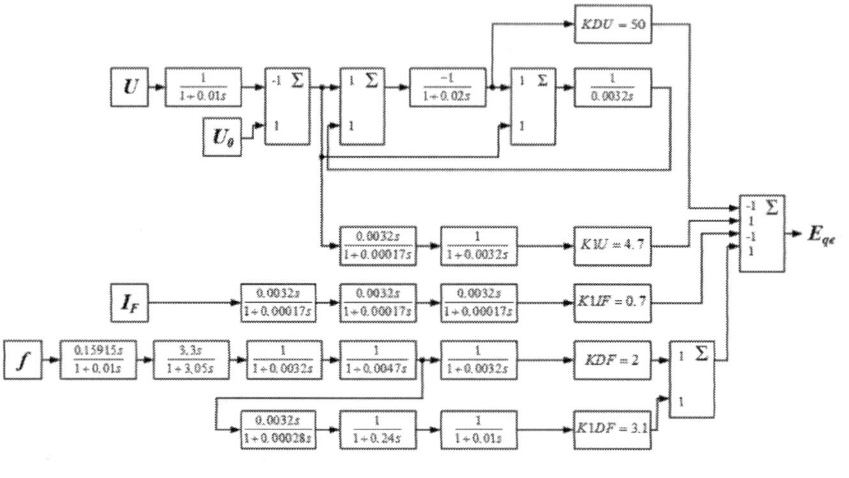

Fig. 1. The mathematical model of AVR, realized in HRTSim.

This structural scheme of AVR mathematical model synthesized on the basis of the analysis of different existing excitation systems and AVR [11, 12], that allows to consider significant factors and elements for the adequate simulation of regulators.

Taking into account the possibility of changing, regulating, implementation of additional channels, resetting all coefficients and parameters in the presented AVR model, this mathematical model allows to adequately and fully simulate the functioning of all types of excitation systems and AVR.

The model of the power system for AVR testing (Fig. 2), presented in [5] and realized in HRTSim, has been used as the basic scheme for researches. As the simulated generator with the testing AVR has been used the generator "*G-1_1*".

DOI: 10.1051/matecconf/201714101038

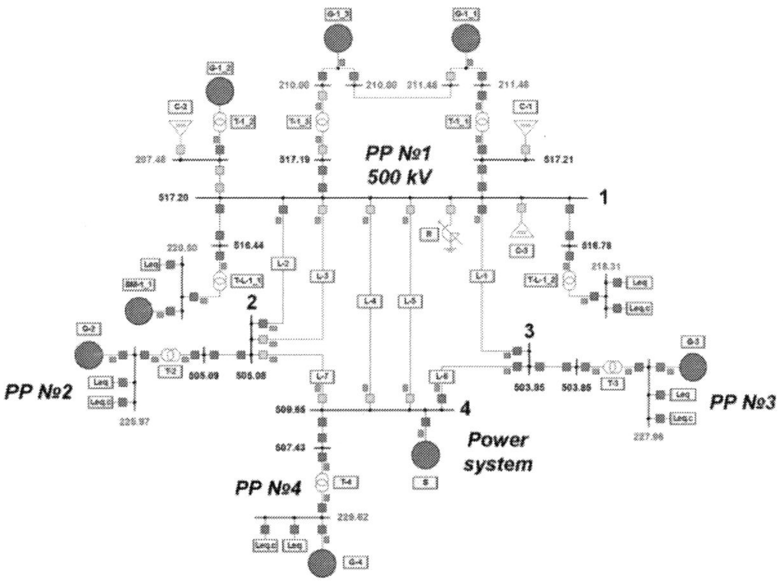

Fig. 2. The model of the power system for AVR testing, realized in HRTSim, according to [5].

As a check of the correct operation of the AVR of synchronous generators has been performed test disturbances, according to the standard "SO UPS" [5]. In this paper presented the results of the next test disturbance – a step changing of voltage on the busbars of power station, where installed testing AVR, by connecting a capacitance *C1* = 188 MVAr to the busbars of this station (the busbars 500 kV PP No.1). Example of the correct operation of AVR at the entrance to the area of minimum excitation limit (MEL) and exit from the area of MEL, according to [5], is shown in Fig. 3a.

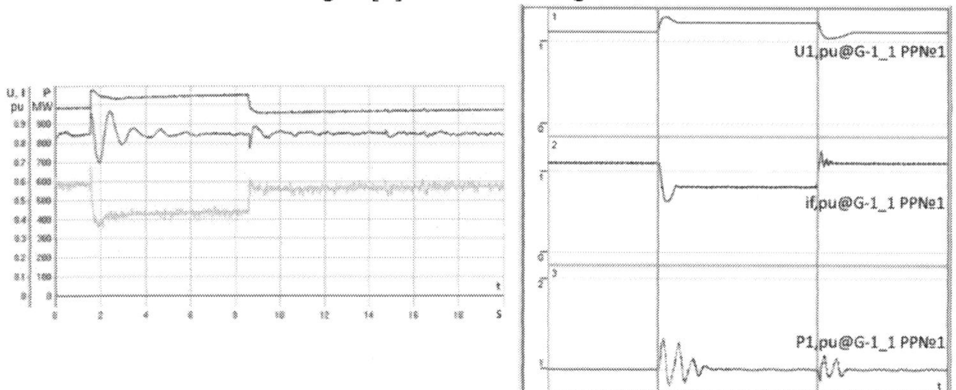

a). *b).*

Fig. 3. Oscillograms of the AVR operation in case of MEL.

In this experiment has been verified the correctness of the AVR operation in the regime of minimum excitation limit. As can be seen from the oscillograms (Fig. 3b), when the balance of reactive power changed under connecting of capacitance, the voltage on the busbars of power plants increased and the excitation current reduced to the minimum value, what ensures the functioning of MEL, which does not allow violation of the static stability of the power system.

4 Analysis of the obtained results

In this article for the first time proposes to use HRTSim for setting and certification of AVR, which combines the advantages of physical and mathematical simulation of power systems.

Further work lies in the development of the communication interface between HRTSim and real AVR devices and also in the modernization of the mobile version of the complex for onsite setting and certification, which in turn would reduce the time and cost of work. This development especially effective for:

– hydropower plants, where the hydro-generators and characteristics of power plant are purely individual;
– nuclear power plants, where full-scale and correct setting of AVR is required for security reasons.

Also the relevance of this work is confirmed by many Russian companies such as JSC "Rosenergoatom Concern", PJSC "RusHydro", PJSC "INTER RAO" [13].

This work was supported by the Ministry of Education and Science of the Russian Federation under the governmental grant "Science" No. 13.5852.2017/BCH (Development of the concept for comprehensive validation of calculating modes and processes in electric power system and tools of its realization).

References

1. A. Suvorov, Y. Borovikov, A. Gusev, A. Sulaymanov, M. Andreev, N. Ruban, R. Ufa, Electr. Eng. J. **99**, 959 (2017), DOI: 10.1007/s00202-016-0464-4
2. N. S. Kumar, J. Gokulakrishnan, Int J. electr. power energy syst. **33**, 1172 (2011)
3. A. Ghorbani, B. Mozafari, S. Soleymani, A.M. Ranjbar, Turk. J. Elec. Eng. Comp. Sci. **24**, 2575 (2016)
4. A. Gusev, A. Suvorov, A. Sulaymanov, MATEC Web Conf. **37**, 01023 (2015), DOI: 10.1051/matecconf/20153701023
5. STO 59012820.29.160.20.001-2012. Requirements to excitation systems and automatic voltage regulators of synchronous generators
6. A. A. Suvorov, A. S. Gusev, A. O. Sulaymanov, O. V. Suslova, MATEC Web Conf. **91**, 01052 (2017), DOI: 10.1051/matecconf/20179101052
7. A. S. Gusev, A. A. Suvorov, A. O. Sulaymanov, IOP Conf. S. **93**, 012016 (2015), DOI: 10.1088/1757-899X/93/1/012016
8. Y. Borovikov, A. Prokhorov, M. Andreev, Proc. 7th Int. Forum Strategic Technol. 341 (2012), DOI: 10.1109/IFOST.2012.6357822
9. A. Prokhorov, Y. Borovikov, M. Andreev, Proc. 7th Int. Forum Strategic Technol. 512 (2012), DOI: 10.1109/IFOST.2012.6357823
10. Y. S. Borovikov, A. Y. Pischulin, R. A. Ufa, IOP Conf. S. **93**, 012018 (2015), DOI: 10.1088/1757-899X/93/1/012018
11. H. Zimmer, B. Niersbach, J. Hanson, Proc. 11th IEEE Int. Conf. CPE-POWERENG, 71 (2017)
12. U. K. Verma, S. Banerjee, A. P. Singh. Proc. 3rd Int. Conf. AEEICB, 299 (2017)
13. M. Andreev, A. Sulaymanov, A. Suvorov, MATEC Web Conf. **37**, 01001 (2015), DOI: 10.1051/matecconf/20153701001

Research of the power plant operational states with block structure

Nina Kosmynina[1,*], and *Victor* Yarmonov[1]

[1]National Research Tomsk Polytechnic University, 634050 Tomsk, Russia

Abstract. In this article the research technique block structure power plant operational states is offered. As an example the operating power plant of OOO Siberian Generation Company with block structure of turbogenerators connection is considered. The choice of the operating power plant has allowed to receive to carry out the analysis real long and emergency states. The offered technique of states identification and the analysis can be used for power plant of other structure after the corresponding correction.

1 Introduction

The main feature of block structure power plants is connection of generators to switchgear by means power transformers or autotransformers. Such schemes are used for nuclear, powerful hydroelectric power plants and thermal power stations. In case of nuclear and thermal power stations generators have the drive from the steam turbine – turbogenerators are used; for hydroelectric power plants – hydrogenerators with the drive of the electric generator from the hydraulic turbine are used. The difference in generators doesn't influence structure and the modes of power plant. In this article the thermal power plant working as a part of Ltd company "The Siberian generation company" is considered. Feature of power plants block diagrams is power development from generators to power transformers and to voltage switchgear [1]. In the figure 1 the block diagram of the studied power plant is shown. The studied power plant has three open voltage switchgear: high voltage switchgear (HV switchgear), the middle voltage switchgear -1 (MV-1 switchgear) and the middle voltage switchgear -2 (MV-2 switchgear). Switchgears have the following classes of voltage: HV switchgear – 500 kV, MV-1 switchgear -220 kV and MV-2 switchgear -110 kV. At the station seven turbogenerators are used: six turbogenerators (G1, G2, G3, G4, G5, G6 with type TVF-165-2У3, and one turbogenerator (G7) with type TGB-500-2. Turbogenerators TVF -165-2У3 the Russian manufacturer with power 165 MW have direct hydrogen cooling of rotor winding and hydrogen indirect cooling of stator winding [2]. The turbogenerator excitement system is static thyristor, according to the scheme of self-excitation. Turbogenerators TGV -500-2 the Kharkiv manufacturer with power 500 MW have direct hydrogen cooling of stator winding and water direct cooling of rotor winding [2]. The turbogenerator excitement system is static thyristor STVB-12B.

* Corresponding author: kosm_nm@tpu.ru

DOI: 10.1051/matecconf/201714101039

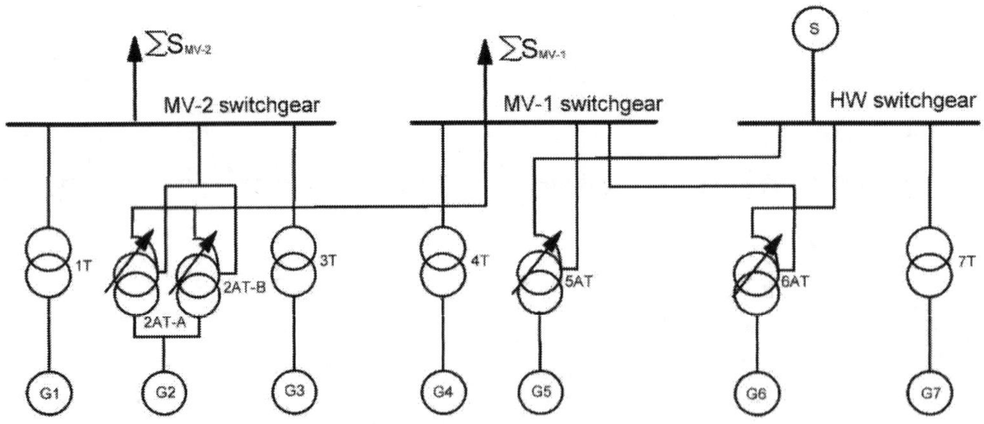

Fig. 1. The block diagram of the studied power plant.

Turbogenerators TVF-165-2У3 the Russian manufacturer with power 165 MW have direct hydrogen cooling of rotor winding and hydrogen indirect cooling of stator winding . [2]. The turbogenerator excitement system is static thyristor, according to the scheme of self-excitation. Turbogenerators type TGB -500-2 the Kharkiv manufacturer with power 500 MW have direct hydrogen cooling of stator winding and water direct cooling of rotor winding [2]. The turbogenerator excitement system is static thyristor, STVB-12B. All turbogenerators work according to the scheme of blocks: the generator – the two winding transformer [1] (G1-1T, G3-3T, G4-4T, G7-7T) with the generating switch; generator - autotransformer (G5-5AT, G6-6AT, G2-2AT-A,2AT-B). Types of block transformers are TDSN-200000/110, TDS -200000/220, 3xORTS -210000/500 [2]; types of block autotransformers are ATDSTN -250000/220/110, ATDSTN -500000/500/220. Power of the ТДЦ-200000/110 is 200 MB*A with higher voltage 110 kV . Power of the TDS - 200000/220 is 200 MB*A with higher voltage 220 kV. Power of the 3xORDS -210000/500 is 630 MB*A with higher voltage 500 kV. Cooling system: compulsory circulation of oil and air, with not directed oil stream. Power of the ATDSTN -250000/220/110 is 250 MB*A, and ATDSTN -500000/500/220 has 500 MB*A. Autotransformer voltage tap of ATDSTN -250000/220/110 are 220 кВ – higher voltage , 110 кВ - middle voltage. Autotransformer voltage tap of ATDSTN -500000/500/220 are 500 kV higher voltage , 220 kV middle voltage [4]. Cooling system: compulsory circulation of oil and air, with not directed oil stream. The autotransformers has the built-in device RPN for an opportunity to carry out adjustment of voltage under loading without shutdown of the autotransformer. All autotransformers carry out the connection between switchgears of power plant. The power supply system (in drawing it is designated as source S) has connection with station on the HV switchgear. The MV -1 switchgear has middle voltage -1 external consumers loading of the total power of SMV-1. The MV -2 switchgear has middle voltage - 2 external consumers loading of the total power of SMV-2.

2 Operational states classification

Depending on external consumer loading, an operational condition of block diagram elements the following operational states of power plant are possible: normal, repair, postemergency. The normal state is characterized by regime parameters admissible values : voltages, currents, capacities. At the same time distinguish the states of the maximum and minimal load of external consumers. For the studied power plant the possible load states described above exist both for MV -1 switchgear and for MV -2 switchgear . The scheme represented in the figure 1 corresponds to these long states.

There are repair states at the scheduled preventive and capital connection transformers repairs. For the studied power plant are possible two repair states. The first state is available at shutdown of one connection autotransformer MV-1 switchgear – MV-2 switchgear. The second state is available at shutdown of one connection autotransformer of MV-2 switchgear – HV switchgear. All possible operational states are considered: change of loading (min., max.) on MV-1 switchgear and MV-2 switchgear.

Only one of postemergency states is considered: shutdown of one block from MV-1 switchgear or from MV-2 switchgear, as emergence two postemergency states is an improbable event. Only one of the repair states is considered: shutdown of connection autotransformers MV-1 switchgear - MV-2 switchgear or MV-2 switchgear - HV switchgear as existence two repair the mode is an improbable event.

3 Operational states analysis

All electrical equipment has work qualitatively in all operational states. Therefore the greatest values of currents or powers flowing through the power plant electrical equipment must be taken into account. The currents and power values in the operational states is carried out on the basis of power balance f or power plant nodes-terminal. So, terminals of each block generator (G1, ... G7); autotransformers neutral and each switchgear are nodes of the studied power plant. The direction and designations of power flows are presented in the figure 2.

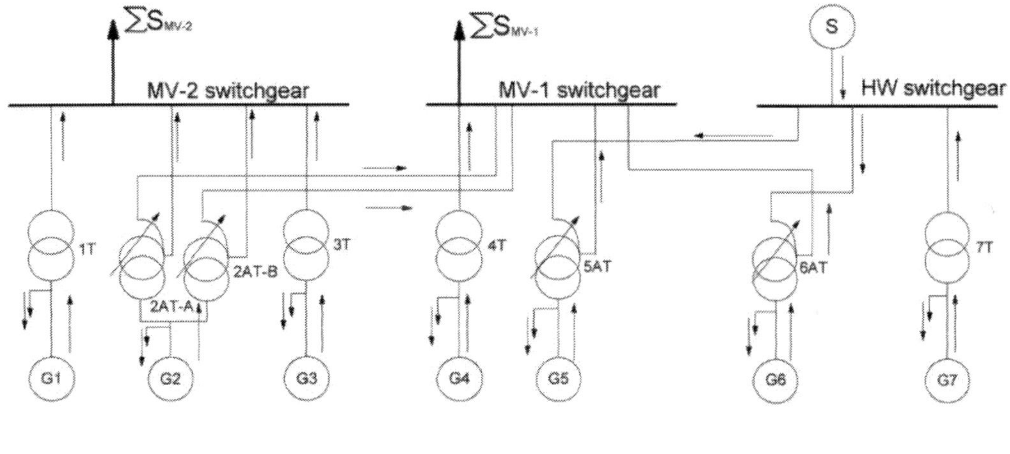

Fig. 2. The power flows block diagram plant.

Principles of the power overflows direction: from generators, to external loading; to common auxiliary (internal loading), from the two winding block transformer to switchgear; randomly for the middle and high terminal of the three winding autotransformer.

Below the example of power flows definition for the connection autotransformer 2AT-B is given. The power flow on the lowest voltage terminal 2AT-B is defined from the power balance equation of the G2 generator node. The power flow on the middle voltage terminal 2AT-B is defined from the power balance equation of the MV-1 switchgear node.

The power flow on the high voltage 2AT-B is defined from the power balance equation of autotransformer neutral.

It should be noted that by deducing the balance equations within power plant losses of electric equipment elements power in aren't considered.

The example of one operational states calculation with using of the "RastrWin3" program without a common auxiliary power is presented in the figure 3.

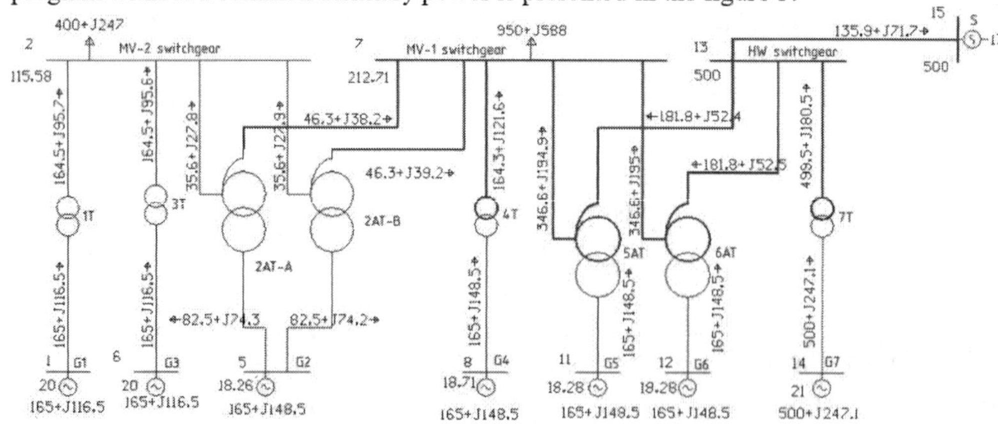

Fig. 3. The one operational states calculation in the "RastrWin3" program.

The RastrWin program complex is intended for the solution of tasks of calculation, the analysis and optimization of the electrical systems states. In Russia the main users of RastrWin are System Operator of the Unified Energy System (SO UES) and its branches.

The programs allowing to make a research of the different power plants operational states are developed for master preparation at the Tomsk Polytechnic University. The program is constructed by the linear principle: the data input windows of generators and loadings; the original structure window (it is presented in the figure 4), the window of the operational states choice, the results window [3].

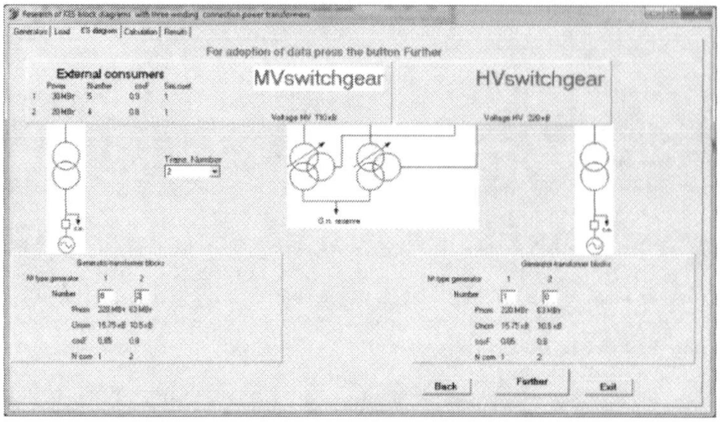

Fig. 4. Window of the power plant block diagram formation.

Power flows via transformers are a basis for determination of the required rated power of the equipment and its choice. In case of block two winding transformers the required rated power is equal to the block generator power. The required rated power for three winding the transformer $S_{RATE} = \max(S_{LW}, S_{MV}, S_{HV})$, S_{LW} – power flow on a lower voltage winding; S_{MV} – power flow on a middle voltage winding; S_{HV} –power flow on a higher voltage winding. In case of the autotransformer $S_{RATE} >= \max(S_O, S_S, S_{LW}) / Kco\_ratio$; S_O - power flow on the common winding; S_S - power flow on the series winding; S_{LW}- power flow on the lower-voltage winding; Kco_ratio – co-ratio of an autotransformer coefficient is defined as Kco_ratio $=(U_{HV}-U_{MV})/U_{HV}$. Besides, the selected autotransformer needs to be checked on loading of a lower-voltage winding: $S_{LV} \leq S_{LV}$_rate.

The program for the autotransformer choice is also developed for students of the Tomsk Polytechnic University [4].

It is provided in the program: the entered data verification; the definition of an autotransformer operational states (the autotransformer, transformer, combined with power transfer to the middle voltage, combined with power transfer to the higher voltage); the calculation of the required autotransformer rated power.

3 Conclusions

In article the technique of operational states identification for power plant is developed: load, repair and postemergency.

For the operating block structure power plant the table of the possible operational states has been created.

The algorithm of power flows calculation via power transformers and autotransformers is offered.

The calculation protocol for one of the operational states is provided in the RastWin program.

The specialized program developed and used for operational states calculation at the Tomsk Polytechnic University is described.

Use of operational states calculations results for determination of the power transformers and autotransformers required rated is shown.

References

1. B.A. Starshinov, M.I.Piratorov, M.A.Kozinov, *Electric part of power plants and substations* (MEI, Moscaw, 2015)
2. B.N. Neklepayev, I.P. Kryuchkov, *Electric part of power plants and substations. Reference materials for course and degree design* (Petersburg: BHV-St. Petersburg, 2014)
3. N. M. Kosmynina, T. T. Valiev. MATEC Web Conf. **91** (2017)
4. I. S. Tsoy, N. M. Kosmynina, IOP Conf. Series: Earth Environ. Sci. **27** (2015)

MATEC Web of Conferences **141**, 01040 (2017) DOI: 10.1051/matecconf/201714101040
Smart Grids 2017

Mathematical simulation of analog-to-digital converter

Mikhail Andreev[1,*], *Aleksey* Suvorov[1], and *Aleksey* Khlebov[1]

[1]National Research Tomsk Polytechnic University, 634050 Tomsk, Russia

Abstract. Numerical relay protection devices (NRP) are a promising area for the development of protection equipment for electric power systems. For its development, research and, in some cases, setting mathematical models are used. At the same time, the converting part of the NRP, which realizes the analog-to-digital conversion (ADC) of the analog input signal to the digital one, with which the microprocessor of the NRP works directly, is often ignored in models. The purpose of this work is to study the characteristics and parameters of the ADC, as well as to evaluate its impact on the operation of the ADC itself and need to take into account those or other characteristics in the simulation. To develop the ADC model, the C programming language was used. The ADC model is takes into account the cutoff frequency of the analog signal, sampling period, and number of samples required. The characteristics and parameters of the ADC are studied, their effect on the operation of the ADC is estimated. The result of this study – the ADC must be modeled as completely as possible, since its operation depends on the correct functioning of the NRP algorithm.

1 Introduction

Relay protection devices play a significant role in ensuring the manageability and reliability of power systems. Until recently, the main share of the relay protection devices utilized was made up of electromechanical and microelectronic devices [1]. Technical means of relay protection morally and physically "grow old". These circumstances indicate that in this area, modernization of relay protection is needed, based on the use of microprocessor systems and use of hardware and software systems.

Modern relay protection devices use logic circuits. For their implementation it is necessary to use a discrete binary information transfer system. The signals necessary for monitoring the state of the provided system come from the transformers in analog form. In order to convert the analog signal into a binary code, use an ADC (analog-to-digital converter). The goal of this work is to simulate the algorithm of the ADC operation in the MathCAD software package.

* Corresponding author: andreevmv@tpu.ru

2 Mathematical model of ADC

The basic principle by which the analog signal is converted into a discrete one consists in sampling the values from an analog signal after a certain period of time and translating these values into binary code. These values repeat the basic form of the analog signal.

The analog signal to the ADC passes through a low-pass filter that passes the frequency spectrum below a certain frequency. The cutoff frequency usually is 50 Hz. The program can specify any number of samples to be performed. In the course of the algorithm, all the numbers are rounded up to larger ones with the help of the function *ceil(x)*, since working with non-integer values when using the binary system is complex and in most cases not advisable. During the operation of the ADC, the simulation program only analyzes the measured value modules. After selecting the initial and associated conditions, we create a matrix of sample values, consisting of twenty-five elements, the number of elements can be very easily changed. The function *num2str(x)* convert numbers to a string. This is necessary to further convert the decimal number to binary. A sampling matrix is obtained consisting of 25 elements of whole decimal numbers written in the form of a string. Next, you need to translate the numbers in the matrix from the decimal to binary, for this a function is created, the form of which is presented in Fig.1.

```
base2base(tal, base_in, base_out) ≡ | NBD ← "0123456789ABCDEF"                                   (1)
                                    | InputNumberLength ← strlen(tal)                          (2)
                                    | NBD_base ← substr(NBD, 0, base_in)                        (3)
                                    | DecimalValue ← 0                                         (4)
                                    | for j ∈ 0 .. InputNumberLength − 1                       (5)
                                    |   | k ← substr(tal, j, 1)                                (6)
                                    |   | cif ← search(NBD_base, k, 0)                         (7)
                                    |   | error("Found a digit in the number that isn't in this base") if cif = −1   (8)
                                    |   | DecimalValue ← DecimalValue + floor[cif·(base_in)^(InputNumberLength−j−1) + 1/2]   (9)
                                    | Out ← ""                                                 (10)
                                    | while DecimalValue > 0                                   (11)
                                    |   | X ← floor(mod(DecimalValue, base_out))               (12)
                                    |   | Out ← concat(substr(NBD, X, 1), Out)                 (13)
                                    |   | DecimalValue ← floor(DecimalValue / base_out)        (14)
                                    | Out                                                      (15)
```

Fig. 1. Operating algorithm of ADC in MathCAD.

Description of the functions of each line in the program is following: (1) – the name of the function and enumeration of its arguments, creating a string that includes all the digits; (2) – counting the number of digits in the entered number; (3) – substring obtained from the NBD string by selecting 10 characters starting at position 0 in the row NBD; (4) – assigns a variable to a value of 0; (5) – beginning of a cycle executed from 0 to a value equal to the number of digits in the entered number -1; (6) – select a specific number in the entered number; (7) – search for the selected digit in NBD and select its serial number; (8) – if there is no digit in the list that is assigned the value -1; (9) – an increment to some variable of an integer, depending on the number of passed stages of the cycle; (10) – assigning an empty string to a variable; (11) – beginning of the loop, while the variable is still greater than zero; (12) – finding the remainder of dividing by 2 and rounding up to a smaller integer; (13) – adding the resulting residue to a string of final value; (14) – decreasing the variable

DOI: 10.1051/matecconf/201714101040

number, then how many times it can be reduced determines the number of values in the response line; (15) – output of the final answer.

3 Research results

To test the algorithm, the ADC input signal is generated by the formula (1):

$$U(t) = 10 \cdot (\sin(314 \cdot t - 60) + \sin(157 \cdot t + 60) + \sin(78 \cdot t + 150)) \tag{1}$$

here $U(t)$ – the dependence of the voltage applied to the analog-to-digital converter on the time; t – time, during which an analog-to-digital converter is supplied with voltage.

In Fig. 2a is a diagram of the obtained sample values in accordance with the ADC algorithm (Fig. 1). Also, for clarity, the original analytical curve with the converted curve of discrete sampling is compatible (Fig. 2b). It can be seen from the graph that the sample almost perfectly repeats the outlines of the original analog signal with a small error in some places, which is caused by rounding of the selected numbers.

Let us analyze the behavior of a rectified discrete curve with increasing or decreasing sampling period within acceptable limits without changing the cutoff frequency of the input signal and its shape. First, consider the change in the discrete signal with a decrease in the sampling period to 0.0025 s. Below is a graph of a combined rectified discrete signal and an analytical signal (Fig. 3a). The Fig. 3b shows a combined graph of voltage changing in time as the sampling period increases to 0.005 s. It can be seen that with a decrease in the sampling period, the output signal is closer to the original analytical form. At the same time, with an increase in the sampling period, the output signal more and more diverges with the original signal, the measurement error increases.

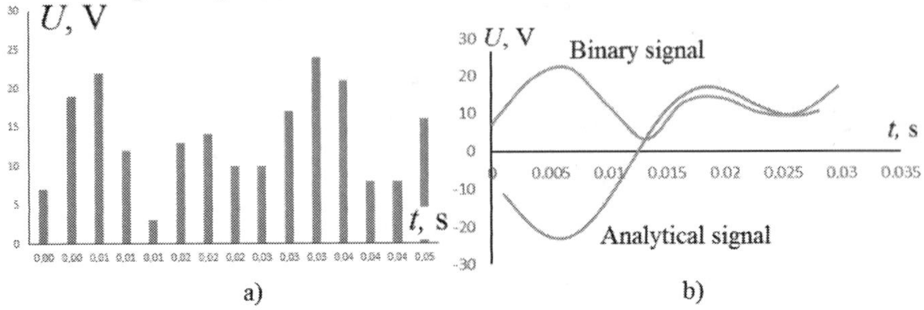

Fig. 2. a) The diagram of values of the received sample in decimal form; b) a combined voltage changing in time.

4 Conclusions

1. In the work done, the operation of the analog-digital converter was investigated. With the use of software complexes, an algorithm for the operation of the ADC has been created. The process of modeling the converter in its general form is described in full.

2. Some binary values are obtained for the received analog signal. The operation of the algorithm is considered for a specific example, but if the input signal is changed, the algorithm will continue to work with observance of the maximum frequency conditions.

3. The effect of sampling period selection is investigated: if the less the period is selected, the more accurate results will be obtained at the output of the ADC.

4. The developed algorithm can be used in the simulation of relay protection [2-8], including power system simulator [9-12]. The implementation of the algorithm will allow

detailed analysis of the operation of the ADC at each of its stages, while varying its parameters.

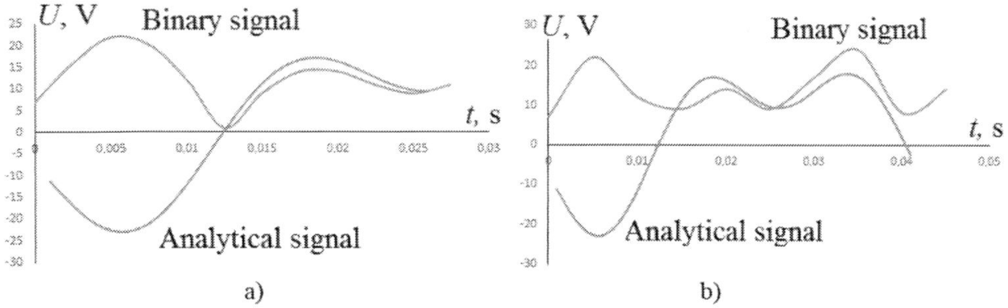

Fig. 3. a) Combined graph of voltage changing in time with decreasing sampling period; b) combined graph of voltage changing in time with increasing sampling period.

The reported study was funded by RFBR, according to the research project No. 16-38-60043 mol_a_dk.

References

1. G. Ziegler, *Numerical Differential Protection* (Erlangen: Publicis Publishing, 2012)
2. A.M. Carter, D.C. Humphreys, A. Dysko, J.R. McDonald, DPSP, 391 (2001)
3. A. Dysko, J.R. McDonal, G.M. Burt, J. Goody, B. Gwyn, DPSP, 381 (1987)
4. A. Dysko, J.R. McDonal, G.M. Burt, J. Goody, B. Gwyn, IEEE Transmission Distribution Conf., 406 (1999)
5. C.-S. Chen, C.-W. Liu, J.-A. Jiang, IEEE Transactions on Power Delivery, **21** (2), 619 (2006)
6. A. Suvorov, Y. Borovikov, A. Gusev, A. Sulaymanov, M. Andreev, N. Ruban, R. Ufa, Electrical Engineering, **99**, 3 (2017), DOI: 10.1007/s00202-016-0464-4
7. N.Yu. Ruban, A.S. Gusev, A.Yu. Pishchulin, MATEC Web Conf. **37** (2015), DOI: 10.1051/matecconf/20153701046
8. N.Y. Ruban, A.O. Sulaymanov, R.A. Ufa, I.A. Razzhivin, MATEC Web Conf. **91** (2016), DOI: 10.1051/matecconf/20179101050
9. N.Y. Ruban, Y.S. Borovikov, A.O. Sulaymanov, A.S. Gusev, IFOST, 311 (2016), DOI: 10.1109/IFOST.2016.7884255
10. R.A. Ufa, A.S. Vasilev, A.S. Gusev, A.A. Suvorov, IFOST, 113 (2016), DOI: 10.1109/IFOST.2016.7884204
11. R.A. Ufa, V.A. Sulaymanova, Y.S. Borovikov, MATEC Web Conf. **37** (2015), DOI: 10.1051/matecconf/20153701061
12. Y.S. Borovikov, A.Y. Pischulin, R.A. Ufa, IOP Conf. Series: Mater. Sci. Eng. **93** (2015), DOI: 10.1088/1757-899X/93/1/012018

MATEC Web of Conferences **141**, 01041 (2017) DOI: 10.1051/matecconf/201714101041
Smart Grids 2017

Approach in the modeling of wind turbines for power system stability studies and evaluation of their impact on electric power systems

Igor Razzhivin[1,*], *Almaz* Sulaimanov[1], and *Sergey* Stavitsky[1]

[1]National Research Tomsk Polytechnic University, 634050 Tomsk, Russia

Abstract. Wind power - one the fastest evolving in renewable energy industry. From year to year the share of wind energy in power systems are increasing as well as increasing the size of wind farms. A large-scale wind farms that produce hundreds of megawatts of electricity have a serious impact on the power grid. This paper discusses proposed approaches to modeling wind farms using the Hybrid Real-Time Power System Simulator (HRTSim), to study stability and transient processes in electric power systems.

1 Introduction

A renewable energy sources play an increasingly important role in energy supply as a result of the constant increase prices on fuel and environmental problems in world. According to experts for 2020, expectations for renewable energy to 280 GW will be installed in Europe, which is 64% of its total planned capacity, and the installed capacity of wind farms will be 136 GW (59%). Large wind farms in Europe, China and the United States consist of hundreds of wind turbines with an installed capacity exceeding 1,000 MW. For example, the wind farm in Gansu Province (China) has more than 5,000 turbines with more than 5 GW a total capacity, and is underway to expand it [1-2].

The growing scale of large wind farms and penetration level of wind power, consider the chaotic nature of its production would consequently exert serious impacts on the operation and control of the electric power systems (EPS), that wind power is integrated. Issues of stability of EPS, regulation of reactive power, voltage, active power, frequency, etc. remain today not fully investigated today are not fully investigated. It is therefore very important to study the interaction between large-scale wind farms and power system connected, and the impacts caused by wind power integration. To fulfil this purpose, adequate wind farm models are required.

Conceptually, there are two approaches to modeling wind farms: detailed modeling and aggregated modeling. In detail, all wind turbines in the wind farm and their interconnection are modeled. For a wind farm with a high number of wind turbines the detailed model is a complex system and leads to a long calculation time of the models in the hardware and software simulation tools. Aggregated modeling uses one or more equivalent wind turbines

* Corresponding author: lionrash@tpu.ru

and is most often used to study transient processes and the stability of a power plant with a high proportion of wind energy [3].

In this article, it is proposed to apply HRTSim as a tool for modeling wind farms in EPS, taking into account all the advantages of this software-hardware real time system.

2 Wind farm modeling

The accuracy of the model of wind turbines has a decisive influence on the results of wind park modeling. Setting the goal of research in studying the complex reaction of wind farms, the impact on power stability when connecting high-power wind farms to the power system, the interaction of the wind farm with other power plants, etc., it is necessary to use accurate and reliable models of all components of the wind farm.

Using a detailed approach in modeling leads to a high complexity of modeling and long calculations and exhausting computation time, especially for wind farms with a large number of wind turbines. Hardware and software simulation tools appropriate for modeling failing with this problem in virtue of the limited mathematical calculations, and the impossibility of solving large systems of differential equations describing processes in EPS. Moreover, when solving the systems of differential equations by numerical methods, the computational error accumulates, which reduces the reliability of the simulation results [4-6]. Various scientific community and research laboratories when modeling large wind parks are necessity to apply certain limitations and assumptions, applying equivalent models in calculations. Therefore, to solve the above-mentioned problems and obtain the most accurate results in the simulation of wind farms the HRTSim was chosen as the modeling instrument [7-8].

Modeling of wind turbine systems requires interdisciplinary knowledge. The wind turbine has a complex device and combines an aerodynamic model, a mechanical model and an electric model, as well as a control system model. In Fig. 1 shows the general structure of the wind turbine model and the input parameters.

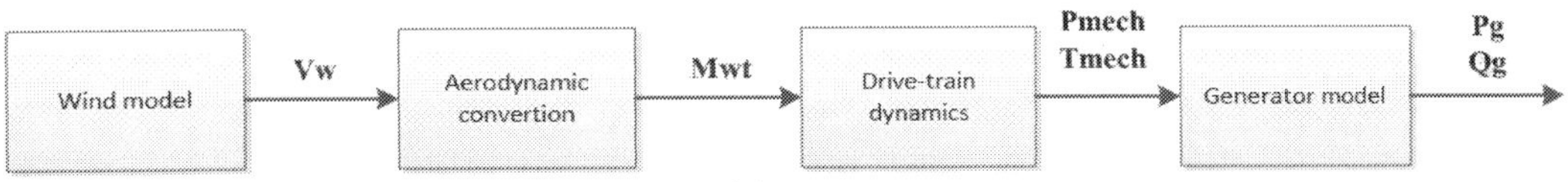

Fig. 1. The basic structure of the overall wind turbine system.

2.1 Wind model

In modeling the wind usually take a constant value equal to the average wind speed or wind define the mathematical model through a sine function which is not quite correct assumption in large wind farms assess the impact on the EPS.

To simulate the dynamic component of wind speed in the HRTSim there is provided to implement the spectral model of Kaimal and Davenport recommended by the international standard [9-11].

$$V(t) = \bar{V} + \Delta V(t), \tag{1}$$

where: $\bar{V}$ is the average wind speed; $\Delta V\ (t)$ - deviation of wind speed (dynamic component of speed)

The average wind speed is given as a function of time. The statistical characteristics of the dynamic component $\Delta V\ (t)$ are specified using the spectral density function $S\ (f)$.

To calculate the dynamic component of wind speed, the spectral density model of Kaimal is described:

$$\frac{f \cdot S(f)}{\sigma^2} = \frac{4 \cdot f \cdot L/\bar{V}}{(1+6 \cdot f \cdot L/\bar{V}'} \tag{2}$$

where: f - the frequency of the harmonic component of the wind speed signal, Hz; $S(f)$ - one-sided spectrum of horizontal component of wind speed; σ - root-mean-square deviation of the horizontal component of wind speed; L - the turbulence coefficient.

Using Fourier Theorem function of the horizontal component of the wind speed may be represented as a series:

$$V(t) = A_0 + \sum_{k=1}^{\infty} A_k \cdot Cos(\omega_k t + \varphi_k), \tag{3}$$

where Ak - the amplitude of the k-th wind speed harmonics; ω_k - the circular frequency of the corresponding harmonic; φ_k - the phase shift of the k-th vibration; A_0 - the probability.

Then the amplitude of the k-th harmonic is equal to:

$$A_k = \sqrt{S(f_k)\Delta f}, \tag{4}$$

where Δf - the interval between neighboring frequencies.

Substituting the corresponding values, we get that the horizontal wind speed in the time interval T is equal to:

$$V(t) = \bar{V} + \sum_{k=1}^{N} A_k \cdot Cos(\omega_k t + \varphi_k), \tag{5}$$

where $\bar{V}$ is the wind speed averaged over a given time interval.

To implement the aerodynamic model of the turbine A mathematical model is proposed that allows to reproduce the torque of the wind turbine M_{wt} in accordance with the equation used in their design and reliably verified practice of its application:

$$M_{wt} = M\pi R^3 \frac{pV^2}{2}, \tag{6}$$

in which R is the radius of the wind wheel, V is the wind speed, and M is determined by the structural features and parameters of the wind turbine wheel:

$$M = \frac{4e}{Z} \cdot \frac{1-e}{1+e}\left[\left(1 - \frac{r_0^2}{R^2}\right) - 2\mu\left(\frac{Z_U}{3} + \frac{1-\frac{r_0}{R}}{Z_0} - \frac{1-\frac{r_0^2}{R^2}}{Z}\right)\right] \tag{7}$$

where $e = V_1/V$ - coefficient of deceleration of the wind wheel, and $V_1 = 1/3 \cdot V$ - loss of wind speed in the plane of the wind wheel; $Z = (\omega_w R)/V$ - speed of the wind turbine, and ωw - angular velocity of the wind wheel; Z_0 - distance from the axis of the wind wheel to the beginning of the blade; $\mu = Cx/Cy$ is the reverse quality of the wind wheel blade, determined from the Cx and Cy diagrams for specific blade profiles; $Z_U = \omega R/(V-V_1)$ - the relative number of modules for the end of the blade [11].

Simulation of the drive and the generators is suggested using the developed and most frequently used mathematical models of EPS [10].

3 Real Time Hybrid Simulation of Electrical Power Systems

HRTSim is based on the concept of hybrid simulation and has the form of parallel, multiprocessor, hybrid software-hardware real time system combining adaptable complex of three-phase specialized hybrid processors (SHP) of all simulated power system elements and data management system (DMS) (Fig 2).

DOI: 10.1051/matecconf/201714101041

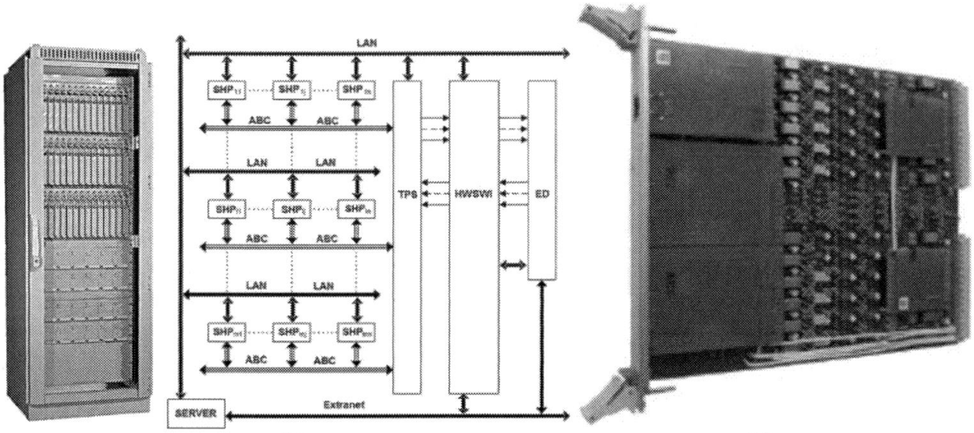

(a) (b)

Fig 2. (a) HRTSim and its operational diagram: SHP - specialized hybrid processors; LAN – local area network; TPS – three-phase switch; HWSWI – hardware/software interface; ED – external device; ABC – three-phase input/output of SHP; (b) SHP of power line.

The hybrid simulation technology developed at the Tomsk Polytechnic University is analogue modeling, digital and physical. Each approach is used to solve individual subtasks, which solve the general problem of EPS modeling. The key task is to avoid a methodological error in solving differential equations [12-13]. The analogue part of the HRTSim provides continuous implicit integration, the digital part - provides flexibility, rapid and accurate control of model parameters and visualization of results; the existence of a physical layer allows us to unite the models of power system elements by converting mathematical values into physical voltages and currents of the model while minimizing the volume of digital information exchange [14].

3 Conclusions

In this paper, a detailed method for modeling a wind farm for analyzing the system stability studies and transient processes was proposed. In this model, individual wind turbines within the wind farm could receive different incoming winds. And also a new model of a wind turbine was proposed.

Also in this article we propose the use of HRTSim for detailed modeling of a large wind power plant, the use of which solves an important problem, given the number of wind turbines, various wind farm models and wind distribution changes. The calculations of mathematical models of the wind and wind turbine, even with detailed modeling, will have high-precision results, and the calculation speed will be rapid due to the features of the HRTSim of the EPS and its advantages in comparison with other Hardware and software simulation tools.

This work was supported by the Ministry of Education and Science of the Russian Federation under the governmental grant "Science" No. 13.5852.2017/БЧ (Development of the concept for comprehensive validation of calculating modes and processes in electric power system and tools of its realization).

References

1. GWEC. Global Wind Statistics, 2015.2.10, Brussels, Belgium (2014)
2. Hongzhi Liu, Zhe Chen, IEEE APPEEC 978-1-4577-0547-2/12/ (2012)

3. R. Ufa, A. Vasilev, A. Gusev, A. Suvorov, IFOST, 113 (2016), DOI: 10.1109/IFOST.2016.7884204
4. M. Andreev, Y. Borovikov, N. Ruban, A Khlebov, MATEC Web Conf. **91**, 01048, (2016), DOI:10.1051/matecconf/20179101048
5. A. Gusev, A. Suvorov, A. Sulaymanov, ICIEAM 127470, 7911415, (2016), DOI:10.1109/ICIEAM.2016.7911415
6. I. Tsoy., N. Kosmynina, IOP Conf. Series, **27** (1), 012065 (2015), DOI:10.1088/1755-1315/27/1/012065
7. A. Askarov, M. Kondrashov, N.Kosmynina, MATEC Web Conf. **37**, 01004 (2015), DOI: 10.1051/matecconf/20153701004
8. S. Obukhov, I. Plotnikov, E. Sarsekeev, Mod. problems sci. educ. **5**, 137 (2013)
9. IEC 61400-1 Third edition (2005)
10. D. Hansen, DIgSILENT technical documentation (2010)
11. N. Ruban, A. Gusev, A. Pishchulin, MATEC Web Conf. **37**, 01046 (2015), DOI:10.1051/matecconf/20153701046
12. A. Suvorov, A. Gusev, A. Sulaymanov, O. Suslova, MATEC Web Conf. **91**, 01052 (2017), DOI:10.1051/matecconf/20179101052
13. N. Ruban, M. Andreev, Y. Borovikov, ICIEAM 127470, 7910994, (2016), DOI:10.1109/ICIEAM.2016.7910994

DOI: 10.1051/matecconf/201714101042

Software and hardware platform for testing of Automatic Generation Control algorithms

Alexey Vasiliev[1], *Artem* Petrov[1,*], and *Sergey* Svechkarev[1]

[1]National Research Tomsk Polytechnic University, 634050 Tomsk, Russia

Abstract. Development and implementation of new Automatic Generation Control (AGC) algorithms requires testing them on a model that adequately simulates primary energetic, information and control processes. In this article an implementation of a test platform based on HRTSim (Hybrid Real Time Simulator) and SCADA CK-2007 (which is widely used by the System Operator of Russia) is proposed. Testing of AGC algorithms on the test platform based on the same SCADA system that is used in operation allows to exclude errors associated with the transfer of AGC algorithms and settings from the test platform to a real power system. A power system including relay protection, automatic control systems and emergency control automatics can be accurately simulated on HRTSim. Besides the information commonly used by conventional AGC systems HRTSim is able to provide a resemblance of Phasor Measurement Unit (PMU) measurements (information about rotor angles, magnitudes and phase angles of currents and voltages etc.). The additional information significantly expands the number of possible AGC algorithms so the test platform is useful in modern AGC system developing. The obtained test results confirm that the proposed system is applicable for the tasks mentioned above.

1 Introduction

AGC systems must be constantly developed due to changes in market conditions, increasing degree of automation in the energy sector, development of computing systems, increasing number of Phasor Measurement Units (PMU), necessity of improving the reliability of power systems and changes in network configuration.

At the moment the AGC system of Russia operates on base of CK-2007 which is the main SCADA system used by the System Operator of the United Power System of Russia. Making changes to the settings and adding new algorithms requires testing the AGC system on adequate power system model. One of the complexes that allow real-time closed-loop testing of AGC system in composition with SCADA system is HRTSim, which was developed at the Tomsk Polytechnic University.

* Corresponding author: apsapetrov@gmail.com

1.1 HRTSim description

The concept of HRTSim (Hybrid Real Time Simulator) simulation is based on the use of three modeling approaches: analog, digital and physical, each of which achieves maximum efficiency in solving individual subtasks. A detailed description of the concepts and its hardware application is presented in [1–3]. Modelling of primary processes in power equipment is done by analog computational units that solve differential equations describing this power equipment. Power system automatics are simulated on microcontrollers in hard real-time. Switching of equipment and interconnection of units into a general model of a power system is carried out on the physical level. Such a solution ensures continuous simulation of primary processes.

Adjusting of power system and power equipment parameters and automation settings as well as the continuous collection and processing of simulation results is carried out by specialized software – VmkServer.

VmkClient software allows to control operating conditions by changing the automatic setpoints values, switching devices positions and parameters of the generation equipment and the load. Also, it is possible to control the operating conditions according to the predetermined algorithms – the mode scenarios.

Interaction with third-party software or SCADA systems is done through the OffsiteClient component.

All the processes are simulated in real time, which allows to solve such problems as training of dispatch personnel, closed-loop testing of relay protection devices and local emergency control automatics, data exchange with SCADA systems for scientific, practical and educational purposes and much more.

1.2 SCADA CK-2007 features

Operational and informational software complex (OIC) CK-2007 is a high-performance real-time SCADA platform which is designed for receiving, processing, storing, and transmitting telemetric, reporting and scheduled information on the operation of energy facilities, networks and systems and provide flexible access to different users and external automated systems.The complex is widely used in Russian Federation as the basis for the construction of operational technological control and management systems in situational centers, data centers and control centers in the power industry for generating, network, marketing companies and system operators.

The complex is delivered with EMS, DMS and MMS applications in various configurations and also has Application Programming Interface (API) and Data Access Component (DAC) library which allows to implement additional modules to solve specific technological tasks [4-5].

2 Implementation of AGC systems test platform

The AGC test platform is composed of two complexes HRTSim and OIC CK-2007. The first plays a role of a controlled object – a power system, and provides an interface for reading/writing of operating parameters in real time via a specialized software interface and a user application. The second complex is the basis for building operational and technological management systems in the electric power industry, it collects state information from control objects, and provides all the necessary tools for implementing third-party modules for further processing.

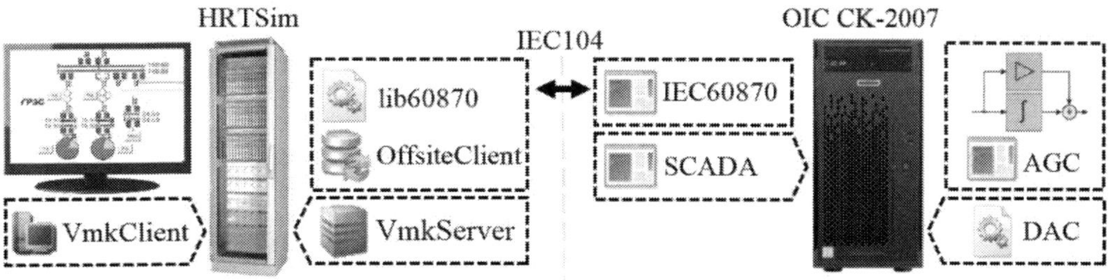

Fig. 1. Automatic Generation Control systems test platform

On the base of OffsiteClient two-way information exchange between HRTSim and OIC has been set up according to the IEC 60870-5-104 protocol, which makes it possible to transmit state information to OIC and to implement remote control of the generators that are modelled in the HRTSim.

Support for the IEC 870-5-104 protocol was implemented using the lib60870.NET library written in C#. The library is free to use for testing, evaluation, educational purposes and open-source projects under the terms of the GPLv3 license, for commercial projects commercial license is needed [6]. The library allows to implement both a client and a server which can exchange telemetry data, interrogation, time synchronization and control commands. Also a C version of the library is available.

The telemetry data is transmitted once a second or by interrogation command from a controlling station (OIC). The data is being filtered so the repeating values are not re-transmitted. It is possible to configure dead band in order to reduce excessive traffic between HRTSim and OIC and to test the impact of accuracy of telemetry data coming to SCADA system on AGC system operation. Datasets for transmission and reception are configured using the graphical user interface shown in Figure 2.

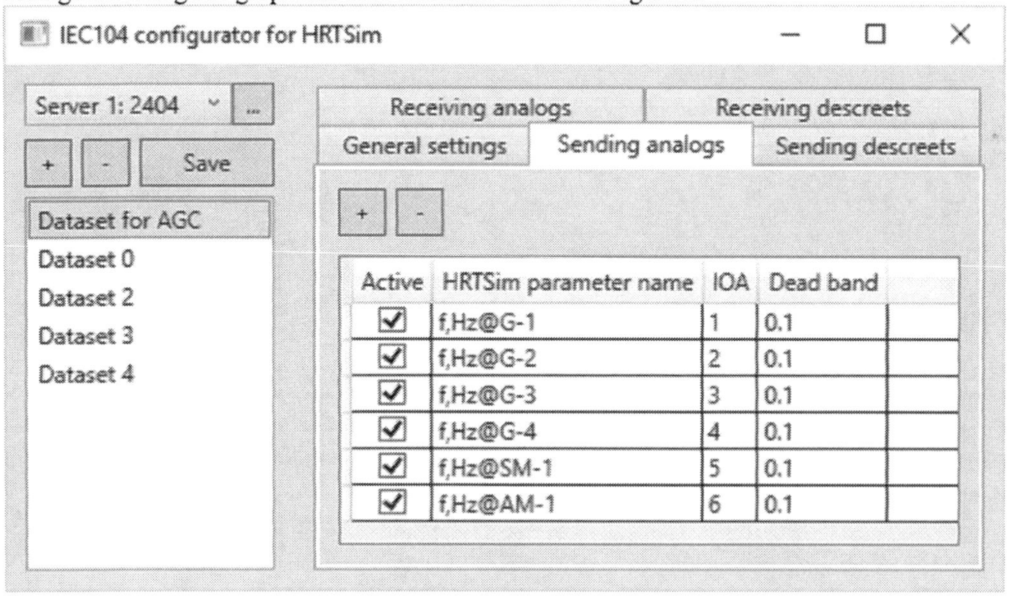

Fig 2. IEC 60870-5-104 protocol configurator for HRTSim

An AGC system testing process is carried out in two stages: preparatory and basic. At the first stage, the test object and testing facilities are being prepared:

- The power system model configuration;
- Development and verification of testing scenarios;

- Preparation of the AGC system (transferring an existing AGC system to the test platform and adding of new parameters and algorithms);
- Configuration of two-way communication protocol;
- Setting up of OIC tools to record state variables and AGC system control actions.

At the second stage, all test scenarios are run sequentially, the results are analyzed, and a conclusion about the readiness and applicability of the AGC system are made.

3 Results

As was already mentioned above, in addition to practical tasks related to the interaction with the unified AGC system of Russia, research tasks can be solved. As an example of a research task, an external program for OIC CK-2007 has been developed that partially reproduces the functionality of the AGC system. The program implements the PI-regulator algorithm that calculates the change of active power generation that is needed to maintain the active balance at the frequency of 50 Hz. The obtained value is distributed between the generators under AGC according to the given proportional coefficients and recorded in the CK-2007. When a variable in CK-2007 corresponding to the output of any of the generators updates, its value automatically sends via the IEC 60870-5-104 protocol and records in the HRTSim through OffsiteClient.

The AGC program is written in C# using the .NET Framework 4.6.1. The reading and writing of information from the OIC is carried out using the Data Access Component library (DAC), which is included in CK-2007 complex. All necessary regulatory parameters involved in the process of operation of the AGC system are put on the form of the program, which allows you to customize the operation of the system for the needs of a particular task, and also use the developed program for scientific and educational purposes.

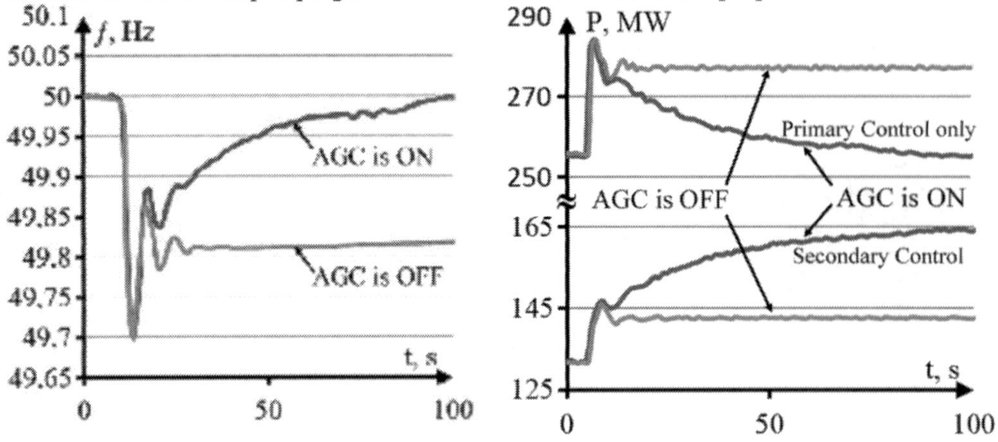

Fig. 3. Frequency and active power response on load switching of generators participating in the primary and secondary control

For example, in Figure 3 graphs of frequency and active power response on load switching of generators participating in the primary and secondary control are shown. The experiment was done twice – with AGC system turned on and off. As can be seen from the graphs, the developed program successfully restores the frequency of the power system that is modelled in the HRTSim to a value set by the setpoint, the processes occurring during the primary and secondary frequency control are fairly accurately reproduced. The value of the frequency exceeded the permissible limit of 50 ± 0.2 Hz in the Russia, but recovered to an allowable value of 50 ± 0.05 Hz in about 40 seconds. The results are explained by the small time constant of the AGC and the low capacity of the modelling power system (even a small change in the load power creates a significant relative imbalance), which does not allow to fully apply the requirements described in State Standard R 55890-2013 [7] to the power system being modeled in HRTSim.

4 Conclusion

The use of software and hardware tools mentioned above for modeling real-time processes in power systems allows testing control algorithms of AGC systems in a closed loop (including the AGC system that is used in operation in Russia). Testing of AGC systems together with the SCADA that is used in operation allows to simplify the transfer of AGC parameters from the test server to the real one and vice versa by automating the process and avoiding transfer errors.

Available in HRTSim extended set of primary state variables, in comparison to the real power system, allows to perform research tasks. CK-2007 Application Program Interface allows to implement third-party software applications not only of AGC systems, but also of other centralized automation systems, for example, Centralized Automatic Emergency Control or Centralized Voltage Control systems. Interaction via standard protocols and the availability to remotely launch VmkClient allows to territorially distribute the system and remove the function of technical support of HRTSim from the end user of the system.

References

1. A. Prokhorov, Yu. Borovikov, A. Gusev, Int. J. Smart Grid Clean Energy, **1** (1), 67 (2012)
2. N.Y. Ruban, A.O. Sulaymanov, R.A. Ufa, I.A. Razzhivin, MATEC Web Conf. **91**, 1 (2016)
3. M.V. Andreev, Y.S. Borovikov, N.Y. Ruban, 11th Int. Forum Strategic Technol. 187 (2017)
4. JSC "Monitor Electric", official website. Available at: http://www.monitel.ru/ (accessed: 5.11.2017)
5. Operational and informational software complex CK-2007C, User manual, JSC "Monitor Electric", (2013)
6. IEC 60870-5-104 Libraries, MZ Automation, official website. Available at: http://mz-automation.de/index.php/products/iec-60870-5-104-protocol-libraries/ (accessed: 5.11.2017)
7. GOST R 55890 – 2013. Operativno-dispetcherskoe upravlenie. regulirovanie chastoty i peretokov aktivnoj moshhnosti. Normy i trebovaniya [State Standard R 55890 - 2013. Operational dispatch control. Load frequency control. Norms and requirements.], (2014)

Validation of short-circuit current calculations in electric power system

Aleksey Suvorov[1,*], and *Mikhail* Andreev[1]

[1]National Research Tomsk Polytechnic University, 634050 Tomsk, Russia

Abstract. The determining factor for the selection of electrical equipment and switching devices is the short-circuit currents (SCC), for which the electrodynamics and thermal resistance is checked for the first, and for the second the switching capacity and wear-resistance. Also, the fault current is calculated for setting the relay protection and automation threshold. As is known, different digital simulation tools are used for calculating the SCC, in which inevitably use simplifications and limitations of mathematical models of equipment and EPS in whole. In this connection, the calculated values of SCC comprise, in some cases, significant error, so the validation is needed for used simulation tools. Proposed to use specialized hybrid software and hardware system for validation as the source of data needed to compare the simulation results of verified tools. It helps to determine their actual properties and capabilities, as well as to increase the reliability of decisions made on the basis of the calculated SCC, by understanding the causes and possible level of errors that arise.

1 Introduction

The task of calculating the short-circuit currents (SCC) is one of the most important in the research, design and operation of electric power systems (EPS). The solution of this problem is necessary for the selection and verification of electrical equipment in accordance with the conditions of electrodynamic and thermal stability, the design and setting of relay protection, automation and et al. Therefore, the evaluation of reliability the SCC calculations, especially their maximum values, is extremely important and urgent. The maximum instantaneous value of SCC, called the peak current, is found at the steady-state value of initial symmetrical short-circuit (SC) and the largest value of the dc short circuit component. The value of the dc component is approximately calculating with the help of a constant factor (k) reflecting the damping of the dc component according to an exponential law with a certain time constant and determined by the formula [1]:

$$k = 1,02 + 0,98 \cdot e^{\frac{-3}{\omega \cdot T_{dc}}}, \tag{1}$$

* Corresponding author: suvorovaa@tpu.ru

where ω – synchronous angular frequency of voltage; T_{dc} – the time constant of the damping of dc component of the SCC.

2 The problem of calculation of short circuit currents and an alternative way to solve it

For a branched network, the theoretically rigorous method of calculating the dc component of SC and its time constant in any branch is the application of the Laplace transform. Such a general and rigorous path of solution is possible in principle, but it is obvious that for a complex scheme it is inapplicable. Therefore, an equivalent network time constant is used, for the determination of which there are several methods [2, 3]. However, the determination of the equivalent time constant in real power system with respect to each faults point with using of these methods is an extremely complex task, the solution of which does not provide a significant increase in reliability in the calculation of SCC due to the approximations of the indicated methods. So, for practical calculations, the time constant is not determined, but takes for it some mean value depending on the location of the fault and the corresponding value of the constant factor.

For SCC calculations are used various software tools (CAPE, PSS/E, PowerFactory, etc.). However, despite the existing differences, they are united by a common approach to SCC calculations, namely the calculation of only steady-state emergency modes, since the accuracy of solving algebraic equations is high and does not cause doubts, in contrast to the solution of differential equations. In this case, the changing during the transient process values are taken into account by means of common coefficients, in particular the constant factor. With this approach, single-line equivalent circuits are used for calculating, the calculation of currents and voltages in asymmetric modes is performed by the method of symmetric components, and in order to simplify the EPS model only the main harmonic component is accounted for, the models of synchronous machines (SM) are given in the form of a constant EMF behind the resistance. In this case, the EMF SM (super-transient or transient) and the internal angle are assumed to be unchanged throughout the calculation. They are assumed to be numerically equal to the value at the time preceding the fault and determined by the known formula [4, 5].

For complete replacement of large-scale data, for understandable reasons, the created in the research and development laboratory of Electric power system simulation of Tomsk Polytechnic University Hybrid real time power system simulator (HRTSim) is used to evaluate the reliability of SCC calculations. HRTSim is intended for continuous methodically accurate three phase simulation of quasi-steady state and transient processes in equipment and EPS in whole for all possible operating conditions and any electric disturbance, including overvoltage [6], on an unlimited range and guaranteed instrumental accuracy. The possibility of obtaining complete and reliable information on the full significant spectrum of quasi-steady and transient processes in EPS using the hybrid tool is confirmed by comparing the results of simulation with full-scale data [7, 8], as well as a number of successfully performed applied works [9-12].

3 Experimental studies and discussion

The EPS of Tomsk region was used as a test model for simulation. The initial topology and the mode of the test EPS correspond to the validated dispatching measurements formed on the data of Tomsks EPS supervisory control and data acquisition (SCADA). For carrying out of experimental researches the most widespread in Russia tool for calculation of SCC - ARM SRZA has been chosen. In each simulation tools, the test model of EPS with the identical topology and parameters of the main equipment is modelling within the limits of the capabilities of used in them mathematical models.

Since in the tools in which the possibility of calculating SCC is declared, static mathematical models of equipment and EPS as a whole are used. It excludes the effect of the network on the dc component of SCC, on the reliability of reproduction of which depends mainly the maximum value of the total SCC. For determining the degree of influence of different values and the trend of the dc component, the three-phase faults at the generator are simulated, at which the influence of the network on the peak current from the generator side is eliminated, and at remote faults in the network, when this influence is significant. For determining the peak current in the ARM SRZA the obtained value of the initial current is multiplied by the constant factor, depending on the fault location [13]. For comparison from the oscillograms of the phase SC obtained in the HRTSim, one should choose those at which the phase value of the voltage is maximally close to zero at the moment of short-circuit. Example results in HRTSim simulation is shown in Fig. 1.

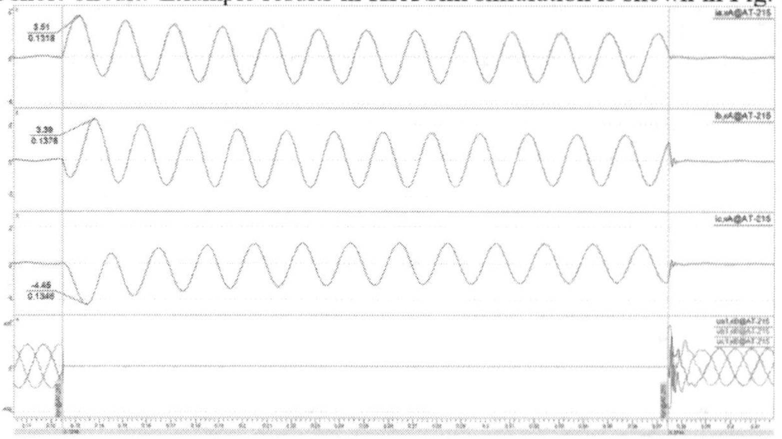

Fig. 1. Oscillogram of instantaneous values of phase currents and voltages of 220 kV AT-215 transmission line with the three-phase short-circuit.

For comparison, the simulation results are summarized in Table 1, in which the error of the peak current (ΔI_p) is determined with respect to HRTSim.

Table 1. Comparison of the peak currents values for three-phase faults obtained in HRTSim and ARM SRZA.

The fault location as the fault point is electrically removed from the sources	**HRTSim**	**ARM SRZA**	**ΔI_p,%**	**k**
	RMS value of peak currents, A			
Faults on the 110 kV bus TPS-3	6872	6449	6.16	1.965
Faults on the 110 kV bus SHC	7714	6386	17.22	1.965
Faults on the 220 kV transmission line AT-215	3144	3259	-3.65	1.78
Faults on the 220 kV transmission line T-211	3005	3186	-6.00	1.78
Faults on the 220 kV transmission line AT-214	4770	4289	10.07	1.78
Faults on the 220 kV transmission line CHP-223	1306	1440	-10.2	1.78
Faults on the 220 kV transmission line TV-221_2	2571	2896	-12.6	1.78
Faults on the 220 kV transmission line T-218	4398	5218	-18.6	1.78
Faults on the 110 kV transmission line C-40	2069	2458	-18.8	1.717
Faults on the 110 kV transmission line C-7B	3137	2223	29.13	1.717
Faults on the 110 kV transmission line C-7A	3409	4711	-38.1	1.717

The following conclusions can be drawn from these results:

1. As the short-circuit point is removed from the sources, the relative error of the peak current increases, which confirms the significant role of the network in the formation of the dc component of SC and the impossibility of an accurate calculation of its magnitude and trend with the help of the common coefficient, since in each specific case the attenuation rate is individual and formed by all elements networks.

2. The results show (in Table 1, the first and second points of the short-circuit), that the error introduced with the assumption of the constant EMF and its calculation by the parameters of the pre-emergency mode can reach significant values and basically depends on the magnitude of the dc component that discarded when applying the principle of constancy of flux linkages.

3. Since real EPS are complex networks, it is absolutely impossible to make an equivalent, and there will always be an error related to this. In addition, the drawback of such transformations is that in the common case, the power of the sources and energy receivers in the real circuit are not equal to the corresponding powers in the equivalent circuit.

4 Conclusions

The conducted experimental researches confirm that the use of the constant factor does not allow to determine reliably the maximum value of the peak current, and in view of the impossibility of accurately calculating the time constant of the dc component for real EPS, the only way to increase the reliability of the calculation of SC is the validation of the software tools. HRTSim can be used as a source of data for the validation of software tools.

The reported study was funded by RFBR, according to the research project No. 16-38-60043 mol_a_dk.

References

1. IEC 60909-0, *Short-circuit currents in three-phase a.c. systems - Part 0: calculation of short-circuit currents* (2001)
2. G. Knight, H. Sieling, IEEE Trans. Ind. Appl. **29** (3), 625 (1993)
3. J. Schlabbach, *Short-circuit currents* (IET Power And Energy Series, United Kingdom, 2008)
4. C. Pierre, IEEE Trans. Ind. Appl. **26** (2), 204 (1990)
5. P. Kundur, *Power System Stability and Control* (McGraw-Hill Education, India, 1994)
6. N. Ruban, A. Sulaymanov, R. Ufa, I. Razzhivin, MATEC Web of Conf. **91**, 01050 (2017), DOI: 10.1051/matecconf/20179101050
7. R. Ufa, A. Vasilev, A. Gusev, A. Suvorov, IFOST 2016, 16760462 (2016), DOI: 10.1109/IFOST.2016.7884204
8. N. Ruban, A. Gusev, A. Pishchulin, MATEC Web Conf. **37**, 01046 (2015), DOI: 10.1051/matecconf/20153701046
9. N. Ruban, Y. Borovikov, A. Sulaymanov, A. Gusev, IFOST 2016, 16760394 (2016), DOI: 10.1109/IFOST.2016.7884255
10. A. Suvorov, Y. Borovikov, A. Gusev, A. Sulaymanov, M. Andreev, N. Ruban, R. Ufa, Electrical Engineering, **99** (3), 959 (2017), DOI: 10.1007/s00202-016-0464-4
11. R. Ufa, A. Sulaymanov, Y. Borovikov, MATEC Web Conf. **37**, 01061 (2015), DOI: 10.1051/matecconf/20153701061

12. Y. Borovikov, A. Pishchulin, R. Ufa, IOP Conf. Series: Materials Science and Engineering, **93**, 012018 (2015), DOI: 10.1088/1757-899X/93/1/012018
13. F. Castelli-Dezza, A. Silvestri, D. Zaninelli, ETEP, **4** (3), 213 (1994)

Research of the surge voltage protection by means of Hybrid Real-Time Power System Simulator

Sergey Stavitsky[1,*], *Yuri* Borovikov[1], and *Yuly* Bay[1]

[1]National Research Tomsk Polytechnic University, 634050 Tomsk, Russia

Abstract. The article considers the simulation of surge voltage protection. A functional diagram of this protection model and the means by which the researches were made are presented. The results are the oscillograms of the surge voltage protection operation for the generator.

1 Introduction

A voltage boost (for example, at the stator winding terminals) can lead to insulation failure and damage in the windings of the generator or transformer of the generator-transformer unit [1]. Dangerous voltage boost for the windings insulation result from the disappearance of the magnetic flux in the stator and a generator speed increase, which occurs when generation shedding is. Therefore, the turbo-generators with a capacity of 160 MW or more are supplied with the surge voltage protection (SVP) [2]. Depending on the insulation state and the degree of overvoltage [3], the SVP can act either to turn off the generator (generator-transformer unit) or to a signal.

Today, there are various types of relay protection (RP) for generators [4] (electromechanical, microelectronic, and microprocessor). Demands are high for all types of protection. Due to the fact that of the above mentioned RP microprocessor devices are the newest and, therefore, the least tested [5]. Additionally, there is a growing tendency of their adoption into electric power systems (EPSs) [6]. As a result the microprocessor protection is of great significance for research in the Hybrid Real-Time Power System Simulator (HRTSim) among the developed RP modeling tools [7].

2 Hybrid Real-Time Power System Simulator as a means of relay protection simulation

Hybrid simulation is the simulation of the processes continuous spectrum in electric power equipment and EPS as a whole with all possible normal and emergency modes, without decomposition into quasi-steady-state processes, electromagnetic and electromechanical transient and supertransient processes [7]. The greatest advantage of HRTSim is the ability of hybrid modeling. HRTSim provides continuous valid three-phase mode modeling in real

* Corresponding author: sas4@tpu.ru

time and on an unlimited duration of processes in all possible operating modes of EPS. Moreover, HRTSim supports the information-control interactions between the created devices of flexible alternating current transmission system (FACTS [8, 9]), RP [10], emergency preventive automatics, etc.

Due to the above mentioned properties and abilities, HRTSim is able to solve reliably and efficiently the following problems of design, research and equipment operation:

- correct setting of various relay protection and emergency preventive automatics types with all possible modes of EPS;
- research and testing of established relay protection and emergency preventive automatics;
- current, retrospective and perspective analysis of real-time processes in equipment, including relay protection and emergency preventive automatics.

Fig. 1 shows the block diagram of information-control interactions of HRTSim with external means of relay protection and emergency preventive automatics

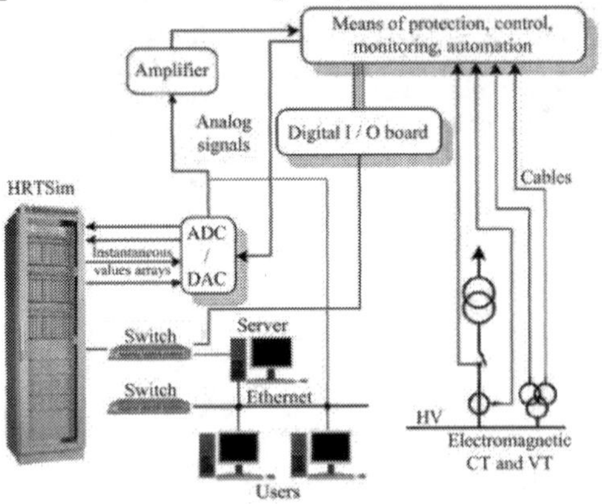

Fig. 1. Block diagram of information-control interactions of HRTSim with external means of relay protection and emergency preventive automatics.

3 Surge voltage protection simulation

In this section we present the results of the simulation of the surge voltage protection implemented in the microprocessor terminal, BMRZ-GR (Russia).

The developed SVP model operates according to the principle of comparison with the U_{svp} (setting of the maximum effective voltage). The U_{svp} setting for the turbo-generators is $1.2*U_{nom}$, and for hydro-generators $(1.5\text{-}1.7)*U_{nom}$. The functional diagram (fig. 2) shows the algorithm for the SVP actions.

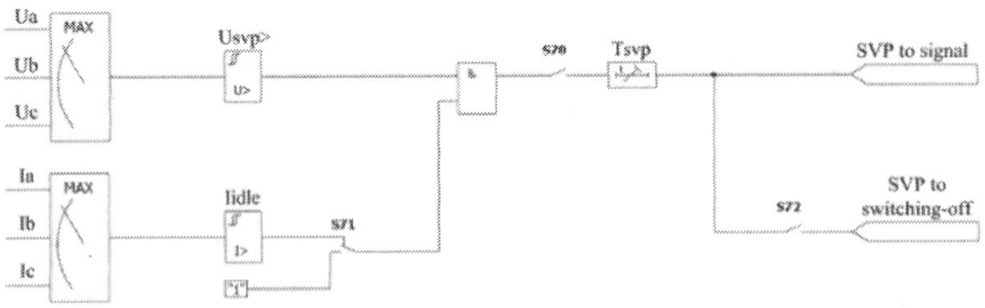

Fig. 2. Functional diagram of the algorithm for the SVP actions.

The SVP model has the ability to select an action to a switching-off or to a signal, similar to a real device. Additionally, the surge voltage protection has an idle current blocking. The blocking is set by default and the SVP only works if the generator is running on load and the maximum current is above the specified value I_{idle} (the generator is not idling).

Triggering of the SVP occurs via a time delay T_{svp}.

Fig. 3 shows the functional diagram of the SVP implemented mathematical model and its dynamic monitoring and control panel.

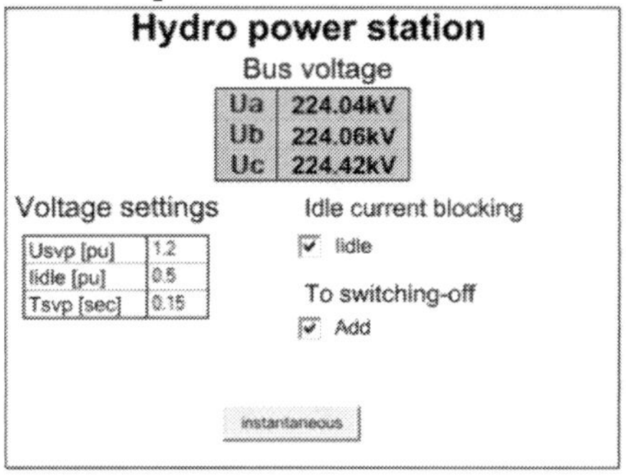

Fig. 3. Dynamic monitoring and control panel of SVP.

Voltage and current settings are set in relative nominal units.

Figure 4 shows oscillograms illustrating the process of increasing the voltage at the generator-transformer unit terminals caused by the load shedding on the 220 kV hydro power station buses and the surge voltage protection model operation, which commands to turn off the generator.

Figure 5 shows oscillograms, according to which the SVP does not operate in a mode similar to the previous one, but with a smaller amount of load shedding.

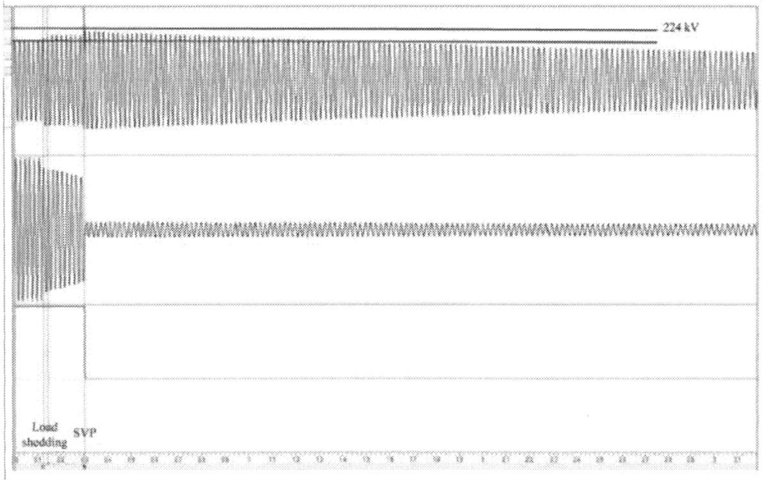

Fig. 4. Oscillograms of phase A voltage and current of the generator-transformer unit with a capacity of 210 MW and a triggering signal of SVP at the load shedding on the 220 kV buses.

 DOI: 10.1051/matecconf/201714101044

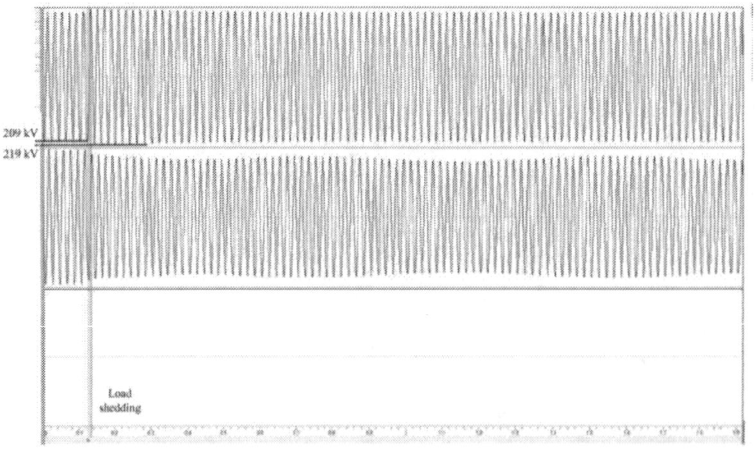

Fig. 5. Oscillograms of phase A voltage and current of the generator-transformer unit with a capacity of 210 MW and the triggering signal absence of SVP at the load shedding on the 220 kV buses.

4 Conclusions

The presented oscillograms confirm the valid operating of the SVP obtained model. The simulation results can be used to properly configure this type of relay protection for a particular object of the modeled EPS.

The reported study was funded by RFBR, according to the research project No. 16-38-60043 mol_a_dk.

References

1. M. Andreev, Y. Borovikov, N. Ruban, A. Khlebov, MATEC Web Conf. **91**, 01048 (2016), DOI: 10.1051/matecconf/20179101048
2. S. Jeong, D. Shin, J. Kim, S. Kim, Asia-Pacific Int. Symp. Electromagnetic Compatibility, 7975464 (2017)
3. L. Chen, L. Yang, Y. Liu, Y. Huang, P.C. Wang, L. Xiao, J. Fusion Energy, **35** (2016)
4. P.M. Vahdati, H. Seyedi, A. Arjmandi, S. Moradi, Int. Trans. Electrical Energy Syst. **25** (2015), DOI: 10.1002/etep.1844
5. T. Bujanovic, P. Ghosh, 4th IEEE Int. Conf. Smart Energy Grid Eng, 7589552 (2016), DOI: 10.1109/SEGE.2016.7589552
6. O.V. Gnana Swathika, K. Karthikeyan, S. Hemamalini, R. Balakrishnan, ARPN J. Eng. Appl. Sci. **11** (2016)
7. M. Andreev, Y. Borovikov, N. Ruban, 11th Int. Forum Strategic Technol. 7884223 (2017), DOI: 10.1109/IFOST.2016.7884223
8. A. Gusev, A. Suvorov, A. Sulaymanov, IOP Conf. Series: Mater. Sci. Eng. **93**, 012016 (2015), DOI: 10.1088/1757-899X/93/1/012016
9. R. Ufa, A. Vasilev, A. Gusev, A. Suvorov, 11th Int. Forum Strategic Technol. 7884204 (2017), DOI: 10.1109/IFOST.2016.7884204
10. N. Ruban, A. Sulaymanov, R. Ufa, I. Razzhivin, MATEC Web Conf. **91**, 01050 (2016)

Comparison of qualities and capabilities of the "EUROSTAG" software and the Hybrid Real Time Power System Simulator

Alexey Maltsev[1,*], *Ruslan* Ufa[1] , and *Nikolay* Ruban[1]

[1]National Research Tomsk Polytechnic University, 634050 Tomsk, Russia

Abstract. To compare the capabilities of the "EUROSTAG" software and the Hybrid Real Time Power System Simulator, simulation of the transient process in both complexes was carried out. The results show which of these complexes is best suited for various tasks.

1 Introduction

Nowadays electric power systems (EPS) become more complex due to the introduction of FACTS devices, HVDC systems, distributed power production technologies and renewable power sources. Actual is, firstly, the need to ensure uninterrupted operation in the context of the complexity of the network, and secondly, to obtain accurate information about the processes in the EPS. The acquisition of sufficiently complete information provides the effective solution of a wide range of research, design and operation tasks of the EPS.

The main analysis tools of the electric power systems are information-measuring complexes (IIC) and simulation systems. The use of IIC has such drawbacks as high laboriousness of analysis of oscillograms of emergency processes with low observability of the EPS, the limited application of results when the configuration of this EPS is changed and in other power systems, in cases of occurrence of previously disturbed effects and a large time expenditure due to the inability to conduct large-scale full-scale tests in EPS [1,2,3]. Thereby, the main analysis tool of the electric power systems' work is simulation. At the same time, special attention is paid to the modeling of dynamic transient processes, since the parameters of transient processes that arise, especially in emergency modes, determine the stability and reliability of the system.

One of the most prevalent modeling tools is "EUROSTAG" software that was developed by TRACTEBEL and ELECTRICITE DE FRANCE [4]. This complex implements a digital approach to modeling.

In Power Grid Simulation Lab the Hybrid Real Time Power System Simulator (HRTSim) was developed which is based on hybrid simulation technology [5, 6].

The purpose of this work is comparison of the simulation results in the "EUROSTAG" and in the Hybrid Real Time Power System Simulator.

* Corresponding author: maltsevaleksey@mail.ru

 DOI: 10.1051/matecconf/201714101045

2 Analysis of models of equipment realized in "EUROSTAG" and in the Hybrid Real Time Power System Simulator

The mathematical model of synchronous machines realized in "EUROSTAG" is based on the classical Park-Gorev theory, and two models are realized:
- Model with the damper winding in the d-axis and two damper windings in the q-axis.
- Model with no damper windings [4].

The load in "EUROSTAG" is modeled by a constant power in the steady-state mode. In the transient mode, a load model with voltage and frequency dependence is realized, a model of the controlled load taking into account the action of the tap changer of the step-down transformer and the distribution network model [4].

As a mathematical model of synchronous machines realized in HRTSim, a complete system of Park-Gorev equations of high accuracy is used, together with equations for the formation of three-phase coordinates a, b, c and equations for the mutual transformation of system variables d, q and a, b, c. Increased accuracy is achieved by modeling a larger number of damper circuits and taking into account the dependence of their parameters on frequency, namely three contours along the d-axis and four contours along the q-axis [3, 7, 8].

The load model realized in HRTSim includes equivalent asynchronous and synchronous motors, the generalized load described by the dependences of the active and reactive power of this load on frequency and voltage, taking into account the voltage regulation range [3, 9, 10].

3 Results of simulation

On the figure 1 there is the scheme of the grid, which includes generator (G), step-up transformer (T), transmission line (TL) and the load (L).

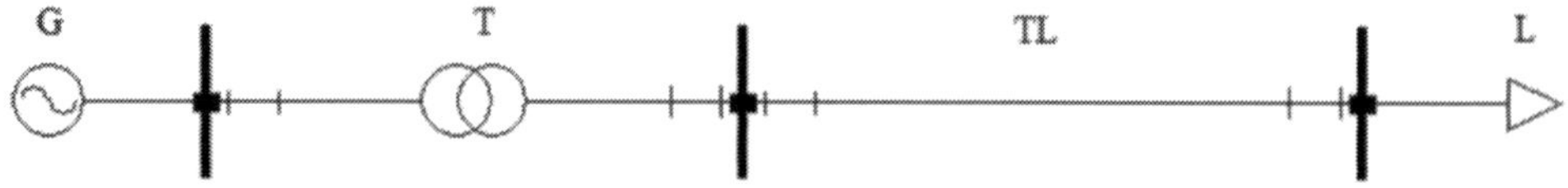

Fig. 1. The scheme of the grid.

In "EUROSTAG", the tests are carried out according to the following scenario: at a time of 10 seconds, a sudden increase in power is produced at the load node up to a time of 100 seconds.

When simulating a synchronous generator, the full model type is used without saturation, with 3 windings (the Park-Gorev equation).The load is modeled by constant resistance. Figure 2 shows the generator voltage and power consumption curves.

Further tests are carried out in a similar scenario in HRTSim. Thus it is possible to take into account fluctuations of parameters of a mode, or to not consider them, having translated the generator in an infinite bus. Figure 3 shows oscillograms of the voltage of generator and infinite bus.

From the graphs obtained, it can be seen that the nature of the process is similar for modeling in the HRTSim and in "EUROSTAG". The voltage curve of the endless power node is close to the voltage curve received in "EUROSTAG", that testifies to the convergence of calculations in these complexes. However, according to the generator voltage plot obtained in the HRTSim, it can be seen that the voltage fluctuations, which are always present in power systems, increase their scope with increasing load. Figure 4 shows a frequency oscillogram in the power system.

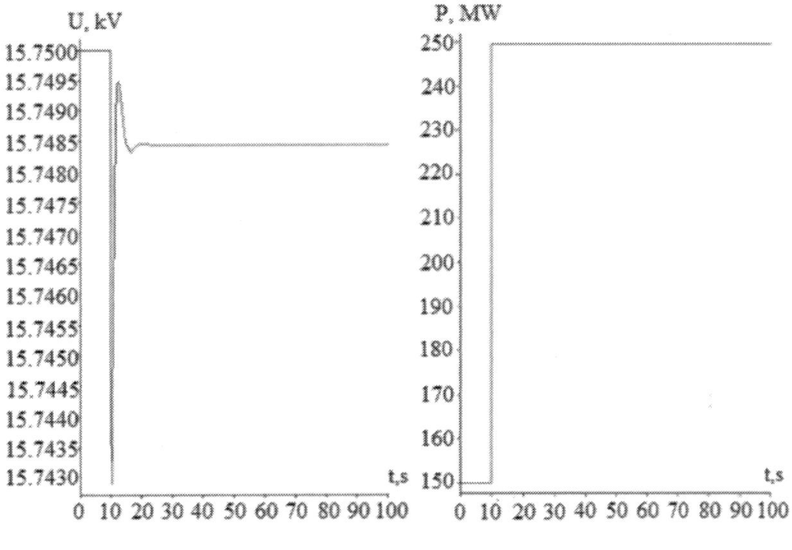

Fig. 2. Generator's voltage and power consumption curves.

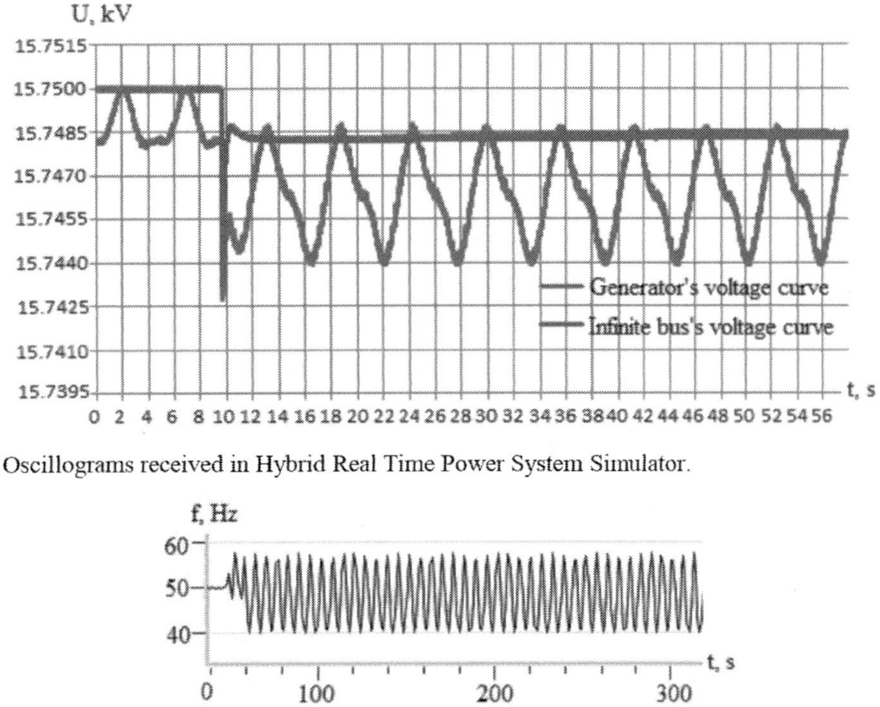

Fig. 3. Oscillograms received in Hybrid Real Time Power System Simulator.

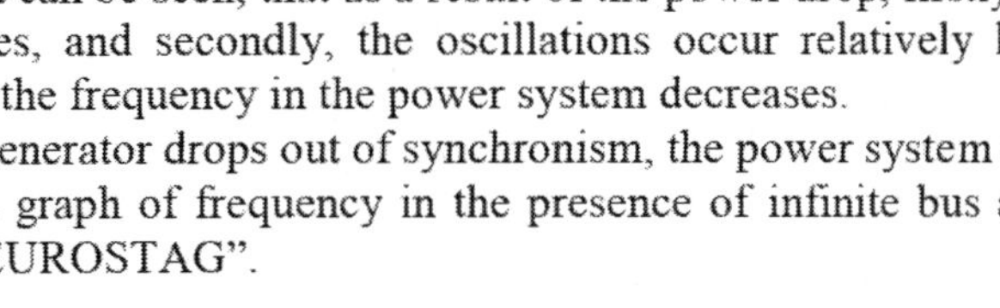

Fig.4. Frequency curve obtained in the Hybrid Real Time Power System Simulator.

From the curve it can be seen, that as a result of the power drop, firstly, the sweep of the fluctuations increases, and secondly, the oscillations occur relatively lower than in the initial mode, that is, the frequency in the power system decreases.

As a result, the generator drops out of synchronism, the power system collapses.

Figure 5 show a graph of frequency in the presence of infinite bus and the frequency graph obtained in "EUROSTAG".

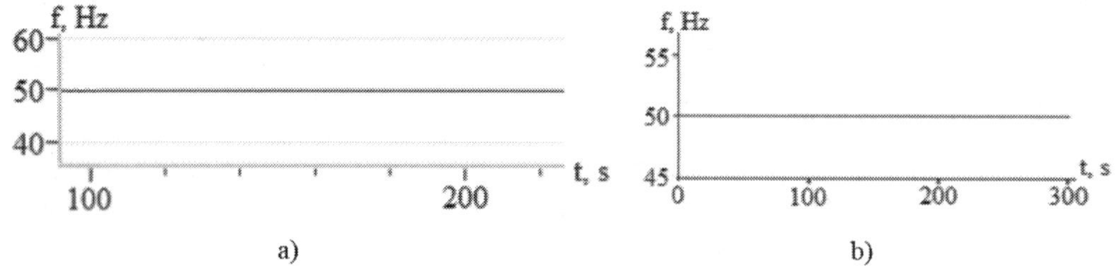

Fig. 5. Graph of frequency in the presence of infinite bus (a) and graph of frequency obtained in "EUROSTAG" (b)

When calculating in "EUROSTAG" and in the presence of infinite bus it is impossible to judge the frequency changes, because infinite bus maintains the frequency value to constant.

As a result of this work it can be seen that with the help of the HRTSim it is possible to obtain more detailed information about the behavior of the power system under the influence of various perturbations. This complex allows us to see not only the nature of the transient process and the new steady-state regime, but also to get the information about the fluctuations of the regime itself and, accordingly, its stability, that is, it gives an idea of the dynamic stability of the energy system and its static stability. "EUROSTAG" models separately steady-state modes, not taking into account the oscillations occurring in the power system, it calculates directly the transients, gives the information only about the dynamic stability.

This work was supported by the Non-profit Partnership "Global Energy" under the project No. MG- 2017/04/1 "Developing a hardware and software platform for all-regimes real-time simulation of intelligent electric power systems".

References

1. M. Khatir, S. Ahmed Zidi, S. Hadjeri, J. Electrical Eng. **61**, 1 (2010)
2. R.L. Sellick, M. Åkerberg, 10th IET Int. Conf. AC DC Power Transmission. Birmingham (2012)
3. M.V. Andreev, A.O. Sulaymanov, A.A. Suvorov, MATEC Web Conf. **37**, 1 (2015), DOI: 10.1051/matecconf/20153701001
4. M. Stubbe, A. Bihain, J. Deuse, J.C. Baader, IEEE Trans. Power systems, **4** (1988)
5. A.A. Suvorov, Yu.S. Borovikiv, A.S. Gusev, A.O. Sulaymanov, M.V. Andreev, N.Yu. Ruban, R.A. Ufa, Electrical eng. **99**, (2017), DOI: 10.1007/s00202-016-0464-4
6. A.A. Suvorov, A.S. Gusev, A.O. Sulaymanov, O.V. Suslova, MATEC Web Conf. **91**, 1 (2016), DOI: 10.1051/matecconf/20179101052
7. A.S. Gusev, A.A. Suvorov, A.O. Sulaymanov, MATEC Web Conf. **37**, 1 (2015), DOI: 10.1051/matecconf/20153701023
8. Yu.S. Borovikov, A.O. Sulaymanov, M.V. Andreev, IOP Conf. Series: Mater. Sci. Eng. **93**, 5 (2015), DOI: 10.1088/1757-899X/93/1/012001
9. Yu.S. Borovikov, A.V. Prokhorov, M.V. Andreev, 7th Int. Forum Strategic Techol. **2**, 604 (2012), DOI: 10.1109/IFOST.2012.6357822
10. A.S. Gusev, A.A. Suvorov, A.O. Sulaymanov, IOP Conf. Series: Mater. Sci. Eng. **93**, 5 (2015), DOI: 10.1088/1757-899X/93/1/012016

MATEC Web of Conferences **141**, 01046 (2017) DOI: 10.1051/matecconf/201714101046
Smart Grids 2017

Analysis of dynamic behavior of the back-to-back High Voltage Direct Current link model as part of electrical power system

Vladimir Rudnik[1,*], *Nikita* Desyukov[1], and *Ruslan* Ufa[1]

[1]National Research Tomsk Polytechnic University, 634050 Tomsk, Russia

Abstract. One of the main directions of the development of electric power systems is the introduction of technologies based on high-power semiconductor switches, such as FACTS devices and HVDC technologies. These systems effectively solve a number of urgent tasks of EPS, connected with asynchronous connection of EPS, transmission of electricity, improve local and systemic flexibility and reliability of EPS, increasing the capacity of network elements that contains a "weak" connection. However, the implementation and operation of mentioned technologies in the EPS determines the need for a wide range of analysis and research that can only be done with the help of mathematical modeling.

1 Introduction

One of the main directions of development of electric power systems (EPS) is the introduction of devices and technologies based on power semiconductor switches (FACTS devices and HVDC technologies), which are designed to improve flexibility, network capacity and ensure the reliability of electricity transmission [1, 2]. B2B HVDC link is a converter station that used to convert alternative current (AC) to direct current (DC) and then convert DC to AC the same or other frequency. B2B HVDC link solves the following tasks: asynchronous connection of EPS; increasing of local and system flexibility and reliability of EPS; increasing of transmission line capacity. Number of research and analysis of B2B HVDC link operation should be done, before introduction in the EPS [2-4].

2 B2B HVDC link model

Properties and opportunities of B2B HVDC link will be analyzed via the B2B HVDC link model realized in the Matlab Simulink and the Hybrid Real Time Simulator (HRTSim) [5-9].

* Corresponding author: fordlp006@mail.ru

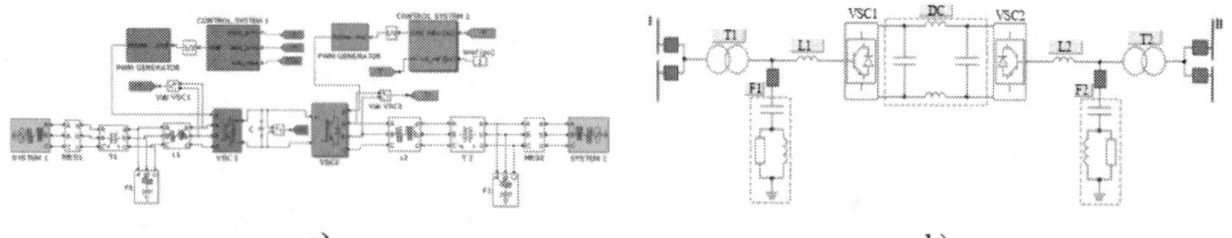

a) b)

Fig. 1. The B2B HVDC link model realized in the Matlab Simulink (a) and HRTSim (b) where *T 1,2* – transformer, *L 1,2* – reactor, *DC* – direct current link, *VSC 1,2* – voltage source converter, *PWM* - pulse-width modulation, *F 1,2* – harmonic filter.

According to the [10, 11] the following scenarios should carry out to confirm the benefits of B2B HVDC link: asynchronous connection of the EPS; short circuit on one side of the EPS; step change and reverse of transmitted power.

2.1 Asynchronous connection of EPS

In Fig. 2 the oscillograms of voltage and currents on the side of the System 1 and System 2, and the voltage in the DC circuit, after the connection of the two power systems, are presented.

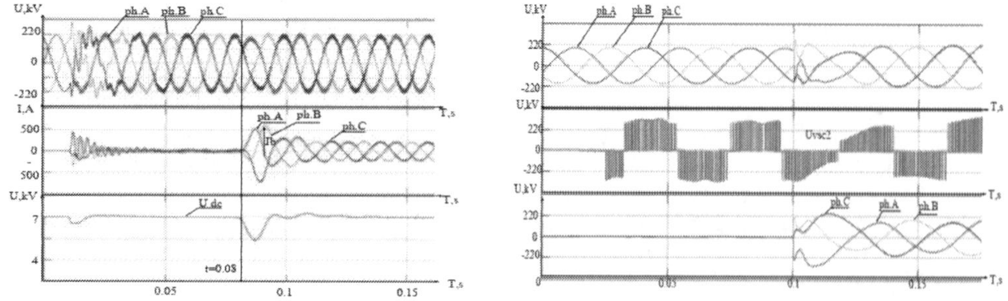

a) b)

Fig. 2. Oscillograms of the voltage in the DC circuit, voltages and currents on the System 1 (a) and and System 2 (b).

According to the oscillograms shown in Figure 3, the voltage of phase A on the side the System 1 and System 2 after the connection of the two power systems (0.08 s.) the same frequency is observed in both power systems. This confirms the correct functioning of the B2B HVDC link, however a higher harmonics, which are generated by VSC, is occurred. To eliminate the higher harmonics possible with the correct tuning of the higher-order harmonic filter.

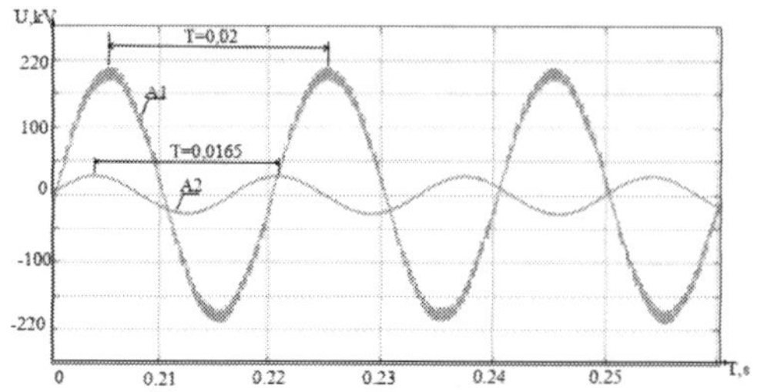

Fig. 3. Oscillograms voltage of the phase A on the side the EPS 1 and the EPS 2 after the connection of the two power systems.

2.2 Short circuit on side the EPS 2

At the initial time of the EPS 1 transmits power to the EPS 2, then, at the 0.1 s. there is a short circuit on the side of the power system 2. Figure 4a shows oscillograms of voltages and currents of power systems using a power line.

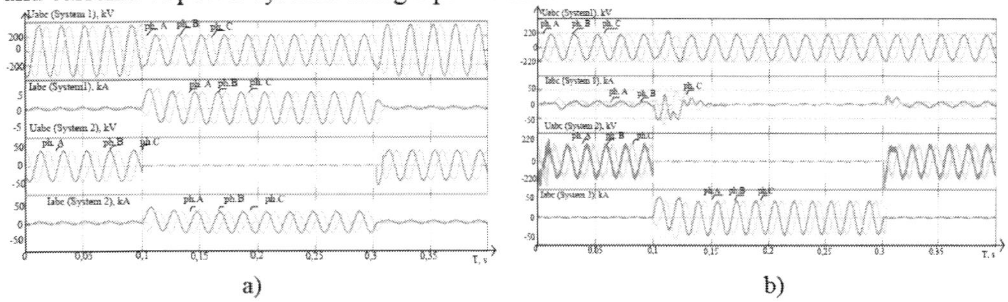

Fig. 4. Oscillograms of currents and voltages under AC short circuit without (a) and with (b) B2B HVDC link realized in the Matlab Simulink.

According to the oscillograms shown in Fig. 4b, when there is a short circuit on the side of the EPS 2, there is an increase in the currents of the two power systems

When power systems are connected with the B2B HVDC link implemented in the Matlab Simulink (Fig. 4b), it is seen that in the time interval 0.1-0.2 s. There was a sharp jump in currents of up to 30 kA, as well as a drawdown of voltages on the side of the EPS 2, while the currents and voltages of the EPS 1 remained practically unchanged.

When connecting the EPS with the HVDC implemented in the HRTSim (Fig. 5), the following scenario. At time 0,5, a three-phase earth fault occurs on the side of the power system 1, then within 0.1 s. the relay protection system disconnects the EPS 1 and at the instant 0,19 the automatic restart of the System 1 is triggered. Fig. 5 shows that during the short-circuit the System 2 operates in normal mode.

The results of the research show that with the use of the B2B HVDC link it is possible to make a complete decoupling of the power systems, that is, an accident that occurred on one power system does not affect the operation of another power system.

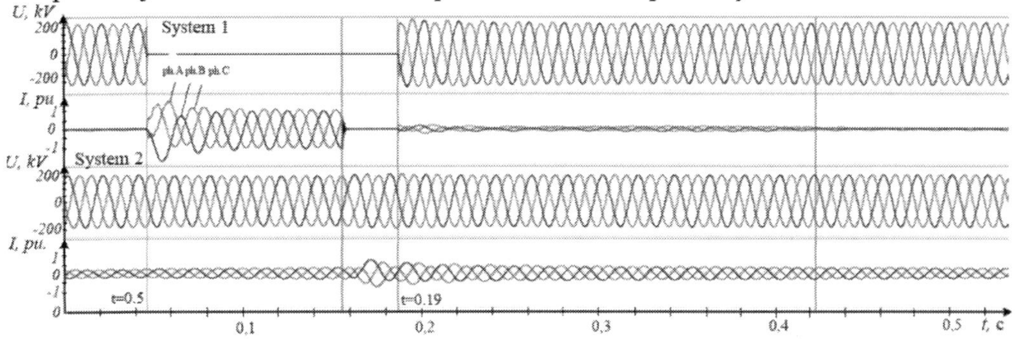

Fig. 5. Oscillograms of currents and voltages under AC short circuit with B2B HVDC link realized in the HRTSim.

3 Conclusion

As a result of the research, an analysis was made of the interaction of the B2B HVDC link with the adjacent power system. One of the ways to carry out such an analysis was modeling, which was carried out with of the Matlab Simulink and the HRTSim. Three situations were considered in connection two asynchronous operating power systems:

frequency differences 50 and 60 Hz; the appearance of a fault; step change and reverse of transmitted power of the B2B HVDC link. The analysis showed that it is possible to connection two asynchronous operating power systems without any failure situations. It is also worth noting that when performing research using the Matlab Simulink and the HRTSim, there is a general trend in the results. This suggests that the HRTSim correctly performs simulation.

This work was supported by the Non-profit Partnership "Global Energy' under the project No. MG- 2017/04/1 "Developing a hardware and software platform for all-regimes real-time simulation of intelligent electric power systems".

References

1. K. Vijay, Sud. HVDC and Controllers: applications of static converters in power systems (2009)
2. R.L. Sellick, M. Åkerberg, 10th IET Int. Conf. AC DC Power Transmission. Birmingham (2012)
3. A. Suvorov, Y. Borovikov, A. Gusev, A. Sulaymanov, M. Andreev, N. Ruban, R. Ufa, Electrical eng. **99** (2017), DOI: 10.1007/s00202-016-0464-4
4. Yu. Borovikov, A. Prokhorov, M. Andreev, 7th Int. Forum Strategic Techol. **2** (2012), 10.1109/IFOST.2012.6357823
5. M. Andreev, A. Sulaymanov, A. Suvorov, MATEC Web Conf. **37** (2015), DOI: 10.1051/matecconf/20153701001
6. A.A. Suvorov, A.S. Gusev, A.O. Sulaymanov, O.V. Suslova, MATEC Web Conf. **91** (2016), DOI: 10.1051/matecconf/20179101052
7. N.Y. Ruban, Y.S. Borovikov, A.O. Sulaymanov, A.S. Gusev, IFOST (2016), DOI: 10.1109/IFOST.2016.7884255
8. N.Yu. Ruban, A.S. Gusev, A.Yu. Pishchulin, MATEC Web Conf. **37** (2015), DOI: 10.1051/matecconf/20153701046
9. A. Gusev, A. Suvorov, A. Sulaymanov, MATEC Web Conf. **37** (2015), DOI: 10.1051/matecconf/20153701023
10. M. Khatir, S. Ahmed Zidi, S.Hadjeri, J. Electrical Eng. **61**, **1** (2010)
11. Xu Lie, B.R. Andersen, P. Cartwright, Transmission Distribution Conf. Exposition, IEEE PES, **2** (2003)

Energy-financial flows of an enterprise

Galina Klimova[1,*], and *Veronika* Morozova[1]

[1]National Research Tomsk Polytechnic University, 634050 Tomsk, Russia

Abstract: Energy, financial, material and other interrelation set common presentation formed on enterprises can be sum in energy-financial flows. Such balances composition and analysis regular practice provides during several flows forming determine contradictions in time, put value of energy efficiency indexes. And, in case of their increasing quickly make management decision. In this way energy-financial balance is enterprise operating management.

Fuel and energy balances are important and sustainable device of energy facility exploitation's efficiency control and analysis. The most interesting and less spread apart energy balances that are in monetary value [1].

Such balance can be used for energy industry's analysis and planning, for quality control of energy-saving and energy resources utilization efficiency improvement measures, for production development programme energy efficiency values estimation. Balance shows energy resources production and consumption volumes of reasonable time, money resources income and expense, fuel storage, energy resources plant demands and utility networks losses, energy resources consumption volumes of organization units according to uses [1 – 7].

There can be designed a model for common whole set of energy, financial, material and other interrelations performance between energy resources packagers and consumers, where these interrelations are presented as flows simulated by network graphs [8].

Energy flows form energy balance that remains as base for lots infrastructural solutions connected with development designing, planning and on-line controlling of areal economy.

Energy balance, because of its physical nature severity, radically can be formed for any arbitrary short period and any arbitrary small site.

Financial flows incidental with energy flows and expressing bought and sold energy resources cash compensation are formed differently. It is hugely complicated to build financial balance for short periods and small sites. Some energy flows do not have responsive financial flow, and some financial flows are generated not by energy flows, but by their realization conditions. Financial flows are greatly varying in time, and small sites financial flows generally do not have recordkeeping. Today standard coasts division into semi-variable (in power industry – fuel, energy, water and etc.) and semi-constant (amortization, salary, taxes and etc.) essentially does not change temporary inequality. Because of it supposed energy and financial flows spread pattern accordance generally in fact is out of a place.

* Corresponding author: gariki@tpu.ru

Energy-financial balance scheme represents graph, branches of which are company interrelationships (resources flows) with objects (graph nodes). Such scheme analytic image is incidence matrix. This matrix elements are formed following way: signals about energy, financial, material and service flows availability and sign are put into matrix columns. Objects having relationship with company (contractors) are put into matrix rows. At the intersection of relevant column and row should be put 1 or 0 in dependence of availability or no-flow condition. Flow direction form its sign: if "plus" –flow directed to company, and if "minus" – flow directed to opposite side.

Matrix fuel elements detailing provides to form fuel balance, and financial elements – financial balance.

Simulated energy company indexes are presented in the tables for visual reference. Table 1 shows some process statements.

Table 1. Annual rate of company activity.

Ser.No.	Characteristics	Targets	Actuals	Notes
1	Energy generation, mil. kWh	3600	3410	
2	Energy purchase at wholesale market, mil. kWh	1400	1599	
3	Electricity consumption for auxiliaries, mil. kWh	288	273	
4	Energy networks losses, mil. kWh	528	530	
5	Productive supply, mil. kWh	4184	4206	
6	Fuel purchase volume, toe	13.0	12,28	1.364 mil. tfoe.
7	Maximum load utilization time, hour	5500	5470	

There is got information about relationships with separate consumers, several levels budgets and markets by elements of matrix rows.

Company budget divides received electrical energy sale proceeds into planning and reporting proceeds items. Table 2 shows an example.

Table 2. Company expenditures, mil. RUR.

Ser.No.	Characteristics	Targets	Actuals	Notes
1	Fuel payment including stock reserve forming	280.2	3410	0.789
2	Purchased electrical energy payment	188.9	215.8	0.05
3	Salary	120.8	100.3	0.02
4	Amortization payments	55.7	55.4	0.01
5	Operating costs	403.5	382.2	0.09
6	Taxes and fees	48.6	46.0	0.02
7	Other	120.8	114.4	0.03\4324.1

Costs some aspects influence on consumers energy production cost is determined by aggregates:

- ✓ Average consumers energy production coast;
- ✓ Specific fuel equivalent consumption for electric generation;
- ✓ Fuel factor of energy production coast.

Relationships with region electricity company all levels budgets are built on principle of salary and taxes payment priority relative to others. Company pay taxes:

- ✓ for profits;
- ✓ for added coast;
- ✓ of personal income tax;
- ✓ for land;
- ✓ for mineral resources extraction and natural resources using;

✓ of road faund;

Company budgetary payments are compensation for public services that mean services having public significance and supplied in terms of legistrated conditions.

By this way energy and financial flows model based on incidence matrix forming provides to estimate wide range of current and integrated indexes. For sure, energy-financial balance is of the utmost interest because it consists information about cash flow connected with energy resources production and using.

Firstly, energy balance is formed in tangible form of each utilizable resource, then in monetary units. Synthetical balance in monetary units provides to outline energy efficiency upgrading ways – consumption reduction, replacing some energy resources to others or resources consumption cost reduction. Boxes, districts, subdivisions, installations analysis of balance expense side helps to find out losses centers and provide means to decrease them.

Considering relative energy resources replaceability, current cost changing and each resource quantity balance can be optimized not only by losses decreasing but also by the way of structural adjustment.

Energy-financial flows formed model using example can be forecast changes of contributing factors on electricity company production and realization cost. In this case, research facility resides in significant model portability in comparison with company actual financial system.

- fuel tariff,
- energy consumption rate,
- electricity sales tariff,
- purchased electricity tariff,
- taxes changes rate,
- electricity distribution and transport cost,
- distribution services cost,
- average salary changes rate,
- repair, connection, transport and other services tariff,
- system operator service tariff,

Investigation of factors changing influence on energy production cost is done by planning method providing upgrade simultaneously large number of variables (factors) and to find target interrelationships.

For a wide range of factors, each influence research by turns leads to large number of experiments. With that, synergies between factors are not explored. Average (basic) factors value set to level achieved in base year, and variation intervals make values corresponding factors changing in range of ±5 % from base level.

Conclusion:

Common spatiotemporal energy and financial flows of energy company visualization indicates about structural observable processes determinism, it provides to use universal mechanisms during operational and strategic decisions. Analysis mechanisms versatility provides to solve many control problems.

Energy-financial balances using provides to look over financial flows values and directions in front of energy flows, it forms competitive position.

References

1. GOST 27322-87. Power plant balance. Main principles [in Russian]

2. GOST P 51379-99. Energy performance certificate of fuel and energy resources power consumer [in Russian]
3. G.N. Klimova, V.V. Litvak, *Seven problems and seven keys of energy saving* (Red banner, Tomsk, 2013) [in Russian]
4. A.S. Nekrasov, Yu.V. Sinyak, *Power plant engineering control* (Energiya, Moscow, 1979) [in Russian]
5. Yu.N. Savenko, E.O. Steinhaus, *Energy budget* (Energiya, Moscow, 1971) [in Russian]
6. S.M. Romanov, G.F. Alekseev, Mountainous research and information bulletin, **1**, 12 (2010) [in Russian]
7. V.V. Litvak, *Zonal energy saving basis* (STT, Tomsk, 2002) [in Russian]
8. V.V. Litvak, G.Z. Markman, M.I. Yavorskiy, Industrial energetics, **5** (2004) [in Russian]
9. E.I. Borisov, I.M. Lyskov, D.Yu. Alexandrov, E.E. Borisova, Electric power industry economy and finances, **9** (2003)
10. Retail (consumers) market electrical (thermal) power regulated tariffs and costs calculation methodology instructions. Approved by the RF Federal Energy Commission 31.07.2002 [in Russian]
11. A.V. Fedchenko, Economics and finance of electric power industry, **7**, 151 (2004) [in Russian]
12. G.N. Klimova, V.V. Litvak, *Origins of energy saving* (STT, Tomsk, 2016) [in Russian]

MATEC Web of Conferences **141**, 01048 (2017)
Smart Grids 2017
DOI: 10.1051/matecconf/201714101048

Development of the heat and mass transfer model for the study of the temperature traces water droplets in a flame

Dmitry Antonov[1], and *Petr* Kostomarov[1,*]

[1]National Research Tomsk Polytechnic University, 634050 Tomsk, Russia

Abstract. The heat and mass transfer model is developed with using Ansys Fluent. The typical temperature of gases in the trace of water droplets is determined (initial temperature of gases 1170 K). Several types for the location of water droplets are studied: two successive water droplets; two parallel water droplets; five water droplets in checkerboard order. The hypothesis about gas temperature reduction in the trace of a moving liquid is confirmed.

1 Introduction

At the present time the development of new oil and gas accumulations is in progress. Oil and gas are highly flammable matters. Because of this, there are special requirements for the development and operation of oil and gas accumulations. The main emphasis is given to ensuring the fire safety [1 - 4]. The most common methods of fire extinguishing are water mist fire suppression system [1, 3] and fire extinguishing with using big water mass [2, 4].

Over the last years researches are conducted [1 - 3]. The main objectives of these researches aimed at improving the efficiency of means and methods of atomization. Unfortunately, the fact is not taken account that the water droplets during motion in high-temperature gases reserve temperature traces which reduce the temperature in the combustion zone due to the evaporation of water droplets.

The goal of the present work is development of heat and mass transfer model for the study of the temperature traces water droplets in a flame with using Ansys Fluent.

2 Physical model of heat and mass transfer

When formulating heat and mass transfer problem, we assumed that the initial temperature of water droplets is T_w=300 K. Temperature of high-temperature gases is T_g=1170 K. Water droplets are heated by conduction in the flow of high-temperature gases. It was assumed that the thermal characteristics of the materials are not dependent on temperature.

* Corresponding author: laz@tpu.ru

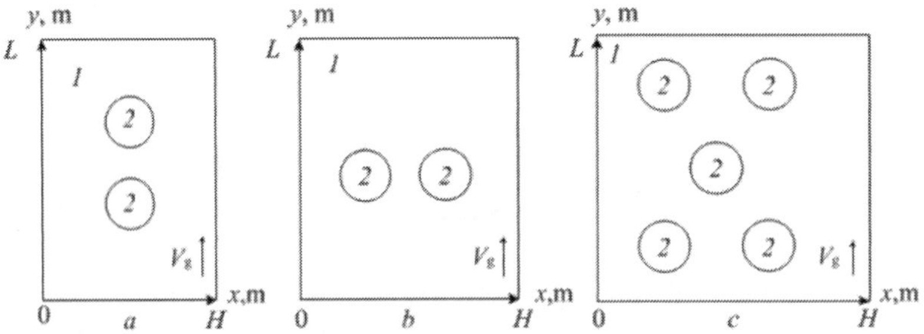

Fig. 1. Scheme of solution area for two successive water droplets (*a*); two parallel water droplets (*b*); five water droplets (*c*): *1* – high-temperature gases, *2* – water droplets

3 Mathematical model and decision methods

The processes under study are described by the system (Fig. 1) of non-stationary partial differential equations.

For high-temperature gases ($0<x<H$, $0<y<L$):

continuity equation:

$$\frac{\partial^2\psi}{\partial x^2}+\frac{\partial^2\psi}{\partial y^2}=-\omega,\ \text{at}\ \omega=\frac{\partial v}{\partial x}-\frac{\partial u}{\partial y}$$

motion equation:

$$\frac{\partial\omega}{\partial t}+u\frac{\partial\omega}{\partial x}+v\frac{\partial\omega}{\partial y}=\gamma(\frac{\partial^2\omega}{\partial x^2}+\frac{\partial^2\omega}{\partial y^2})+\beta g_y\frac{\partial T_1}{\partial x},\ \text{at}\ u=\frac{\partial\psi}{\partial y},\ v=\frac{\partial\psi}{\partial x}$$

energy equation:

$$(\frac{\partial T_1}{\partial t}+u\frac{\partial T_1}{\partial x}+w\frac{\partial T_1}{\partial y})=a_1(\frac{\partial^2 T_1}{\partial x^2}+\frac{\partial^2 T_1}{\partial y^2})$$

water vapor diffusion equation:

$$\rho_2(\frac{\partial C_w}{\partial t}+u\frac{\partial C_w}{\partial x}+w\frac{\partial C_w}{\partial y})=\rho_2 D_2(\frac{\partial^2 C_w}{\partial x^2}+\frac{\partial^2 C_w}{\partial y^2})$$

balance equation:

$$C_g+C_w=1$$

For water droplets ($0<r<R_d$, $0<\varphi<2\pi$):

thermal conductivity equation

$$\frac{\partial T_3}{\partial t}=a_3\left[\frac{1}{r^2}\frac{\partial}{\partial r}(r^2\frac{\partial T_3}{\partial r})+\frac{1}{r^2\sin(\varphi)}\frac{\partial}{\partial\varphi}(\sin(\varphi)\frac{\partial T_3}{\partial\varphi})\right]$$

Initial conditions (t=0): T=T_0, C_w=0, C_g=1, ψ=0, ω=0 at 0<x<H, 0<y<L; T=T_0 at 0<r<R_d, 0<φ<2π.

Boundary conditions at t >0:

$$T=T_g,\ C_w=0,\ C_g=1,\ \frac{\partial \psi}{\partial y}=V_g \text{ at } y=0,\ 0<x<H;$$

$$\frac{\partial^2 T}{\partial y^2}=0,\ \frac{\partial^2 C_w}{\partial y^2}=0,\ \frac{\partial^2 C_g}{\partial y^2}=0,\ \frac{\partial \psi}{\partial y}=0 \text{ at } y=L,\ 0<x<H;$$

$$\frac{\partial T}{\partial t}=0,\ C_w=0,\ C_g=0,\ \frac{\partial \psi}{\partial x}=0 \text{ at } x=0,\ 0<y<L;$$

$$\frac{\partial T}{\partial t}=0,\ C_w=0,\ C_g=0,\ \frac{\partial \psi}{\partial x}=0 \text{ at } x=H,\ 0<y<L$$

Boundary conditions for liquid – gas were set taking into account vaporization [5]:

$$R=R_1,\ 0<\varphi<2\pi,\ \lambda_3\frac{\partial T_3}{\partial R}=\lambda_2\frac{\partial T_2}{\partial R}-W_e Q_e,\ \rho_2 D_2\frac{\partial C_w}{\partial r}=W_e$$

Nomenclature: x, y – coordinates of the Cartesian coordinate system, m; u, v – x and y components of convection velocity of the gas-vapor mixture, m/s; ψ – flow function, m^2/s; ω – rotation vector velocity, s^{-1}; t – time, s; γ – kinematic viscosity, m^2/s; β – thermal expansion coefficient, K^{-1}; g_y – acceleration of gravity, m/s^2; T – temperature, K; a – thermal conductivity, m^2/s; C_w – concentration of water vapor; D – diffusion coefficient, m^2/s; C_g – concentration gases (0<C_g<1); r, φ – coordinate of the spherical systems; R_d – radius of droplet, m; H, L – dimensions of solution area, m; T_0 – initial temperature, K; T_g – temperature of gases, K; T^* – temperature of gases in the trace of droplets, K; Q_e – water vaporization heat, J/kg; W_e – evaporation rate, kg/(m^2·s); V_g – velocity of high-temperature gases, m/s; symbols: *1* – high-temperature gases, *2* – water vapor, *3* – water liquid.Equations should be centred and should be numbered with the number on the right-hand side.

4 Results and discussion

Fig. 2 shows temperature and velocity fields of high-temperature gases for systems (Fig. 1). The resulting temperature fields illustrate that over time the temperature of gases in the trace of droplets changes. This is primarily due to intensive evaporation of water droplets and a corresponding decrease in the droplet size. The water evaporation area is maximized when water droplets are moving in the beginning.

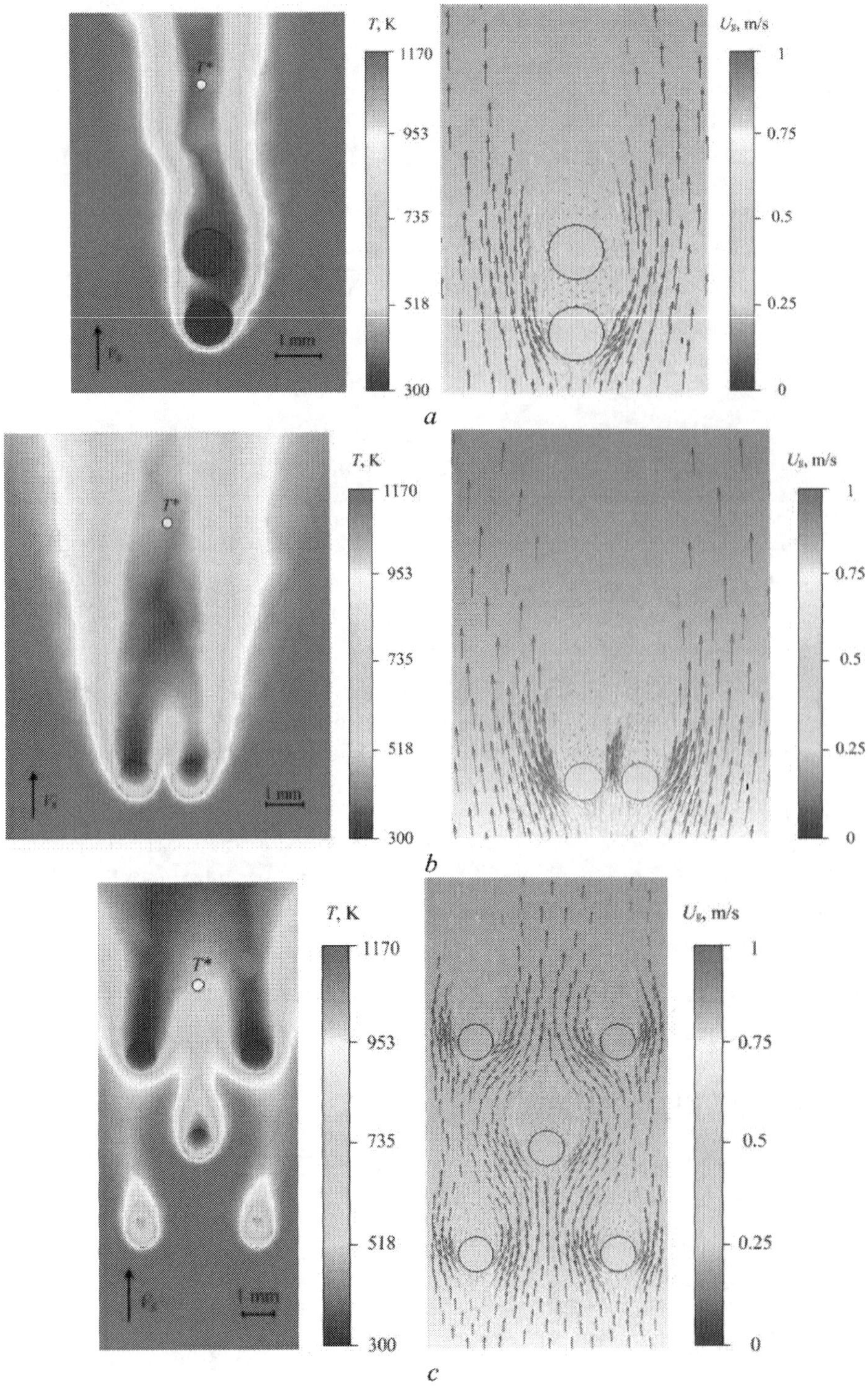

Fig. 2. Temperature fields and velocity fields with two successive water droplets (*a*); two parallel water droplets (*b*); five water droplets (*c*).

The resulting temperature fields prove the hypothesis that aerodynamic and temperature traces of an aerosol and non-sprayed water are significantly different. Moreover, the transverse dimensions of the temperature traces of droplets are considerably different, especially the areas with a temperature which is substantially lower than the initial one (by several hundred degrees).

Moreover, the deceleration and turning of the evaporating droplets influence the temperature trace. In particular, sizes of droplets are reduced when moving through the counter flow of combustion products. Then after deceleration and turning, droplets begin to move in the same direction with the flow and again have a significant influence on the temperature in the trace due to evaporation. As a result, the lifetimes of the temperature trace are several times greater than for the massif and an individual droplet.

The mathematical simulation also illustrates different causes and mechanisms of a temperature reduction in the trace of sprayed and non-sprayed water. In particular, in the case of an aerosol, evaporation plays a decisive role. Aerosol droplets are heated up to the maximum possible temperatures corresponding to the intense vaporization. In the case of the large droplets and water massifs, the intensive vaporization was not observed. This is because the evaporation area is small. Thus, the liquid is heated at relatively low rate due to the large sizes of massifs. As a consequence, in the case of massifs and large droplets, the temperature is decreased significantly only in a small vicinity of their trajectory. This is due to the fact that water accumulates gas energy.

5 Results and discussion

As a result of numerical simulation models are designed for different schemes of heat and mass transfer location of several water droplets to vary the distance between them. The results are important for the development of fire extinguishing technologies in the fields of oil and gas industry.

This work was supported by the Russian Science Foundation (No. 14-39-00003).

References

1. S.S. Sazhin, W.A. Abdelghaffar, E.M. Sazhina, E.M. Heikal, J. Therm. Sci. **44**, 610 (2005)
2. D.V. Antonov, I.S. Voitkov, P.A. Strizhak, J. EPJ, **110**, 1 (2016)
3. Y.V. Lyulin, J.A. Kabov, Appl. Phys. Lett. **38**, 88 (2013)
4. R.S. Volkov, G.V. Kuznetsov, P.A. Strizhak, Int. J. Therm. Sci. **88**, 193 (2015)
5. W.H. Lee, Technical ReportLA-UR, **79**, 975 (1979)

MATEC Web of Conferences **141**, 01049 (2017) DOI: 10.1051/matecconf/201714101049
Smart Grids 2017

Artificial neural networks as the data ruling components of future intelligent energy systems

Anna Buran[1,*], and *Maksim* Denisov[1]

[1]National Research Tomsk Polytechnic University, 634050 Tomsk, Russia

Abstract. The main target of this work is to study artificial neural networks and their role in the future intelligent energy systems. Artificial neural networks in the near future will be a very effective tool for data analysis and very perspective for energy consumers. The accelerating growth in the field of software engineering research will prove appearing of neural networks in intelligent energy systems. This study is expected to be important for researchers and engineers studying artificial neural networks (ANNs) in a wide range of domains.

1 Introduction

Researchers and scientists study artificial neural networks (ANNs) in different spheres of our life where neural networks can be used, from economics, business, computer games to induction motors. The artificial neural network was designed as a computational model based on the brain to solve particular problems. In the modern world ANNs take the form of different calculating, data-analyzing, computing algorithms which can already be used in our life, for example, by any mobile phone owner who desires for fun to change his appearance in the photo – it can be provided by such mobile apps as MSQRD and Prisma, based on neural algorithms. Some scientists try to predict the accuracy of components produced by Wire Electrical Discharge Machining (WEMD) by using an Elman-based Layer Recurrent Neural Network (LRNN) [1]. Several studies discuss a novel conceptual framework: Fractional Hopfield neural networks (FHNN), neural network-based command filtered control for induction motors with input saturation, design of a neural network based distributed power flow controller (DPFC) for power system stability, while others observe the feedback principle [2, 3, 4, 5]. Deisseroth and Schnitzer study engineering approaches to illuminating brain structure and dynamics. Electrical engineering has long influenced neuroscience, having contributions in cable theory [6, 7] and radioelectronics [7]. This study is aimed at investigating the role of future intellectual energy systems in the field of neuronal creation. First, the principles on which neuronets are based should be considered.

* Corresponding author: abuand@tpu.ru

1.1 Neurons, synaptic connections, synapse weight, activation functions

We can suggest that neural networks are inspired by the biological neural networks of animal and, correspondingly, human brains. Heaton studied the influence of artificial intelligence for humans [8]. One of the main advantages of these algorithms – progressive self learning process based on considering examples and learning from mistakes; the same, but much more complicated logics is developed by nature in human brains. When we refer to artificial neural networks, we deal with the following definitions which came from biology. 1) Neuron – in ANN is a mathematical function, or computing unit which receives information (in the form of signal), calculates it and transmits further. There are three main types of neurons: input, covered and output. Input neurons usually sum signals of another quantity of neurons and using activation functions output information; covered work with the information received from other covered or input neurons; output neurons give the final answer to indignation of this information system.

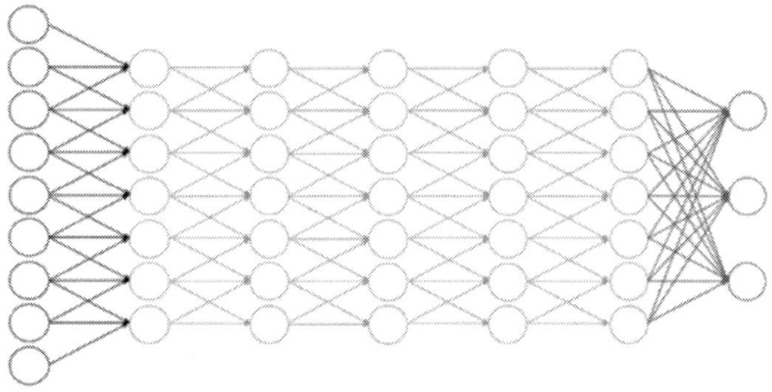

Fig. 1. Different types of neurons.

Synaptic connection exists between two neurons. Synapses have one important parameter – synapses weight. It's responsible for processing the information, and on the whole all the synaptic weights of ANNs composes "the brain" of the system. The synapse weight can be demonstrated with the help of colour mix (fig. 1) – mixing colours in the right proportion (due to their "weight" in new composed colour) creates a new colour.

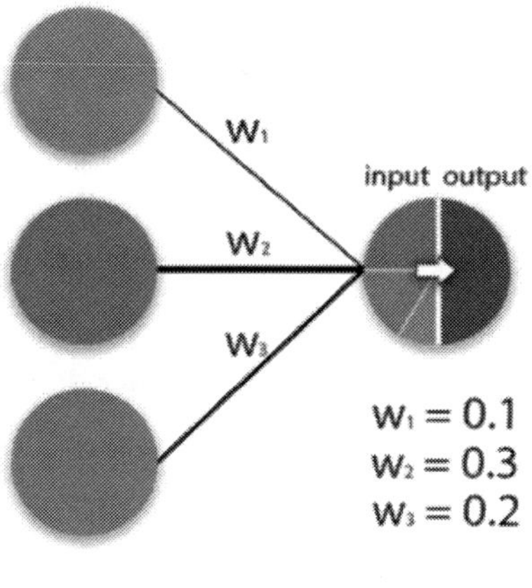

Fig. 2. Synapses weight demonstration on the mix of colours. W – weight.

The following figure shows input and output data producing, where I1, I2 are input neurons, H1 is covered (output) neuron, w1, w2 are synaptic weights.

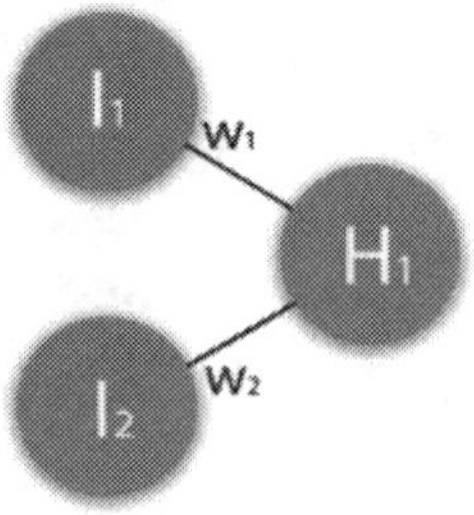

$$1)\ H_{1_{input}} = (I_1 * w_1) + (I_2 * w_2)$$

$$2)\ H_{1_{output}} = f_{activation}(H_{1_{input}})$$

Fig. 3. Input and output data producing. I1, I2 – input neurons, H1 – covered (output) neuron, w1, w2 – synaptic weights.

Output information is a function of neuron activation. Activation function is the function of data normalization, in other words – how the quantity of signals which come to income neurons turn into one signal. Once the ANN's ready for activation, it begins repeating data analysis. After several attempts to calculate/compute right answer, it makes mistakes which are factored into the next calculating attempt. If the system calculates the right answer, it will always be able to work with this type of work without any mistake.

1.2 ANN in intellectual energy systems

Artificial neural networks in smart grids can input data, analyzing programs effectively and responding the needs of certain energy consumers.

What are the characteristics of the future smart grids which require ANN abilities?

1) Flexibility of different energy source usage, including all types of fossil fuels and renewable sources allowable for certain energy consumer;

2) Stability in the energy system of consumer which requires "smart" decisions close in the compartment with decisions made by human;

3) Increasing of energy efficiency connected directly with the information about the energy source supply, which in turn is connected with global net information (as, for example, the weather forecast connected with the energy supply of a certain detached house).

What sources of energy and information will be the input information for ANN?

1) ANN will control the quantity of allowable renewable sources – wind, solar, geothermal energy which is not constant during, for example, one period of usage (day, month, year); 2) ANN will control the usage of the main grid sources – electricity got from electrical grid of country region;

3) Information directly connected with the renewable source generation (like weather forecast) will be getting from Web sources, so ANN will build energy consumption plans for the day/month.

But if we consider ANNs, we have to answer the next question: which way is ANN going to learn how to rule allowed resources and how to manage information about their existence. There are several foreseeable decisions to this problem.

1. ANN's actions before its utilization in real energy systems are going to be "trained" in special conditions by experienced people on the base of computer centers. Training will be similar to the training of modern ANN's with its repeated question-answer-mistake circle. Then trained ANN will be implemented into a real energy system with its conditions similar to conditions ANN learnt in.

2. ANN's background will be fully built in special computer centers, but the real studying of ANN has already occurred in the consumer's energy system. The studying will be provided by active dialogue with consumers on the base of human interface connection with question-answer logics.

Conclusion

Artificial neural networks in the foreseeable future will be able to become a very effective tool for data analysis, for self studying and very perspective for smart grids of energy consumers from detached houses to little industries. The accelerating growth in the field of software engineering research will prove appearing of ANNs very soon.

References

1. A. Conde, A. Arriandiaga, J. A.Sanchez, E. Portillo, S. Plaza, I. Cabanes, Robot Comput Integr Manuf. (to be published)
2. Y.-F. Pu, Z. Yi, J.-L. Zhou, IEEE Trans. Neural Netw. Learn. Syst. **28** (2017)
3. C. Fu, L. Zhao, J. Yu, H. Yu, C. Lin, IET Control Theory Appl. **11** (2017)
4. G.M. Rao, V.A. Kumar, B.V.S. Ram, *International Conference on Signal Processing, Communication, Power and Embedded System, SCOPES 2016 – Proceedings 7955509* (2017)
5. A. L. Buran, V.E. Kozlov, A.V. Filyukov, MATEC Web Conf. **110,** 01021 (2017)
6. T. V. Susakova, A.M. Evseeva, I.A. Chesnokova, MATEC Web Conf., **110,** 01077 (2017)
7. K. Deisseroth, M. J. Schnitzer, NEURON, **80** (3) (2013)
8. J. Heaton, *Artificial Intelligence for Humans. Volume 1: Fundamental Algorithms* (CreateSpace Independent Publishing, North Charleston, 2013)

MATEC Web of Conferences **141**, 01050 (2017) DOI: 10.1051/matecconf/201714101050
Smart Grids 2017

Methods of reducing power losses in distribution systems

Yakov Wilms[1,*], *Sergey* Fedorovich, and *Nikolai* A. Kachalov

[1]National Research Tomsk Polytechnic University, 634050 Tomsk Russia

Abstract. The object of research is plot of an urban electricity distribution network of 10/0.4 kV of Tomsk city. The purpose of work – research and development of scientific and technical solutions to optimize the plot operation modes of an urban power supply system on the basis of modern software and hardware. The study found ways to reduce the unbalance of currents and voltages. Applications: urban and rural distribution networks 10/0.4 kV. Economic efficiency: the measures proposed can reduce the power loss. In the course of the research calculation and analyzing of the operation mode in the programs were carried out.

1 Introduction

Russia is one of the most energy-intensive economies in the world. This is due to climatic conditions, as well as the location of industry and population in its vast territory. According to the conducted research there is a significant potential for energy saving in Russia and this will reduce electricity use by half. There is the problem inefficient use of electricity and energy equipment in distribution networks as a result of unbalanced phase load, etc.

Thus, the strategic aim of this research is determination of the optimal operating mode in distribution networks and development of measures to reduce power losses.

There are different methods for reduce power losses and mainly these methods are grouped in three groups: organizational, technical and methods for improvement an accounting electricity system. Organizational events consist of:

1. finding of disjunction locations with duplicate electric power supply;
2. reducing an unbalanced phase load;
3. optimization of loading of power transformers.

Technical events mean using special equipment, for example: to reduce power losses, a shunt capacitor bank can be installed on distribution primary feeders, also to reduce an unbalanced phase load; a harmonic filter can be installed. Methods for improving an accounting electricity system is due with an application of automatic counting systems of electricity and an application of high accuracy level electric meters.

The object of research is plot an urban electricity distribution network of 10/0.4 kV in Tomsk city. The main disadvantage of this part is inefficient equipment loading (load factor; most of transformers have load factor below than 45%). Since this system is in working order, consideration of organizational events is preferable because they don't need reconstruction

* Corresponding author: yawilms@yandex.ru

and also they don`t demand a significant investment. In addition, in the work of I.I. Elfimov [1] considered the possibility of installing capacitor banks and replacing transformers with zigzag transformers. As a result, reduction of the current asymmetry and power losses were reported, but this result is not significantly.

This paper provides the next events:

1. finding of disjunction locations with duplicate electric power supply;
2. disconnection of equipment for minimization power losses.

2 Modes of operation of electrical networks

The mode of operation of electrical networks is the state of the system, which is characterized by indicators that quantify its operation. The operating parameters include: values of power, voltage, current, frequency, etc.

The operating modes of electrical networks are divided into steady-state and transient ones, connected with the studies of dynamic stability [6]. In turn, the steady-state modes are comprised of:

1. normal mode;
2. emergency state;
3. forced state [7].

A Load curve changes depending on the seasons of the year, in connection with it there are the maximum load mode (winter maximum) and the minimum load mode (summer minimum).

Calculations of the modes of the electric network are performed to determine:

1. loading of network elements, matching the network capacity to expected power flows;
2. cross-sections of wires and power cables of transformers and autotransformers;
3. the level of voltage in the nodes of the electrical network and measures ensuring the maintenance of voltage within the permissible limits;
4. loss of power and electrical energy;
5. the capacity of the electrical network, etc.

3 Finding of disjunction locations with duplicate electric power supply

The existing methods of selecting the optimal places for disjunction electric network can be divided into 2 groups. The methods of the first group imply a completely closed position of circuit breakers at first, but methods of the second group are completely opened originally. Then, circuit breakers operate (opening/closing) occur until minimum power losses develop in the network. There were considered both groups, but most visible was the first group method. In general form, the algorithm of the first group method is shown in figure 1.

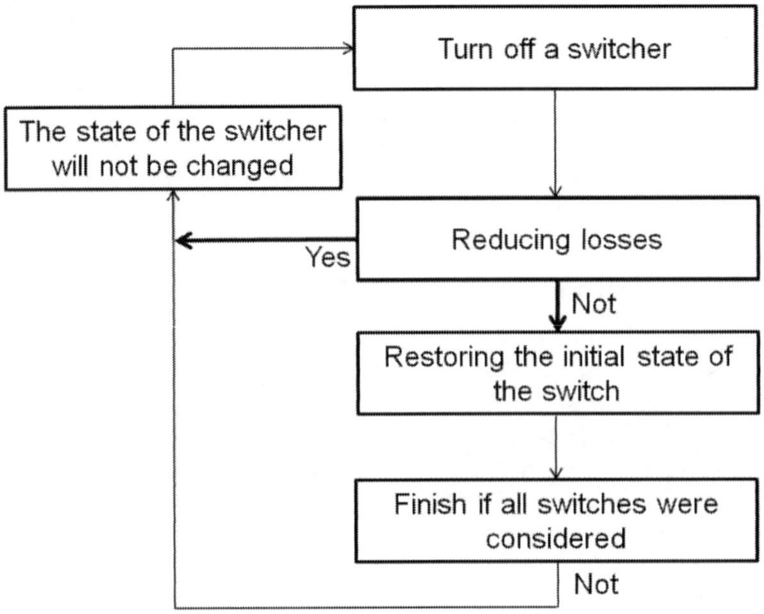

Fig. 1. Finding of disjunction locations algorithm.

Both of these methods have the same results, but the use of them changed nothing for this plot, because the resulting state of the network resembled the original state.

4 Disconnection part of equipment (Switching off part of the equipment)

Such as the main feature of this plot is the low load of power grid equipment, it is worth considering the possibility of its disconnection. There are transformers and cable lines in this plot.

There are no-load losses (rotational losses) and load losses (variable losses) in transformers, and the greater the power of the transformer, the greater the no-load losses. Power losses are calculated by the formula (for one transformer):

$$\Delta P_{trans} = \Delta P_{N-LL} + \Delta P_{LL} \cdot \frac{S^2}{S_{nom}^2} \tag{1}$$

Therefore, disconnecting one of the two transformers on the substation will lead to a reduction of no-load losses. For two transformers, power losses are calculated by the formula (Parallel):

$$\Delta P_{trans} = \Delta P_{N-LL} \cdot n + \frac{1}{n} \cdot \Delta P_{LL} \cdot \frac{S^2}{S_{nom}^2} \tag{2}$$

In that load factors of each transformer on substations are different, there is need to examine parallel and separate modes of the transformers work. To determine the value of the load power at which the operation of one transformer is economically more feasible than the parallel operation of two transformers we can use the following formula:

$$S_{optimal} \geq S_{nom} \cdot \sqrt{\frac{2 \cdot \Delta P_{N-LL}}{\Delta P_{LL}}} \tag{3}$$

DOI: 10.1051/matecconf/201714101050

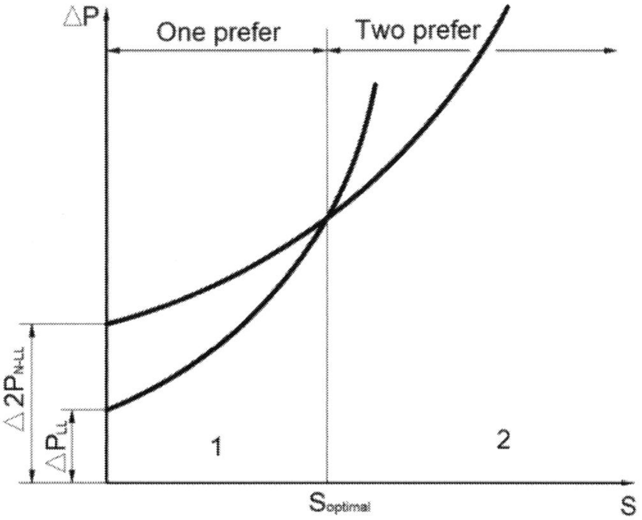

Fig. 2. Suitable operating modes of transformers.

According to the calculation results, usage of one transformer is preferred for all substations (figure 2, plot 1). However, the simulation of this mode showed the opposite, because cable lines have only load losses, and for disconnection of the first transformer followed the disconnection the first cable line on the substation, and also loading of the second cable line was accompanied by an increase in losses. In the issue of the power losses in transformers decreased and the power losses in cable lines increased by the same value. Nevertheless there was a stable voltage profile (level voltage) originally, but a working mode with disconnected second transformers on a substation have an under voltage level, and therefore disconnection parts of energy equipment is ineffective for this plot of distribution network.

There is a problem to reduce losses in this plot of power distribution power system, but this was considering the on-peak (LP winter case) conditions and this case is idealized. It is necessary to consider the real case – daily load curve. As the result of that investigation found (figure 3), there is a way to reduce power losses with increased load power (specific losses). Proceeding from the aforesaid the next part of our investigation is putting the system out of balance – modeling of load power increase.

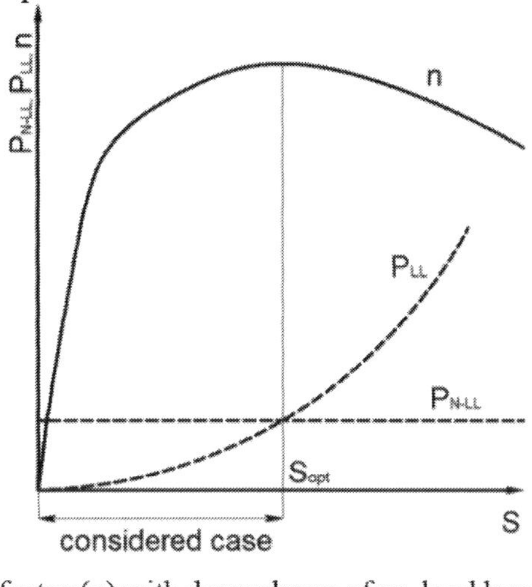

Fig. 3. Curve of efficiency factor (n) with dependence of no-load losses ($P_{N\text{-}LL}$) and load losses (P_{LL}).

References

1. I.I. Elfimov, E. A. Shutov, *Optimization of operating modes of electrical networks* (TPU, Tomsk, 2016)
2. G.N. Klimova, *Energy saving on industrial enterprises* (TPU, Tomsk, 2014)
3. D.E. Dulepov, T.E. Tyundina, Vestnik NGIEI, **4** (2015)
4. M.I. Fursanov, Energetika, **60** (2014)
5. A.L. Trushnikov, V.N. Radkevich, GGTU, **1** (2006)
6. E.S. Tarasova , I.B. Daminov, EPJ Web Conf. **110** (2016)
7. E.S. Tarasova, I.B. Daminov, MATEC Web Conf. **37** (2015)

MATEC Web of Conferences **141**, 01051 (2017)
Smart Grids 2017
DOI: 10.1051/matecconf/201714101051

Onomatological analysis of the technonyms in the sphere of energetics (based on the Russian, English and Chinese languages)

Tatiana S. Plutalova[1,*], *Juri* V. Kobenko[2], and *Elena* S. Riabova[3]

[1]National Research Tomsk Polytechnic University, 634050 Tomsk, Russia
[2]National Research Tomsk State University, 634050 Tomsk, Russia
[3]Samara University, 443086 Samara, Russia

Abstract. The article deals with the results of the onomatological analysis of the names of renewable energy facilities in the Russian, English and Chinese languages. The relevance of the study is determined by the trends in the development of these languages and their onomasiological foundation, which require focused attention of linguists in the context of an avalanche-type enlargement of terminological spheres. The results of the research can be useful for a wide range of specialists in humanities, including interpreters working with terminological thesauri in the sphere of renewable and resource energy.

1 Introduction

The modern world is in the state of a continuous development: the emergence and spread of various technologies in the industrial sectors cause a continuous creation of new technological objects. In connection with the current ecological situation on the planet, renewable energy is becoming an increasingly important alternative to raw energy, replacing fossil energy sources with renewable ones. Thus, the number of new energy facilities is constantly growing all over the world, each of them is called in a certain way that is typical for the nominative traditions of a particular language. Interpreters in the field of professional communication have to adequately and equivalently interpret national technonyms in the context of intercultural transfer, which requires a high level of proficiency in terminological thesauri, especially in such a demanded branch of the economics as renewable energy. The creation of the thesaurus needs a logical, motivated, consistent classification of the technonyms, which will allow us to trace certain patterns in the name of objects in the field of renewable energy in the Chinese, English and Russian languages [1]. This article presents the results of the onomatological analysis (the term of A.K. Matveev [4]), the aim of which is to identify the peculiarities of the alternative energy objects (technonyms) naming in the English, Russian and Chinese languages. Technonyms

* Corresponding author: plutalova_tat@mail.ru

in onomastics include the names of technology and electronics, and in the context of our research the names of energy objects are also included [2]. Being a part of a single energy sector, the names "Sayano-Shushenskaya HPP" and "Baikhetan Daba 白鹤滩大坝" (translated as: "White Crane Coast" Dam), "Churchill Waterfall" look completely different and incomparable, and they represent the multi-structural names of the different group languages but not single-word onyms for which the word-formation analysis would be applicable [3]. In this regard, it seems relevant to carry out an onomatological analysis of Russian, Chinese and English nominations in order to identify onomastic features of technonyms naming in the field of renewable energy in languages mentioned above.
The object of onomatological analysis was onomastic segments of technonyms in the field of renewable energy in the English, Russian and Chinese languages. While compiling the trilingual corpus of nominations with the help of a continuous sample, 124 units were singled out, that contained 23 nominations in Chinese, 42 in English, 59 in Russian. Technonyms in the field of energetics in different languages as well as onomastic vocabulary in general are strikingly different: in Chinese, for example, technonyms of toponymic character prevail (toponym is the name of one's own part of space [5]), for example: HPP "Three Gorges 三峡" is the largest hydroelectric power station in the world, named after the territory on which it is located, the name derives from the name of the region (oronim – the relief notion), and in the English language there is a tendency to form technonyms of anthroponymic character (in the honor of outstanding personalities, for example: HPP "*W. A. C. Bennett Dam*" named after British Columbia's 25th Canadian Prime Minister, William Bennett).

2 The results of the onomatological analysis of the technonyms

It should be emphasized that very often the aim of onomatological analysis is the identification of subclasses of onyms that form a definite onomastic segment of the vocabulary [6]. Analysis data of the technonyms in the English language are presented in Table 1

Table 1. English technonyms in the sphere of energetics.

subclasses of technonyms		**naming units**
toponyms	hydronym	Manapouri Power Station
	hydronym	Tumut Hydroelectric Power Station
	hydronym	Gordon Dam
	hydronym	Murray Hydroelectric Power Station
toponyms	oronym	Snowy Mountains Scheme
	oronym	Kárahnjúkar
	oronym	Grand Coulee Dam
	oronym	Mojave
	oronym	Solar Ranch
toponyms	oikonym	Guthega
	oikonym	Mica Dam
	oikonym	Revelstoke Dam
anthroponyms		Robert-Bourassa generating station
		Churchill Falls
		W. A. C. Bennett Dam
		Daniel-Johnson Dam
		Chief Joseph Dam
		Hoover Dam

	Robert Moses Niagara Power Plant
	Solana
	Clyde Dam
ethnonym	Ivanpah
astronym	Star
phytonym	Mesquite

It is important to note that the technonyms naming of the English language has a tendency to use the names of famous personalities in the technonyms formation. Nominators often refer to the image of outstanding public figures and use the idea of these people as charismatic leaders, emphasizing the unique character of the object.

Most of the technonyms in the Chinese language are represented by toponyms (technonyms that include the geographical name of the territorial object), oronyms in particular (technonyms that include the name of any crustal relief object) and oikonyms (nominations that include the name of the locality). The analysis of the Chinese onomastic technology segment in the energetics sector is presented in Table 2.

Table 2. Chinese technonyms in the sphere of energetics.

subclasses of technonyms		**naming units**
toponyms	oronym	HPP "Three Gorges" 三峡大坝
	oikonym	HPP Silodu 奚落度
	oikonym	HPP Jinping 锦屏一级水电站
	oikonym	HPP Huyaosia "虎跳峡大坝"(Tiger Leaping Gorge
	oikonym	Liujiaxia Dam 刘家峡大坝
	oikonym	Dabancheng Wind Farm 达坂城区
	oikonym	Gansu Wind Farm (also called Jiuquan Wind Power Base) 酒泉市
	oikonym	Longyangxia Dam 龙羊峡水库

Onomatological analysis allows us to conclude that the names of the Chinese language are most marked by poetry and metaphoricity, which is easily explained by the cultural pecularities of the country, characterized by aesthetic onomastic space. According to the definition given by T.V. Shmelyova in the textbook "Onomastics" (the term was proposed by A.V. Superanskaya in a slightly different interpretation), the onomastic space is a part of reality marked by proper names [5]. The frequency of use of the metaphorical names in the Chinese language is explained by the richness of the tropes in the language itself: *白鹤滩大坝 – the dam of the "White Crane Shore", 二滩大坝 – the dam of the "Two Shores",龙滩大坝 – the dam of "The dragons' Shore"*. Among the studied technonyms of the Chinese language, metaphors of various types are used, see Table 3.

Table 3. Types of metaphors of Chinese technonyms.

types of metaphors	**naming units**
geographical metaphors	HPP Xiangjiaaba 向家坝 "Home-oriented"

chronotope metaphor	HPP Gezhouba 长江葛洲坝水利枢纽工程 “The Edge of the Ge Kingdom”
portrait metaphor	HPP Baichatan 白鹤滩大坝 “White Crane Shore”
landscape metaphor	HPP Luntan 龙滩大坝 “Dragon Shore”
synesthesia	HPP Manvan 漫湾大坝 “Quiet Bay”
technical metaphor	HPP Banciao 板桥水库 “Wooden Bridge”

The symbols of natural components are associated with the philosophy of “Feng Shui”, which aims at harmonizing the relationship between the nature and human society. T.V. Shmelyova states that it is important to be able to explain the meaning of the onym, to understand the purpose of a certain name for a particular object, to identify a nominative typology in order to assess the degree of originality and expressiveness of the name [5]. Thus, the onomatological analysis of object names allows us to draw some conclusions about the language environment in which these names are formed. Hence, the historical and sociological factor can play a significant role in the formation of the names of certain technical objects, which at first glance seem completely remote and unrelated to the cultural realities of the language environment. “Dry” names of the Soviet era with no trace of artistic style – even they are able to reflect the specific character of the environment in which they were formed.

The analysis of the power facilities names of the Russian-language register revealed that the genus proximum and the differentia specifica of most of the names have a standardized, “non-poetic” character aimed at excluding any misunderstanding. Data analysis of the technonyms in the Russian language are presented in Table 4.

Table 4. Russian technonyms in the sphere of energetics.

subclasses of technonyms	naming units
toponyms	Perevolotskaya SPP
	Buribaevskaya SPP (Baymak SPP)
	Irkutsk HPP
	Saratovskaya HPP
	Bratsk HPP
	Volzhskaya HPP
	Sayano-Shushenskaya HPP
	Zelenograd WPS
	Yeisk WPS
	Perevolotskaya SPP
astronyms	SPP Solar
	SPP Orion Solar
	SPP Orion Solar
ornithonyms	SPP Ospriy Solar

Among the 59 Russian names, 3 astronyms were recorded; 5 names of power facilities of the post-Soviet era formed by the letters of the Greek and Latin alphabets; 1 ornithonym; 49 toponyms. The clear predominance of toponymic technonyms in the Russian language register indicates strictness, clear standardization, stylistically neutral nature of the formation of the power engineering objects names that refer to the realities of the Soviet era, when the naming unit of industrial facilities did not imply any allegory, cf.: *Sayano-Shushenskaya HPP, Donuzlav WPS, Zelenograd WPP etc.* Only a few names of modern

SPP (*Crimean SPP*, built after 2010) stand out of strictness and restraint of Russian names: thus, zoomorphic metaphors appear, cf.: *Osprey Solar* (*Osprey* - Scop bird of prey).

3 Conclusion

Thus, when composing the trilingual corpus of the names of renewable energy facilities, 124 technonyms were received. While determining the theoretical foundations of technonyms in the sphere of renewable energy, it was found out that the majority of the studied technonyms are toponymic in nature, i.e. onyms were mostly formed from the names of geographical objects where the power engineering objects were installed. Onomatological analysis also shows that now in Russia and China there is a tendency to call objects using Anglicisms, which indicates a trend towards the internationalization of the technology [7].

Nowadays the branch of renewable energy is dynamically developing, power engineering facilities continuously appear one by one, thus, the onomastic space of languages is continually enriched with new unique names. The study of the peculiarities of the naming units makes it possible to learn more about the cultural realities and unique characteristics of the language environment in which the names of objects are formed, as they are a direct reflection of native speakers' way of thinking in a particular language.

References

1. A.A. Kohanskaya, Baltic Humanities Magazine, **134**, 24 (2013) [in Russian]
2. N.V. Podolskaya, The dictionary of Russian onomastic terminology, **192** (1988) [in Russian]
3. O.I. Kopach, The grammatical structure of the Belarusian language, ways of historical development and current trends, 267 (2007) [in Russian]
4. A.K. Matveev, Onomastic questions, 9 (2005) [in Russian]
5. T.V. Shmeleva, Onomastics, **161** (2013) [in Russian]
6. T.I. Meremkulova, Onomatological analysis of units of the borrowed Gallic formation in modern German literary language, Philological Sciences. Questions of theory and practice, **10**, 139 (2017) [in Russian]
7. D.S. Poendaeva, Yu.V. Kobenko, E.S. Riabova, Matec Web Conf. **110** (2017)

MATEC Web of Conferences **141**, 01052 (2017) DOI: 10.1051/matecconf/201714101052
Smart Grids 2017

Grounding systems for power supply facilities

Vladimir Konusarov[1,*], and *Ekaterina* Tarasova[2]

[1]National Research Tomsk Polytechnic University, 634050 Tomsk, Russia

Abstract This study aims to define some recommendations for choice of grounding system type for power supply facility. Operating modes of 0.4 kV networks approved by International Electrotechnical Commission (IEC) were considered and their advantages and disadvantages were identified. The comparison criteria were based on the conditions of human safety from electric shock in the breakdown of insulation on the body of electrical equipment, possibilities of expanding the network and uninterruptible power supply. The real application of the neutral modes and the possibility of installing protective devices were also taken into account. All the recommendations made is planned to apply for the modernization of the urban networks and the calculation of the earthing arrangement by using special software.

1 Introduction

Grounding is one of the most important technological methods used to protect against an electrical shock. Reliable protection of power supply objects from various types of overvoltage requires the creation of an effective system for grounding electrical installations and equalizing the potentials on them. This study work is devoted to systems of grounding of elements of city distributive networks. In these networks, recently there have been increasing trends to the separation of protective and zero conductors, this transition makes it possible to protect against impulse over voltages, as well as it is more reliable in terms of electrical safety. [1] In addition to the grounding and equalizing the potentials, a significant influence is given to protective devices whose switching circuits depend on the mode of operation of the neutrals. We consider in more detail the earthing methods of urban distribution networks of 0.4 kV and supply substation 10/0.4 kV.

2.1 Neutral operation modes in 0.4 kV networks

Technical features of electrical installations and their power supply networks cause the application of various types of grounding systems. In this case type of grounding system is a position relative to the ground of a substation's neutral and electrical installations' neutral. IEC provides application of three neutrals' and a conductive parts' work modes.

* Corresponding author: Konusarov@mail.ru

 DOI: 10.1051/matecconf/201714101052

Table 1. Modes of neutral's work.

Type of grounding system	Description
TN	Substation's neutral is solidly grounded. Electrical installations cases are connected to neutral conductor.
TT	Substation's neutral and Electrical installations cases' neutrals are solidly grounded and unconnected.
IT	Substation's neutral is isolated. Electrical installations cases' neutrals are dead grounded.

The neutral grounding mode TN is divided into 3 types, table 2.

Table 2. Types of grounding modes of TN neutrals.

Type of grounding system	Description
TN-C	Zero protective and neutral conductors are united in one conductor throughout it's all length
TN-S	Zero protective and neutral conductors are separated
TN-C-S	Zero protective and neutral conductors are united at the head, and then separated

The method of grounding the neutral largely determines:

- Conditions of safety in electrical networks, protection from the risk of electric shock;
- Method of limiting overvoltage;
- Electromagnetic compatibility in normal and emergency modes;
- Fire safety;
- Currents with single-phase faults, damageability, selection of electrical equipment;
- Continuity of power supply;
- Network design and operation.

It will be noted the advantages and disadvantages of the existing neutral operation modes in 0.4 kV networks, having chosen the criteria listed above as comparison criteria.

2.1.1 TN–C grid

Networks with this neutral mode of operation are most common in Russia. The network diagram is shown in Fig. 1.

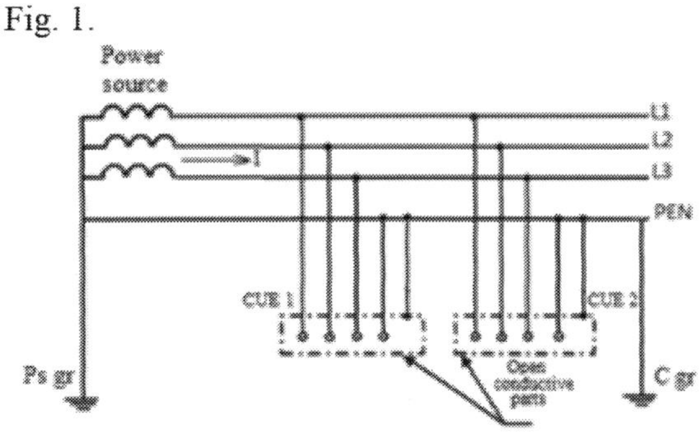

Fig. 1 Diagram of an electrical network with TN-C neutral grounding system.

Electrical safety with indirect contact in this case is provided by switching off the fuse or circuit breaker. If the fault is remote from the source, its elimination time increases in this case the danger of electric shock to persons increases. To ensure electrical safety, short-circuit protection must be switched off within a time less than 0,2 s, which is provided with fuses and circuit breakers only in case

 DOI: 10.1051/matecconf/201714101052

$$I_{fault} = (5\text{-}6)\ I_{nom} \tag{1}$$

Thus, with an indirect contact in remote faults, the TN-C neutral is not safe. It should be noted that designing this type of networks demands to measure or calculate the resistances of all connections and phase-zero loops for their protection settings, and when the network parameters are changing, it is necessary to do a recalculation to ensure the reliability of the protections.

The biggest disadvantage of the TN-C network is the inability of the residual current devices (RCD) to function [2]. The fire safety of this network is low. This is due to the significant currents of single-phase fault and as mentioned above with the weak sensitivity of the protection for remote damage. For TN-C networks, the appearance of electromagnetic disturbances is typical, even in normal mode a voltage drop occurs in the zero wire [2].

This grounding system was used in the Soviet Union and now it can be found in houses belonging to the old buildings. Today it is also used in the networks of street lighting, where the degree of risk is minimal.

2.1.2 TN–S grid

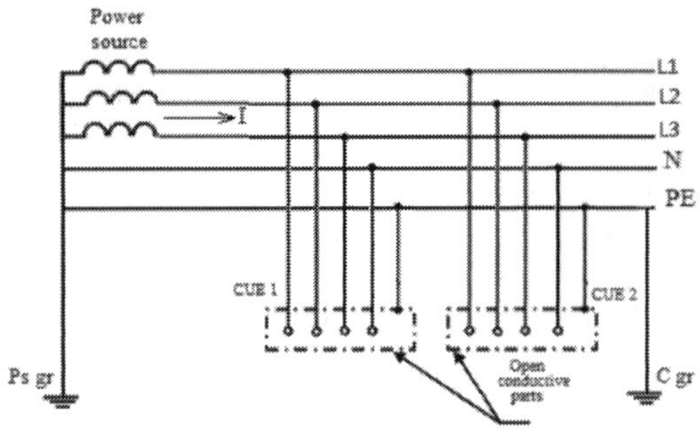

Fig. 2 Diagram of an electrical network with TN-S neutral grounding system.

Separation of the working and protective zero conductors do not provide electrical safety with an indirect contact similar to the TN-C network. The advantage of this network is the ability to use an RCD, it enhances the electrical network. Fire safety of TN-S network, due to the action of RCD is significantly higher in comparison with the TN-C networks. With regard to the continuity of the electricity supply network are similar.

In terms of designing, configuring and maintaining protection, TN-S networks do not have significant advantages over TN-C networks, besides they are significantly more expensive due to the installation of an RCD and the presence of a fifth wire. This grounding TN-S first appeared in Europe and is still used there. In Russia TN-S networks are used in the construction of multi-storey buildings.

2.1.3 TN–C–S grid

The TN-C-S network is a combination of TN-C and TN-S networks. The TN-C-S network is characterized by all the advantages and disadvantages noted above of the two previous networks. The TN-C-S grounding system is used in urban buildings; the conductors are separated in the basement of the building and running separately in the rises.

DOI: 10.1051/matecconf/201714101052

2.1.4 TT grid

The TT network is characterized by the fact that the neutral of the power source and the conductive parts of the electrical receivers are dead-earth grounded, through unconnected conductors, the circuit is shown in Fig. 4.

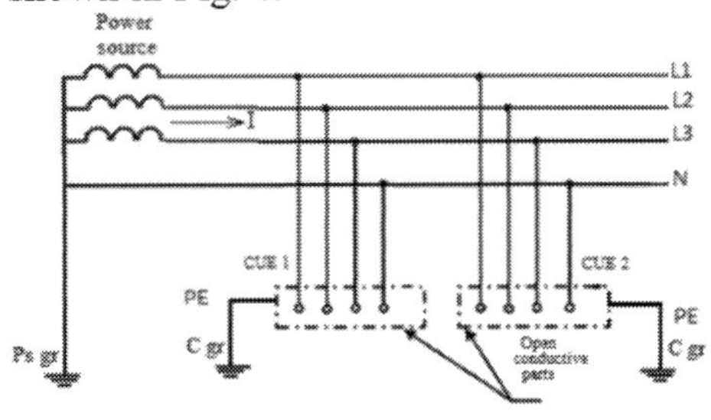

Fig. 4 Diagram of an electrical network with TT neutral grounding system.

With regard to the electrical network, the single-phase fault on the conductive body make the voltage on it determined from the ratio of the grounding resistor of supply substation and local grounding.

$$\frac{U_{fault}}{U_{instulation\ body}} = \frac{R_{substation}}{R_{local\ grounding}} \tag{2}$$

TT grounding system does not provide safety against indirect contact, so the use of RCD is necessary. Fire safety of networks of this type is higher than in networks TN-C. This is due to the lower values of the single-phase fault current and the use of an RCD. Uninterrupted power supply is not ensured. Electromagnetic disturbances are much lower; the amount of damage in the emergency mode is insignificant. [3]

In terms of design, the use of an RCD eliminates the problem of limiting the length of the PE conductor, there is no need to know the impedance of the phase-zero loop, it is possible to expand the network without recalculation of short-circuit currents. TT networks are widespread in rural areas, because of poor quality of power transmission line supports; they are also used in private buildings and in urban conditions to electrify temporary consumers.

2.1.5 IT grid

The IT network, in other words, can be called a network with isolated neutral. The scheme of the grounding system of the IT network is shown in Fig. 5.

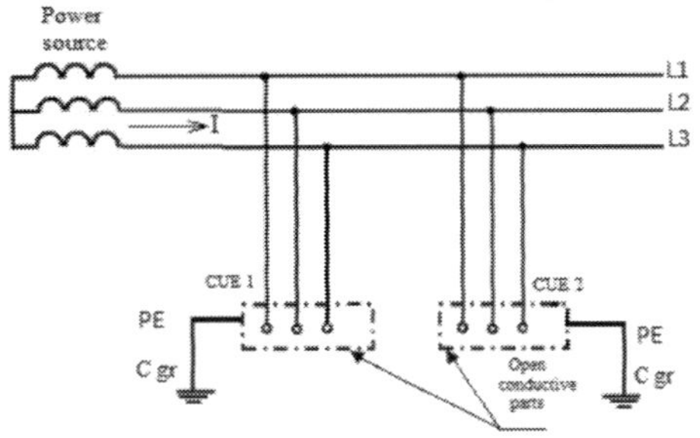

Fig. 5 Diagram of an electrical network with IT neutral grounding system.

IT networks are most electrically safe; voltage with indirect contact is practically nonexistent. Fire these networks as compared to the one discussed above is much higher; it is once again connected with small values of the fault current. IT networks are characterized

DOI: 10.1051/matecconf/201714101052

by uninterrupted operation, single-phase fault to ground or conductive equipment housing does not require immediate shutdown.

It should be noted that during the operation of the network, difficulties may occur, related to the location of the fault. Today they are widely used in dangerous industries, where electrical and fire safety is very important. Also, IT networks are often used in the construction of private houses.

3 Recommendations for choosing the type of grid

The following recommendations for choosing type of grid can be made:

As recommendations for choosing a network, the following conclusions can be drawn:
1. TN-S and TN-C-S networks are characterized by a low level of electrical safety, fire safety and significant electromagnetic disturbances.

2. TN-S networks can be recommended for selection in the case it's not planning to change it.

3. TT networks should be used to supply temporary or variable electrical installations.

4. IT networks are recommended to use, if it's absolutely necessary to ensure uninterruptible power supply.

4 Conclusion

Each of the above modes of neutral grounding is not universal. Therefore, designing a network should be based on advantages and disadvantages. It's also should be carried out a technical and economical comparison and make a choice in favor of a certain grounding regime.

There are also cases of combined use of networks, for example, when a part of consumers are powered from the TN-S network, and the other part is connected via a separation transformer to the IT network.

References

1. A.V. Kabyshev, *Lightning protection of electrical installation systems* (TPU, Tomsk, 2006) [in Russian]
2. Rules of Electrical Facilities Maintenance (RF Ministry of Energy, 2003) [in Russian]
3. S.V. Petukhov, S.V. Butakov, V.V. Radyushin, *Calculation of the grounding device* (MSU, Arkhangelsk, 2011) [in Russian]

Analysis of risk assessment algorithm from power supply interruption due to transmission line fault

Alina Andrukhova[1,*], *Natalia* Batseva[1], and *Arina* Evseeva[1]

[1]National Research Tomsk Polytechnic University, 634050 Tomsk, Russia

Abstract. The guaranteed reliable and uninterrupted supply of electrical energy to consumers is an important task. It can be noted that currently there is no well-structured system for ensuring reliability or security and preventing risk occurrence. In this report, risk assessment algorithms in case of accidents on transmission line are developed and tested by means of the analytical example.

1 Introduction

Reliability of consumer or consumer groups electricity supply is related to the technical state and operating conditions of external distribution networks, through which they are fed.

Decreasing of reliability from transmission line faults leads to a shortage of electricity to consumers, which is called "damage from electricity sacrifice" in cash-equivalent term, thus the study of risks from transmission line faults is relevant. The problem of reliability and risk is discussed by Billington R., Allan R., Beehler M, Guk Yu.B., [2, 3, 6]. In addition to this, modern authors such as Syrri A.L.A., Mancarella P., Junlakarn S., Ciapessoni E., consider this problem in their articles [9, 8, 5].

Nowadays, requirements that regulate the risk assessment and reliability calculation are either absent or obsolete. Consequently, the utility practice requires the theory and methods for reliability calculation taking into account the risks from faults.

2 What is Reliability?

Reliability refers to reliability connected with electrical mode (henceforth referred to as reliability) or service security. [7, 8]

Generally, reliability is the ability of a power system to hold over a new state due to disturbances without violation of power system stability. It characterizes the emergency procedures by the criterion (*n-i*), where *n* – is the total number of components and *i* – is the number of suddenly disconnected system components [6].

* Corresponding author: ava40@tpu.ru

Reliability of consumer energy supply can be defined as the difference between electrical supply with full consumer demand (W_C) and electricity sacrifice due to accidents and other events (ΔW) to W_C [4]:

$$H_W = \frac{W_C - \Delta W}{W_C} \cdot 100\%. \tag{1}$$

Service security is determined by the level of physical reliability of individual components and component connection layout.

The main indicators of service security are:

– failure flow parameter [ω, failure per year] called SAIFI in [3] is the ratio of component number that failed per a time unit to the total number of the same tested components;

– recovery time [T_r, hour] – the ratio of total recovery time of technical devices for the selected fixed period to the number of its failures for the same fixed period.

The main reliability indicators of systems with various connections of recoverable components are calculated according to the serial and parallel connection of components [2].

3 Calculation of the damage from electricity sacrifice

Estimation of the damage from electricity sacrifice costs can be calculated as

$$D_s = d_0 \cdot W_s, \tag{2}$$

where W_s is the expected load at the delivery point, kWh; d_0 , is the specific cost for energy which is not supplied, $/kWh [1].

4 Risk Assessments from power supply interruption

Qualitative analysis corresponds to the stage of risk identification, quantitative analysis allows determining the size of individual risks numerically and risks in general. [9]

4.1 Quantitative method

The calculation of this indicator is carried out by the assumption that production costs can be divided into conditionally variable $C_{con-var}$ (depending on the volume of production B_p) and conditionally constant $C_{con-fix}$ (independent of the volume of production).

$$V_{cr} = \frac{C_{con-fix}}{B_p - C_{con-var}} \cdot 100\%, \tag{3}$$

where V_{cr} – the critical volume of transferred electricity, %.

To calculate the break-even point, the planning interval, for example, the year is chosen and full production capacity is achieved. The project is considered sustainable if V_{cr} <75 - 80% from normal level. Advantages of calculation of the break-even point are the simplicity and clarity. [5]

4.2 Qualitative method

For a qualitative risk assessment of electricity supply, it is necessary to determine the main components of the risk and the severity of fault consequences (Table 1).

Table 1. Scales for establishing the severity category of failure consequences.

Severity category of failure conse-quences	Severity characteristics of failure consequences
IV-catastrophic failure	The failure, which rapidly and with high probability can cause significant damage to the object and / or the environment, death or serious injuries of people, failure of the task
III-critical failure	The failure, which rapidly and with high probability can cause significant damage to the facility itself and/or the environment, disrupts the task, but creates a negli-gible threat to life and health of people
II-not critical failure	The failure that leads to delaying in the fulfillment of the task, decreasing the readiness and efficiency of facilities, but does not pose a threat to the environment, facilities themselves and human health
I-with small consequences	The failure, which leads to decreasing the quality of the function of facilities, does not pose a threat to the environment and human health

To perform this type of analysis, each type of failure must be assigned. The level of occurrence probability is illustrated in Table 2.

Table 2. Levels of occurrence probability.

Level probability of failure	*Description*	*Analysis*
A	Frequent failure. The probability of failure exceeds 0.2	In-depth quantitative analysis of criticality is required
B	Probable failure. The probability of failure from 0.1 to 0.2	Quantitative analysis of criticality is desirable
C	Possible failure. The probability of failure from 0.01 to 0.1	Only qualitative analysis is re-quired
D	Rare failure. The probability of failure is from 0.001 to 0.01	Analysis is not required
E	Remote failure. The probability of failure during a given time is below 0.001	Analysis is not required

The procedure for qualitative analysis of criticality consists of assigning priorities for corrective and compensative actions to the types of failures, depending on the failure category and given level of failure probability.

Types of failures are distributed according to the criticality schedule, where on the horizontal axis the values of failure categories are shown, and on the vertical axis the failure probability levels are illustrated (Figure 1).

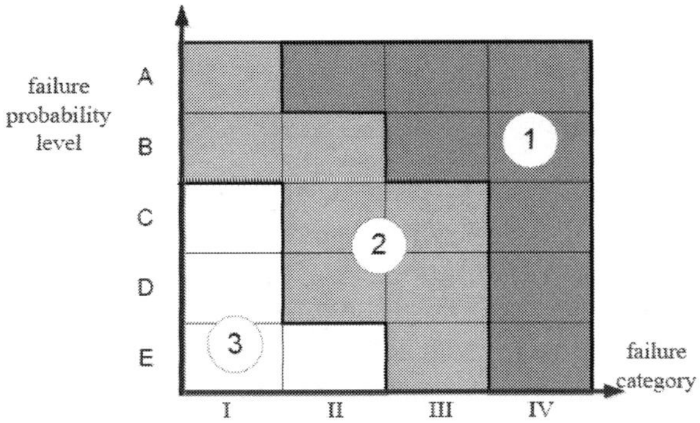

Fig. 1. Characteristics of the severity of failure consequences in a power supply system.

According to the schedule in Figure 1, areas of equal priorities are singled out:
1 – high. A component whose failure requires special attention in the design.
2 – medium. A component whose failures require attention during development.
3 – low. A component that requires careful analysis of species, consequences and criticality of failures to confirm priority values.

The advantage of qualitative analysis is that it is performed in early stages of development, when the design of products is not fully defined and there is no quantitative data on reliability of all components; as a result, it is simple and evident.

5 Calculation of practical objectives

The problem situation: The plant is fed from two power sources: CHPP (Combined Heat and Power Plant) and System (Figure 2). Each transmission line can transfer total power required by the plant. However, CHPP can not produce more than 15.5 MW. Parameters are conventionally accepted and given in Table 3.

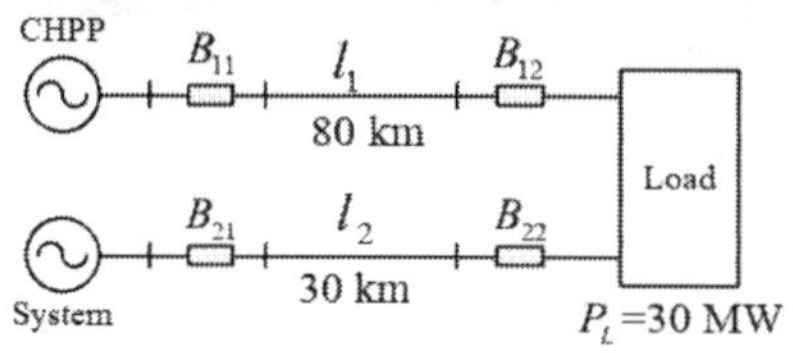

Fig. 2. Scheme of the power supply system.

The task is to define parameters of the failure flow in this system, the probability of failure, damage from the electricity sacrifice in a year, and then identify the risks.

Given: V_{nom}=110 kV.

Lines: r_0=0.118 Om, l_1=80 km, l_2=30 km.

Load: $\cos\varphi = 0.9$, P_l=30 MWh.

Table 3. Power Transmission Reliability Parameters.

Indicators of reliability	System component					
	B_{11}	l_1	B_{12}	B_{21}	l_2	B_{22}
ω_0 ,1/(km per year)	-	0.051	-	-	0.051	-
ω, 1/year	0.02	-	0.02	0.02	-	0.02
T_r , hour	7	8	20	20	7	20

It is assumed that unintentional disconnections of series-connected circuit components are combined in time.

Calculations:

Probability of failure for circuits I and II:

$$q_I = \omega_{B_{11}} \cdot T_{rB_{11}} + \omega_{0l_1} \cdot l_1 \cdot T_{rl_1} + \omega_{B_{21}} \cdot T_{rB_{21}} = (0.02\cdot 7 + 0.051\cdot 80\cdot 8 + 0.02\cdot 7)/8760 = 3.78\cdot 10^{-3};$$

$$q_{II} = \omega_{B_{21}} \cdot T_{rB_{21}} + \omega_{0l_2} \cdot l_2 \cdot T_{rl_2} + \omega_{B_{22}} \cdot T_{rB_{22}} = (0.02\cdot 7 + 0.051\cdot 30\cdot 8 + 0.02\cdot 7)/8760 = 1.437\cdot 10^{-3}.$$

Failure can be classified as category III i.e. the critical failure. The probability of failure of all components is defined separately.

$$q_{B_{11}} = q_{B_{12}} = q_{B_{21}} = q_{B_{22}} = \omega_{B_{11}} \cdot T_{BB_{11}} = 0.02\cdot 7/8760 = 0.00002;$$

$$q_{l_1} = \omega_{l_1} \cdot T_{rl_1} = 0.051\cdot 80\cdot 8/8760 = 0.004;$$

$$q_{l_2} = \omega_{l_2} \cdot T_{rl_2} = 0.051\cdot 30\cdot 8/8760 = 0.0014.$$

Based on these calculations, the breakers can be assigned for group *E*, transmission lines for group *D*, therefore, no analysis is required for the components. Comparing the obtained data according to the graph, shown in Figure 1, it can be concluded that all considered components of the system belong to the second priority of the component classification and it is not necessary to enter control systems and compensate the failure.

In the steady-state mode the risk assessment is defined by quantitative method (break-even point) from the electricity sacrifice.

Total investment: $I_{\Sigma} = 1921000$ \$.

Conditional-fixed costs: $C_{con-fix}$=94090 \$.

Conditional-variable costs:

$$C_{con-var} = \Delta W_{год} \cdot T_{trans} = 8100 \cdot 10^3 \cdot 0.013 = 108100 \,\$,$$

where T_{trans} – tariff for electricity transmission, \$/kWh;

ΔW – losses of energy per year, kWh.

The volume of production:

$$B_p = W_{year} \cdot T_{trans} = 30 \cdot 10^3 \cdot 8760 \cdot 0.013 = 3508000 \,\$.$$

Breakeven point:

$$V_{cr} = \frac{C_{con-fix}}{B_p - C_{con-var}} \cdot 100\% = \frac{94090}{3508000 - 108100} \cdot 100\% = 2.78\% .$$

The scheme is reliable since V_{cr} <75 - 80% from the normal level.

When an accident on transmission line occurs the risk assessment is calculated by quantitative method.

$$H_W = \frac{W_C - \Delta W}{W_C} \cdot 100\% = \frac{30 - 15}{30} \cdot 100\% = 50\% .$$

Taking into account the average recovery time of the line, the estimated damage from electricity sacrifice costs:

$$D_s = d_0 \cdot W_s = 0.313 \cdot 15 \cdot 8 \cdot 10^3 = 37720 \,\$.$$

The proceeds from the sale of products are changed

$$B'_p = B_p - D_s = 3508000 - 37720 = 3470000 \,\$.$$

Break-even point:

$$V_{cr} = \frac{C_{con-fix}}{B'_p - C_{con-var}} \cdot 100\% = \frac{94090}{3470000 - 108100} \cdot 100\% = 3\% .$$

Based on the calculation results, it can be noticed that the critical sale volume when one line is out of service has slightly increased, the project is remaining reliable.

6 The total algorithm of risk assessment

Based on the analytical problem solution, the algorithm is developed taking into account risks from accidents on transmission lines:

1. Visually determine the sections of the circuit with weak circuit reliability.
2. Modeling electric modes that detect "narrow" places by regime reliability, capturing areas with weak circuit reliability.
3. Define reliability indicators.
4. Determine the probability of failure.
5. Determine the damage from under-supply of electrical energy.

6. Range failure by category: with small consequences, no critical failure, and critical failure. To determine the method of risk research depending on the category.
7. Identify risk by quantitative and/or qualitative method.
8. Compare the damage from the electricity sacrifice with investments for measures to improve reliability.

References

1. A.A. Augusto, M.B. Do Coutto, J.C. Stacchini de Souza, V. Miranda, IET Gener. Transm. Distrib. **10**, 2933 (2016)
2. M.E. Beehler, Transm. Distrib. Conf. **10**, 1023 (1998)
3. R. Billinton, R.N. Allan, *Reliability evaluation of power systems* (Plenum, New York, 1996)
4. G. Celli, E. Ghiani, F. Pilo, G.G. Soma, Electr. Pow. Syst. Res. **104**, 164 (2013)
5. E. Ciapessoni, D. Cirio, G. Kjolle, S. Massucco, A. Pitto, M. Sforna, IEEE Trans. Smart Grid, **6**, 2890 (2016)
6. Yu.B. Guk *The theory of reliability in the electric power industry* (Energoatomizdat, Leningrad, 1990)
7. I. Hernando-Gil, I.S. Ilie, S.Z. Djokic, IET Gener. Transm. Distrib.**10**, 93 (2016)
8. S. Junlakarn, M. Ilic, IEEE Trans. Smart Grid, **5**, 2227 (2014)
9. A.L.A. Syrri, P. Mancarella, Grids and Networks, **7**, 1 (2016)

MATEC Web of Conferences **141**, 01054 (2017)
Smart Grids 2017
DOI: 10.1051/matecconf/201714101054

Approaches to formation of the digital learning estimation system in the Russian Federation

Ivan Nikanorov[1,*], and *Antoniy* Shvindt[2]

[1]Federal Agency for Scientific Organizations, 119334 Moscow, Russian Federation
[2]Ministry of Education and Science of the Russian Federation, 125993 Moscow, Russian Federation

Abstract. The article is devoted to consideration of mechanisms of evaluation of the MOOC in the global context and to the development of the Russian e-learning market. The criteria and mechanism for assessing the quality of online courses by consumers in the development contest of the Russian electronic digital environment are proposed. The approbation of the proposed mechanism was carried out and conclusions were made on the significance of the quality assessment of MOOC with their implication in the educational process.

1 Introduction

The transition to a new technological order and the increase in the flow of information in the social space are actualizing the development of new forms of education that would meet these challenges. Since it is the system of vocational education that requires the development of new forms of development of the educational environment allowing implementation of the principle of life-long learning, which meets the modern requirements of the global labor market. The trend towards the development of digital technologies has already significantly changed the educational space, in particular the higher school, democratizing the process of obtaining education and allowing to speed up the process of updating knowledge and improving the skills of people who already have an education. The understanding of the need for the development of e-learning to be one of the most important ways to overcome the global challenges faced by the education system was reflected in a number of international strategic planning documents, in particular the Icheon Declaration adopted at the World Education Forum in May 2015 and outlined the development priorities for the education systems of countries – participants of UNESCO [1]. The most current form of implementing e-learning in higher and additional professional education is Massive Open Online Course (MOOC) [2, 3].

2 International experience

The appeal to e-learning as one of the necessary elements of a modern vocational education system to address the global challenges of the labor market has revealed a number of

* Corresponding author: nikanorovin@gmail.com

problems related to the need to assess the quality of such technologies, as well as the conditions and results of developing such programs [4, 5]. In this regard, a number of international organizations have proposed their own mechanisms for assessing the quality of e-learning, for example Quality Matters TM [6] and Open ECB Check-Quality [7], which provide a system of criteria and indicators for the evaluation of online courses, and the creation of an expert community for their implementation in e-learning. It should be noted that the listed formats are primarily focused on peer review and suggest some mechanism for certifying courses based on the results of the evaluation.

3 Russian experience

In the Russian Federation, there is currently no common approach to assessing the quality of e-learning. However, there is already a significant reserve associated with its implementation. The effectiveness of the implementation of the MEP in the university's educational environment and the educational programs implemented by it are demonstrated, in particular, by the Higher School of Economics [8] and other successfully developing Russian universities [9].

The priority project "Modern digital educational environment" (the Project passport approved by the Protocol of the meeting of the Presidium of the Council under the President of the Russian Federation for Strategic Development and Priority Projects of October 25, 2016 No. 9) aims to create conditions for a systemic improvement of quality and empowerment by 2018 continuous education for all categories of citizens through the development of the Russian digital educational space and the increase in the number of trained educational organizations, having mastered *x* online courses up to 11 million people by the end of 2025. As one of the results of the project, it is planned to "create a system for assessing the quality of online courses and online resources of general education, combining automatic and peer review, which provides the formation of a rating." As we see, the declared system assumes the creation of a unified system for assessing courses on the territory of the Russian Federation.

4 The proposed evaluation mechanism

Simultaneously with the expert review, a number of Russian organizations consider it appropriate to develop mechanisms for assessing the quality of e-learning and the MOOC by consumers that could be integrated into the national system for assessing the quality of e-learning. To this end, the All-Russian Public Organization "For Quality Education" in cooperation with the Commission for the Quality of Education of the Council of the Russian Ministry of Education and Science for Youth Affairs and the "Internet + Education" Competence Center, autonomous noncommercial organization "Institute for Internet Development" proposed a project to consolidate the interests of consumers of e-learning educational services.

In the framework of this activity, focus groups were held in 10 Russian universities, the result of their work was a single document – "The Consumer Quality Standard for Education Implemented with the Use of E-Learning", organized on the basis of the "Standards and Recommendations for the Quality Assurance of Higher Education in the European Space (ESG)" [10]. It was divided into structural blocks; we omit the descriptive aspect and give the contents of the main sections.

1) Assessment of the educational content quality

This section contains indicators and criteria related to the content of the course. In the section the main requirements of the consumer to the content of the course are considered.

DOI: 10.1051/matecconf/201714101054

2) Evaluation of the results of e-learning and acknowledgment of e-learning by educational organizations and employers

The section contains the criteria and indicators of the consumer to the system for assessing the results of completed courses, as well as to the procedures for verification, identification and setting off the results of courses in higher education organizations, considering them by employers.

3) Assessment of architectural requirements

The section describes the requirements for the design and interface of the courses, as well as the methods of information transfer.

Criteria for assessing the quality by consumers were focused primarily on the availability of materials, digestibility and the convenience of the interface and self-evaluation of learning outcomes. These criteria do not duplicate and do not substitute for professional and expert evaluation, but supplement it, allowing to characterize the needs of consumers, which will allow the manufacturers of the courses to adjust their production in their interests.

5 Approbation of the proposed model

As a result, a mechanism was proposed to complement the system of peer review that was being developed in the Russian Federation by criteria and mechanisms that would allow to see the view of the consumer of online courses and to provide a mechanism for adequate feedback for the platform creators. As an important factor in the effectiveness of its implementation, the mass appeal of users of the MOOC should appear, which will allow to rank the courses effectively. In order to approve the proposed mechanisms within the framework of the special project of the contest of best e-learning practices EdCrunch Award (within the framework of international forum EdCrunch-2017 held annually by National University of Science and Technology "MISiS", I. Rybakov Charitable Foundation and the Ministry of Education and Science of the Russian Federation), a special nomination "Best course from the consumer's point of view (student evaluation)".

More than 100 MOOC consumers and more than 40 courses offered at leading Russian platforms and meeting the basic criteria of the MOOC took part in the approbation. The participation of platforms was not of a declarative nature, but, on the contrary, it was performed by selected students anonymously after passing the course and without informing its creators. This made possible to make the assessment more objective and to guarantee the impartiality of consumers. The high status of the competition was emphasized by the fact that the award was given to the winners by the Deputy Prime Minister of the Government of the Russian Federation A. Dvorkovich and the Deputy Minister of Education and Science of the Russian Federation L. Ogorodova.

The approbation of the consumer assessment of MOOC within the framework of the EdCrunch Award allowed to draw several conclusions.

Consumers show the greatest interest in the most practical courses or soft skills courses; theoretical courses have attracted the least interest. In particular, the most relevant were MOOCs devoted to data analysis in Google Analytics and the implementation of own business idea in the framework of technological entrepreneurship. This allows making a cautious conclusion that consumers are eager to train via e-learning exactly the competencies they often lack in formal education. The approbation of the proposed mechanism was very conditional in the presented sample, but it demonstrates the obvious productivity of attracting consumers to the assessment of the MOOC.

DOI: 10.1051/matecconf/201714101054

6 Conclusions

Global challenges, expressed in the need for the comprehensive development of human capital and the continuous improvement of the labor resources skills in connection with the changing technological order, can be overcome with regard to e-learning only if it is neatly implemented and a multi-level system for assessing its quality. In the Russian Federation favorable conditions were created for the development of e-learning and its implementation in university educational programs. The quality assurance system for such training includes an automated assessment of the course data and an expert evaluation. Its supplementation with the consumer appraisal, in our opinion, would make the system more flexible and enhance its ability to take into account the aspirations of students and their needs.

References

1. Icheon Declaration and Framework for Action for the implementation of Sustainable Development Goal 4 Electronic resource: http://unesdoc.unesco.org/images/0024/002456/245656R.pdf
2. A. McAuley, B. Stewart, G. Siemens, D. Cormier, *The MOOC Model for Digital Practice* (CC Attribution, Charlottetown 2010)
3. L. Mapping Web 2.0 benefits to known best practices in distance education // 2010 January-February . Electronic resource: http://deoracle.org/onlinepedagogy/emergingtechnologies/mapping-newly-identified-web2-benefits.html.
4. J. Zhang, M. Skryabin, X. Song, Distance Education, **37** (3), 270 (2016)
5. L. Stuchlíková, A. Kósa ICETA 2013 - 11th IEEE International Conference on Emerging eLearning Technologies and Applications, Proceedings (2013)
6. Quality Matters TM Electronic resource: https://www.qualitymatters.org/
7. Open ECB Check-Quality Electronic resource: www.ecb-check.net
8. N. Emelyanova, E. Voronina, IRRODL, **15** (1), 272 (2014)
9. K. Mertins, V. Ivanova, M. Abdrashitova, D. Isaeva, MATEC Web Conf. **48**, 06005 (2016)
10. I. Nikanorov, S. Volkova, MATEC Web Conf. **41**, 01022 (2016)

Mechanisms for credit transfer of e-learning results within the framework of educational programs of higher education

Antoniy Shvindt[1,*], and *Ivan* Nikanorov[2]

[1]Ministry of Education and Science of the Russian Federation, 125993 Moscow, Russian Federation
[2]Federal Agency for Scientific Organizations, 119334 Moscow, Russian Federation

Abstract. The evolving mechanisms of e-learning leads to new challenges for traditional institutions of higher education. Different countries adopt various mechanisms for their harmonization and integration of the MOOC to the educational process. The Russian system of higher education is distinguished by rigidity of its standards and the requirements both to the educational process and to the results of education. This had significantly complicated the implementation of the MOOC in the educational programs of Russian universities. Changes in the legislation and the adoption of a strategy for the development of the digital educational environment in the Russian Federation has provided new opportunities to Russian universities in this regard. The article is devoted to various options, mechanisms and practices for the credit transfer of e-learning by Russian universities, which were revealed during the monitoring of their legal acts placed in the open sources.

1 Introduction

Increasing speed of technological progress requires increased flexibility and adaptability of the education system. A key issue that is relevant today to any country claiming technological leadership in the world is how to implement the trends in the national education system for personalization and individualization and to make it adaptable to a rapidly changing world. Different national educational systems offer their own solutions consistent with certain global challenges in one way or another. One of the most important of them has become global digitalization of economy and social services area, which requires rapid changes in the system of professional training of personnel and the environment of higher education [1], [2].

* Corresponding author: anshvindt@gmail.com

2 Global context

One of the answers to this question was the emergence of OER – Open Educational Resources included by the United Nations in the document accepted in 2015 – "Transforming our world: the 2030 Agenda for Sustainable Development". The implementation of its fourth goal of sustainable development (SDG 4) is entrusted to UNESCO. The key document determining the approaches to SDG 4 implementation is the Incheon Declaration accepted at the World Education Forum in 2015 and the "Framework for Action for the implementation of Sustainable Development Goal 4" confirmed in Paris on November,4, 2015 [3].

Also, the priorities identified in the 2016 Global Education Monitoring Report are worth mentioning, which are necessary to achieve SDG 4, including the need for the development of Open Educational Resources (OER) as an innovative method of knowledge sharing, contributing to high-quality education getting throughout life for everybody. To one of the most widespread and gaining popularity of OER during the last 5 years can be attributed Massive Open Online Courses (MOOC) [4]. According to Education International's research, the global education market volume is about $4.5–5 trillion and in the coming years its possible growth is estimated to be $1–2 trillion. The share of online education is about 3%, which is $165 billion. Due to the steady growth dynamics by 2023, the digital part of the industry promises to overcome the mark of 240 billion US dollars, adding more than 5% per year. The next largest region is Southeast Asia (the largest share here is China and India), with a market growth of about 17%. Western Europe (6.8 billion US dollars) lags behind Eastern Europe (1.2 billion US dollars) yet, but it should be noted that Russia in this region is the driver of this direction with an average annual growth of 17–25% [5].

3 Russian development of e-learning environment

However, in Russian MOOC-segment of EdTech factors are worth noting that hamper its development in formal education [6]. Among such factors it is necessary to note a lack of trust in distance education, yet unformed system of implementing individual educational trajectories in general, in particular – the introduction of the MOOC to the implementation of basic educational programs, the lack of elaborated mechanisms for cost recovery as an individual student in the case of developing an online course on an external platform (in another educational organization), and the acquisition of MOOC by one educational organization from another at the expense of the federal budget, in accordance with the control figures for admission as a state task for the implementation of higher professional education programs, the disorganization of mechanisms of conducting state accreditation expertise and the implementation of state supervision and control in the field of education, taking into account the integration of MOOCs established in other educational and (or) other organizations , in the basic educational programs.

Summarizing these factors, we can say that they relate to the regulatory and administrative barriers that exist at two levels: bylaws adopted by federal government bodies and local normative legal acts approved at the level of each educational organization. Let is consider the issue related to federal legislation and bylaws providing the possibility of credit transfer of the taken online courses as part of the module or as a whole discipline in the framework of mastering the basic educational programs of higher education by students.

It is important to note here that the key normative legal act of the Russian Federation in the field of education is the Federal Law No.273-FZ "On Education in the Russian Federation" presupposes, on the one hand, the academic rights exist of students to be certified by an organization that carries out educational activities, mastering students'

DOI: 10.1051/matecconf/201714101055

subjects, courses, disciplines (modules), practice, additional educational programs in other organizations that carry out educational activities, and on the other hand, the educational organization has autonomy, which is understood as independence in the implementation of educational, scientific, administrative, financial and economic activities, the development and adoption of local regulations, as well as its competence includes the right to exercise current monitoring of academic performance and intermediate certification of students, the establishment of its forms and periodicity.

4 Normative grounds for the implementation of e-learning in the system of higher education in Russia

Integration of statutory rights of students for credit transfer and educational organizations for their autonomy is established in paragraph 41 of the Order of the Ministry of Education and Science of Russia No. 301 of April 5, 2017, which was enacted on September 1, 2017, as well as in paragraph 8 of the Order of the Ministry of Education and Science No. 816 of August 23, 2017. In connection with the right of educational organizations to establish the procedure and forms for the studies passing in individual disciplines and (or) individual practices that have been mastered (passed) by students in obtaining secondary vocational education and (or) higher education, as well as additional education interest analysis of the local regulatory legal acts adopted by educational organizations, which provide for the relevant procedure. In accordance with subparagraph e) of paragraph 2 of art. 29 of the Law on Education in the Russian Federation, educational organizations ensure openness and accessibility via the publication of copies of local regulatory legal acts on the basic issues of organization and implementation of educational activities, including regulating the procedure and forms for the exercise of noted right of students for credit transfer on their official website.

5 Variability of practices of implementation by universities of e-learning in educational programs

Analysing local normative legal acts from open sources of 287 educational organizations, it is possible to single out four main models that are implemented today in Russian universities. The first model – the educational organization preliminarily establishes a list of platforms containing online courses or a list of specific online courses (created in Russia or in other countries), and the document on the passage is passed by default, without additional tests. This model is implemented by 37% of organizations. The second model – before the beginning of the semester the educational organization accepts the applications from students with the list of online courses planned for passage in the current semester, forms lists of such students, and during the intermediate certification conducts own tests to confirm the knowledge received on the online course. This model is implemented by 21% of organizations. Such tests are fixed in the local normative legal acts of universities as an examination, differentiated credit, offset or interview. The third model – the educational organization transfers credits for the disciplines, including ones that mastered by the student in the form of an online course, only during the procedure of transfer from one educational organization to another, or the mastering of an educational program on an individual curriculum. This model is implemented by 6% of organizations. It should be noted that 7% of educational organizations do not provide the possibility of credit transfer of the passage of external open online courses, but only their own ones and they can be attributed to the fourth model. Another 29% of universities do not have local normative legal documentation, which provides an appropriate procedure for credit transfer.

Conclusions

Among the reasons why educational organizations today do not fully use the tools of modern educational technologies, such as OER as a whole and MOOC, as the most actively developing forms of their implementation, one can note first of all normative barriers, as well as distrust of quality and conservatism, As a property of any educational system as a whole. At the same time, the implementation of the norms of the Law on Education in the part of the right of students for credit transfer, enforcing secondary regulatory legal acts accepted by the Ministry of Education and Science of the Russian Federation in 2017, the appearance of online courses in these documents and the process of changing of local regulatory legal framework of educational organizations reduce administrative and regulatory barriers for active development of Russian market EdTech in the formal education sector and create opportunities for Russia to implement personalization and individualization of educational trajectories of students [10], as well as to implement to goal 4 in the field of sustainable development "Ensure inclusive and quality education for all and promote lifelong learning", which will allow to unlock fully the potential of human capital.

References

1. K. Donnelly, S. Rizvi, L. Summers, *An avalanche is coming: Higher education and the revolution ahead Research raport* (The Institute for Public Policy Research, London, 2013)
2. L. Shultz, M. Viczko, *Lynette Shultz Assembling and Governing the Higher Education Institution. Democracy, Social Justice and Leadership in Global Higher Education* (Palgrave Macmillan, UK, 2016) DOI 10.1057/978-1-137-52261-0
3. "Transforming our world: the 2030 Agenda for Sustainable Development" Electronic resource: https://documents-dds-ny.un.org/doc/UNDOC/GEN/N15/291/89/PDF/N1529189.pdf Open Element
4. Ichon Declaration and Framework for Action for the implementation of Sustainable Development Goal 4 Electronic resource: http://unesdoc.unesco.org/images/0024/002456/245656R.pdf
5. «Research of the Russian market of online education and educational technologies» Electronic resource: http://edumarket.digital.
6. M. Carnoy, Y. Kuzminov Voprosy obrazovaniya (Educational Studies), **3**, 8 [In Russian] DOI: 10.17323/1814-9545-2015-3-8-43.
7. K. Mertins, V. Ivanova, M. Abdrashitova, D. Isaeva MATEC Web Conf. **48**, 06005 (2016)

MATEC Web of Conferences **141**, 01056 (2017) DOI: 10.1051/matecconf/201714101056
Smart Grids 2017

Soft skills and Moodle

Veronica Ivanova [1*], *Kseniya* Mertins[1], *Mariya* Alexandrova[1], and *Pavel* Baranov[1]

[1]National Research Tomsk Polytechnic University, 634050 Tomsk, Russia

Abstract. At present, technical university graduates are expected to acquire a set of not only hard skills but soft ones as well, which are in the first instance communication skills that can be developed through active and interactive methods during in-class learning. The issue related to communicative skills development is being discussed in different countries throughout the world. This problem is faced by university graduates from Europe, the US and Russia, and in the East. The learning process exploits a variety of electronic platforms, which, on the one hand, significantly increase the pool of students, but, on the other hand, hinder the development of communicative skills. This poses the question about blended learning which combines active and interactive teaching methods with e-learning. Consider a good example of these two types of combined student activity obtained using interactive methods in the Moodle course "Professional training in English" designed for the Bachelor's Degree Programs, Tomsk Polytechnic University (Tomsk, Russia). This paper considers the main types of tasks used in the e-course. It is shown that the teacher's first aim is to be not a scientific knowledge translator, but to foster a creative educational environment by selecting an optimal teaching strategy through modern educational technology; an e-course is a good teaching aid to build hard and soft skills.

1 Introduction

We live in an era of technological revolution, rapid advances in communication technologies. The pace of technological development in the modern world requires the fastest and cheapest techniques of generating and transmitting knowledge. Emerging technologies and associated real situation in the education sector led to the emergence of E-learning, which is currently successfully implemented in corporate universities and higher educational institutions for training of future specialists in many countries [1-4].

The benefits provided for students by E-learning are as follows: interactivity that facilitates understanding; inclusion through a variety of learning styles; learning at one's own pace; around-the-clock access to training materials. This allows students to choose an individual educational trajectory, improves their technical and educational skills and increases responsibility [5,6].

When preparing tasks, the teacher must adhere to conditions, the implementation of which will make the designed tasks diverse and interesting. The tasks should be focused on

* Corresponding author: kurator@tpu.ru

search for solutions in a clear, interesting and creative manner within a short time and the formation of skills in a particular area.

In the disciplines of the basic and professional modules [7], the examples of practical application of the theory and homework assignment should be related to the future professional activity that stimulates the students' motivation.

In addition, tasks implying teamwork that promotes student responsibility, communication and organization skills should be actively used.

The educational technologies of greatest interest for e-learning are those that focus on cooperative learning, group work of students, active learning process and simultaneous work with various sources of information. These technologies include widespread use of research and problem methods, the use of the knowledge gained in cooperative or individual learning activities, the development of independent critical thinking, communication culture, the ability to perform different social roles in team working. These technologies provide the most effective solution to problems of student-centered learning. Students receive real opportunities to achieve certain results in various fields of knowledge, to acquire better understanding of the knowledge gained, and to form a well-reasoned position on different issues of science, technology and society in accordance with an individual educational trajectory.

2 Methodology

Theoretical and empirical research methods are used to identify meaningful patterns and productive achievements, which can be further implemented in educational practice. The used methods are observation, experience study, analysis of pedagogical documentation and study of the products of training and creative activity of students and on-line surveys.

3 Task samples

Exploring theoretical material and testing after completion of the task, which implies reading texts related to the professional area, develop the skill of working with large volumes of information in a foreign language and its structuring.

3.1 Compiling a glossary and doing a crossword puzzle

Compiling a glossary and doing a crossword puzzle (based on previously compiled glossary, teamwork in small groups) increases the level of synthesizing skills, and the information analysis and perception.

This task can be given to students as an interim one after completion of a module and as a midterm task at the end of the semester.

At the beginning of the semester, the students are given the task to compile a glossary in Moodle. The students are supposed to enter unknown words while studying a particular module. Within a week, the student is expected to enter at least 10 new words and definitions. The glossary is arranged so that the terms may reoccur. Thus, at the end of the semester, the student has a glossary that contains at least 160 terms and definitions in English, in addition to those made up in the core glossary by the course teacher. Based on these terms, the students compile crossword puzzles that include no less than 20 terms. This work is performed using the Google environment as well as the Moodle Wiki tool. The students are encouraged to work in small groups of 2–3 people. Access to the environment is permitted for the members of the micro groups only. While the students are compiling a crossword and working in an electronic environment, the teacher has another opportunity to estimate the contribution of each student in the performance of this task. The students put a blank crossword grid and questions in a shared file. Then the teacher distributes the crosswords to do in small groups. After working the crossword puzzles, the authors of those check the results.

When performing the task, on the one hand, the students learn to formulate a definition and make another resort to the terms. On the other hand, they learn to interact with other students, to negotiate and come to an optimum decision in terms of the teamwork. In addition, the students learn concisely express their thoughts explaining the terms to ensure the ambiguity and scientificity of perception.

3.2 The tasks with case-studies

The tasks containing case-studies reproduce a variety of professional situations and help students immerse in these situations, thereby reinforcing the experience in solving professional problems in a foreign language. The preparation stage can be done in Wiki (Moodle) or Google –disk. The teacher can make monitoring and some correction of student activity (if needed). It also helps to make objective assessment by taking into account the contribution of each person.

The stages of preparation of a lesson by the method of business game are presented below:

- Defining the problem, theme, aims and objectives of the game.
- Selecting an object of the business game. A business game is a model or a segment of professional activity; therefore, a part of the production or labor process and a problem-based situation are chosen for simulation; the conditions of typical situations are defined.
- Development of the business game scenario.
- Development of materials for the game.

Development of the business game "from scratch" is a time-consuming process. To facilitate the task, the teacher can use a ready-made scenario, which can be found in the public domain on the Internet, or develop it together with colleagues, other subject teachers being involved in the process as well. Business games are often conducted using a multidisciplinary approach.

The scenario of the business game 'Innovator" playing in an electronic environment developed for this course is as follows:

The preparatory stage of the game (development of criteria to evaluate the activities of the group with respect to the role assigned) is performed in the electronic environment – each team works in individual Google-Docs. The teacher monitors the discussion, evaluates the contribution of each participant and corrects the students' work when necessary.

During the game, the participants – design engineers, bankers, officials and ecologists – fulfilled their role functions in closed network communities "Wiki" of the e-learning course in Moodle. They came up with their ideas, discussed and developed these, then analyzed and exchanged the ideas and materials following the game scenario.

The teacher performed general guidance of the business game on the page of on-line course and summarized the results. The game develops the ability to work in the electronic environment, develops critical and systems thinking, communication competence, etc.

3.3 Writing papers and presenting

Writing papers and presenting them at different conferences develop skills in presentation and protection of the results of complex engineering activities in a foreign language. Here Moodle can be used at the preparation stage as well as for preliminary practicing in writing abstracts, papers and making presentations performed in accordance with the requirements

DOI: 10.1051/matecconf/201714101056

of the international scientific community. The tools such as Forum, Seminar and Test can be used for that case.

The abstract can be evaluated by the following criteria: the strength of the central idea (thesis) that is related to the assignment; clearance, logical organization with well-developed major points that are supported with concrete and specific evidence; the use of effective transitions between ideas; to compose sophisticated sentences; fresh and vivid expression of ideas; absence of mechanical, grammatical, and spelling errors.

For instance, students working on scientific project "Planar fluxgate transducer for magnetic vacuum system of quantum computer" Russian Science Foundation (RSF), Grant Number 17-79-10083 use Moodle tool to prepare two papers. The papers contain the results of their scientific research. These papers were submitted to the 2nd International Conference on Mechatronics and Electrical Systems and International Conference On Electrical Engineering, Control And Robotic 2018, were reviewed and were accepted for presenting and publication.

3.4 Mind maps

Mind mapping is a highly effective way of getting information in and out of the human's brain. It is a creative and logical means of note-taking and note-making that literally "maps out" human's idea.

Mind maps help to structure and organize information, to increase the level of systems thinking, to use the gained language skills to work with information. The Moodle tools such as "Seminar" and "Forum" are appropriated here. The Moodle gives an opportunity to make cross-evaluation of works by students.

Mind maps can be designed using special software (fig.1a) as well as drawn by hand (fig.1 b) and uploaded to the system as a scan copy or photo.

3.5 Structured discussions

The purpose of method implementation is to improve the quality of learning the material of the course by the students through their involvement in active discussion and development of competencies in finding the problem-solution, evaluation of learning outcomes (including those of other students).

Each discussion usually goes through three stages: Orientation, when the talk participants are oriented and adapted to the problem itself, to each other, and to the overall atmosphere. This is how the students are set to address the problem presented. Evaluation is the stage when the participants compare information and different standpoints and generate

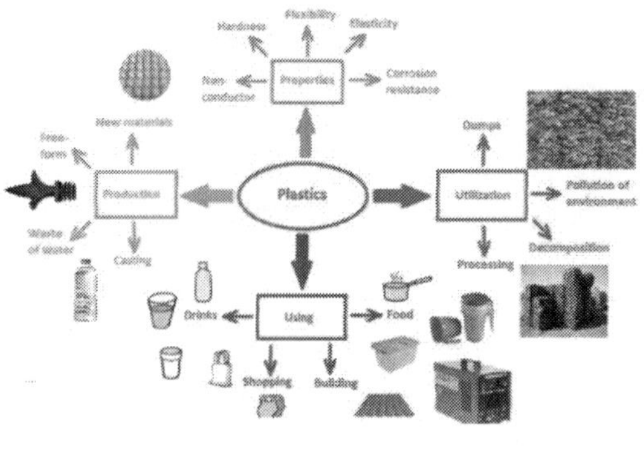

Fig. 1a. Mind map using software.

Fig. 1b. Hand-made mind map.

ideas. Consolidation is the last stage when the participants are expected to develop uniform or compromise decisions, opinions and positions.

Structured discussions held in the form of scientific debates improve students' initiative in finding materials for in-class learning and intensify the activity during debates. It is essential that information sources are diverse with different approaches to problem solving, and the discussion is supported by the teacher. The tool "Forum" can be successfully used for this purpose.

For example, in the discipline "Professional training in English" (module "Actual problems in Instrument Making"), the students were offered 10 innovative discoveries that can change the world for the better from the point of view of scientists. The students were asked to choose three of these discoveries most important from their point of view to discuss with group-mates and to come up with three most important discoveries chosen by the whole group. This activity develops the ability to work with information, teamwork and systems thinking. It should be noted that the task appeared to be rather difficult for Master's degree students. A similar task was simultaneously given to a different group of students in the classroom (in the presence of a teacher) to be done within 90 minutes. Both groups, one in the classroom and another one in the E-learning environment, could not reach a consensus. This result indicates the need to develop communication skills as it does not depend on the environment, electronic media and online communication, in which the participants were to do the task [8].

Conclusion

All the tasks that include the elements of teamwork greatly contribute to soft skills development, teach to interact as a team and help to develop leadership and decision-making skills, and responsibility.

It is shown that nowadays the teacher's first aim is to be not a scientific knowledge translator, but to foster a creative educational environment by selecting an optimal teaching strategy through educational technology.

Thus, it is concluded that an e-course is a good teaching aid to develop not only hard, but also soft skills of graduates and can be used widely in stages of preparation for different active and interactive methods.

The research is funded from Russian Science Foundation (RSF), Grant Number 17-79-10083.

References

1. V. Bošković, T. Gajić, I. Tomić, *SINTEZA 2014*, 480 (2014)
2. J. Lasić-Lazić, T. Ivanjko, I. Grubješi, *MIPRO 2017/CE.* 1021 (2017)
3. I.A.E. Spanjers, K. D. Konings, and other, ERR 15, 59 (2015)
4. L. Deng, N. J. Tavares, C& E, **68,** 167(2013)
5. I. Nikanorov, MATEC Web Conf. (to be published)
6. A. Shvindt, MATEC Web Conf. (to be published)
7. The federal state educational standards of higher professional education. Federal'nyj gosudarstvennyj obrazovatel'nyj standart vysshego obrazovanija, uroven' vysshego obrazovanija - magistratura, napravlenie podgotovki 12.04.01 "Priborostroenie" [in Russian]
8. V. Ivanova, K.Mertins, and others, MATEC Web Conf. **48**, 06005 (2016)

MATEC Web of Conferences **141**, 01057 (2017) DOI: 10.1051/matecconf/201714101057
Smart Grids 2017

Optimization of virtual power plant with a distributed generation

Elena Sosnina[1], *Andrey* Shalukho[1,*], and *Alexander* Kechkin[1]

[1]NNSTU n.a. R.E. Alekseev, 603950 N. Novgorod, Russia

Abstract. The article is devoted to the task of optimizing the structure of virtual power plants (VPP) with distributed generation sources. The solution of the problem is based on the structural topological and probabilistic-statistical analysis of the VPP structure. An algorithm for estimating the structure of VPP has been developed. The optimization criteria for VPP are proposed and investigated. The criteria allow to evaluate VPP structural properties and to take into account the inconstancy of generation from renewable energy sources (RES) in VPP composition.

1 Introduction

An important direction in the development of distributed energy is connected with a merger of various small sources of generation (SSG) with energy storage device and consumers in the VPP [1-3]. In this case the problems of choosing the optimal structure of VPP are modern and research require. For evaluation the efficiency and reliability of systems with SSG widely indications apply: voltage level, loss of power, short circuit currents, SAIDI and SAIFI characteristics. However to constitutive properties of grid and existence of equivocations not enough attention have given and this can lead to the adoption of suboptimal solutions.

The authors investigate the solution of optimization VPP problems which based on structurally topological and probabilistic statistical analysis of VPP structure.

2 Materials and methods

Evaluation algorithm of VPP structure was developed by using the theory of graphs. The algorithm consists of three stages.

Stage 1 - electrotechnical complex of VPP is represented as an oriented graph [4]. It is possible to build several graphs for analysis which reflect the various options for electrical connections in VPP.

Stage 2 - the considered graphs are represented in matrix form, are calculated structurally topological graph characteristics: structural connectivity, structural redundancy, uneven distribution of connections, structural compactness, degree of centralization [5].

Structural connectivity (c):

* Corresponding author: shaluho.andrey@mail.ru

$$c = \sum_{i=1}^{n} \sum_{j=1}^{n} c_{ij}, \quad (1)$$

where $c_{ij} = 1$, if $a_{ij} \geq 1$ and $c_{ij} = 0$, if $a_{ij} = 0$ (a_{ij} is the total number of paths from vertex i to vertex j).

The larger the value of c, the structure is more connected and potentially more reliable.

Structural redundancy (R):

$$R = \left[\sum_{i=1}^{n} \sum_{j=1}^{n} a_{ij} \right] \cdot \frac{1}{n-1} - 1, \quad (2)$$

where a_{ij} is the element of the adjacency matrix of vertices A, n is the number of vertices of the graph.

The minimum value of R indicates the smallest number of excess electrical connections that increase the cost of power lines and equipment.

Uneven distribution of connections (ε^2):

$$\varepsilon^2 = \sum_{i=1}^{n} \left(\rho_i^2 - \frac{4m^2}{n} \right), \quad (3)$$

where ρ_i is the number of edges incident to the i-th vertex of the graph, m is the number of edges of the graph.

The structure with the minimum value ε^2 is considered to be the most effective.

Structural compactness (Q):

$$Q = \sum_{i=1}^{n} \sum_{j=1}^{n} d_{ij}, \quad (4)$$

where d_{ij} - is the minimum path length from vertex i to vertex j.

Structures with a smaller value of Q are characterized by potentially smaller losses and a greater degree of efficiency in management.

Degree of centralization (δ):

$$\delta = (n-1) \frac{2Z_{MAX} - n}{Z_{MAX}(n-2)}, \quad (5)$$

where Z_{MAX} is the value determined from the calculation of structural compactness.

A high value of δ places high demands on the capacity of the central element.

Stage 3 - using the weighted sums of criteria method, an integral evaluation of the VPP structures being compared is performed:

$$K_j = \sum_{i=1}^{M} k_i P_{ij}, i = 1,...M, j = 1,...N, \quad (6)$$

where N is the number of compared structures, M is the number of compared characteristics (c, R, ε^2, Q, δ), P_{ij} is the estimated estimate of the i-th characteristic of the j-th structure, k_i is the coefficient of value of the i-th characteristic.

 DOI: 10.1051/matecconf/201714101057

The most effective structure is considered to be the VPP, to which the largest value of K_j corresponds.

3 Result and discussion

Consider an example of the integration of two IMGs (wind power plant (WPP) and gas piston unit (GPU)) into the power supply system of consumers receiving power from a centralized electrical network.

A simplified single-line diagram and the obtained graph of the distributed electric network are shown in Fig. 1 and Fig. 2. For 13 possible structures structural-topological characteristics (1) - (5) were calculated and their integral estimation was performed.

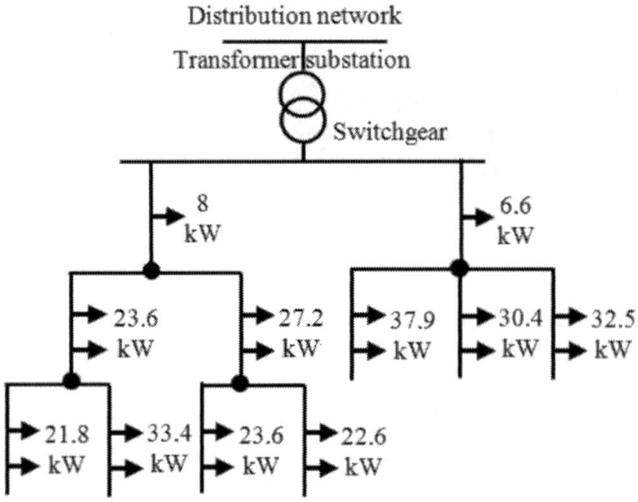

Fig. 1. Simplified single-line diagram of a distributed electrical network.

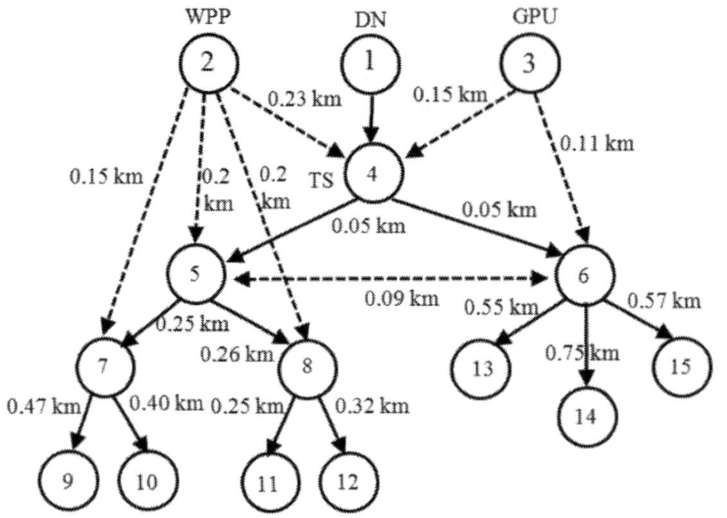

Fig. 2. Graph of a distributed electrical network.

The results of calculations for two variants of the structure with the largest values of K_j are given in Table 1.

 DOI: 10.1051/matecconf/201714101057

Table 1. Results of structural topological analysis.

Option structure	c	R	ε^2	Q	δ	K_j
K1	50	0	37.8	103	1.429	0.958
K2	50	0	37.8	103	1.429	0.958

The WPP is connected to the node 7 in the variant K1, and in the variant K2 to the node 8. The GPU is connected to the node 6 in both versions of the structure.

For a more detailed analysis of the selected version of the structure of the VPP, in addition to calculating technical and economic indicators, it is proposed to use the ranking of its elements. The most significant elements in the structure of the VPP are determined. To assess the significance, the "element rank" parameter can be used [6]:

$$r_i = \sum_{j=1}^{n} a_{ij}^{k} \Big/ \sum_{i=1}^{n}\sum_{j=1}^{n} a_{ij}^{k}, \tag{7}$$

where a^{k}_{ij} is an element of the adjacency matrix, k is the path length from vertex i to vertex j.

The higher the rank of the element, the more severe the consequences will be if its functioning is disrupted. The results of calculating the ranks of the elements for the selected VPP structures are shown in Table 2.

Table 2. Results of calculating the ranks of the elements.

Structure	Elements of the graph and their ranks				
	1	2	3	4	5 - 16
K1	0.64	0	0	0.36	0
K2	0.64	0	0	0.36	0

For both structures, the key elements are nodes 1 and 4. For more complex systems, the uniform distribution of the participation of elements in the formation of the structure is considered to be the best. When renewable energy sources (RES) are part of the VPP structure, it is important to take into account the volatility of their generation. The inconstancy of RES generation is proposed to be assessed using a special criterion - R_{RES}.

Under R_{RES} is the probability that within a certain time interval (for example, a month) the consumer will use the entire amount of RES produced. It is assumed that the distribution of generated electricity corresponds to the normal law. In this case, the R_{RES} value is determined by solving the integral function of the normal distribution [6]:

$$R_{RES} = \int_{-\infty}^{W_{CE}} \frac{1}{\sqrt{2\pi\sigma(W_{RES})}} e^{-\frac{\left(W_{RESi}-\mu(W_{RES})\right)^2}{2\sigma(W_{RES})^2}} dW \tag{8}$$

where W_{CE} is the amount of consumed electricity, W_{RESi} is the generated electricity of RES, $\mu(W_{RES})$ is the mathematical expectation of the renewable electricity produced, $\sigma(W_{RES})$ is the standard deviation of the generated electricity. The calculation of R_{RES} for connection of WPP to nodes 7 and 8 has been performed (Table 3).

Table 3. The probability of consumption of the produced electric power by WPP.

Node	Time of day	Month											
		1	2	3	4	5	6	7	8	9	10	11	12
7	Day	69.5	74.9	65.5	35.6	50.8	46.8	72.6	64.4	79.1	81.3	97.5	53.6
	Night	54.8	58.7	59.5	62.9	99.6	95.4	96.9	85.8	99.0	94.2	92.1	67.4
8	Day	80	85.1	77.3	42.5	59.1	55.2	81.1	73.2	88.3	88.5	99.6	59.9
	Night	53.6	56.8	57.5	57.5	98.6	91.3	92.9	80.8	97.6	92.8	90.5	65.2

From the results it follows that it is necessary to provide for the possibility of transferring electric power from the VPP to neighboring nodes with a load.

For VPP with a large number of aggregated objects, an important criterion for optimization can be an assessment of the chaotic nature of processes. The VPP can be characterized as a dynamic dissipative nonlinear system [7]. It is known that chaotic processes are typical for such systems. The danger of chaotic regimes is that they can lead to the system's exit from a steady state and an accident. Therefore, the chaotic processes in the VPP should be studied with a view to a deeper understanding of the mechanisms for ensuring the sustainability of the VPP.

4 Conclusion

In conclusion, it can be noted that the proposed and investigated optimization criteria for the VPP allow us to evaluate the structural properties of the VPP and take into account the volatility of the generation of renewable energy.

The work was carried out with the financial support of the Ministry of Education and Science of the Russian Federation (state task No. 13/2078/2017/ПЧ).

References

1. E.N. Sosnina, A.V. Shalukho, A.Yu. Kechkin, Vestnik NNSEEI, **4**, 50 (2015)
2. E. Sosnina, I. Lipuzhin, A. Shalukho, A. Kechkin, IJAER, **10**, 44363 (2015)
3. E.N. Sosnina, A.V. Shalukho, I.A. Lipuzhin, A.Yu. Kechkin, Vestnik SSTU Tech. Sci. Series, **2**, 121 (2017)
4. E.N. Sosnina, A.V. Shalukho, A.Yu. Kechkin, N.V. Shumsky, Operative manage. electrical power-engine., **4**, 48 (2017)
5. N.I. Voropay, *The theory of systems for electric power engineering specialists* (Novosibirsk, 2010) [in Russian]
6. E.N. Sosnina, A.V. Shalukho, Power Tech. Engine., **6**, 467 (2013)
7. K.S. Demirchyan, Power Engine. Transp., **3**, 3 (1987)

Using of non-traditional current and voltage sensors for the fault location

Andrey Yablokov[1,*], *Galina* Filatova[1], and *Alexander* Timofeev[1]

[1]Ivanovo State Power Engineering University, 153024 Ivanovo, Russia

Abstract. The article describes the possibilities of using the non-traditional current and voltage sensors to solve the problem of fault location in terms of digital transformers developed at the Ivanovo State Power Engineering University (ISPEU). Frequency and metrological characteristics of current and voltage sensors are given.

1 Introduction

The high-voltage power lines are quite often damaged parts of energy system. Line outage is always accompanied by undersupply of energy, decreasing of reliability, cost and quality of electric supply. That's why one of the important tasks of line maintenance is quick looking for location of the damage and organization of the rehabilitation.

So there are wide-spreading methods and means of remote fault location on the power lines in power industry. Solving of these problems is the one of the most important tasks related with the conception of creating active-adaptive electrical system (Smart Grid). This conception has new such properties as self-diagnosis, self-healing, the ability to identify "weak" parts of the system and change its operating mode, and it helps to prevent the emergence of technological accidents. The development of market relations in the power industry, the requirements to ensure reliable power supply and rapid recovery of power industry connections lead to the need for accurate determination of the fault location and special algorithms.

The development of non-traditional current and voltage sensors (including digital current and voltage transformers) that allow to measure voltage and current without distortion due to lack of enrichment, but with higher accuracy due to exclusion of analogue secondary conduits and isolating transformers, and also having wider frequency ranges [1], involve a significant increase of the accuracy of remote fault location algorithms [2].

Nowadays, with the development of the satellite synchronization technology the interest in pulsed methods of remote fault location based on measurement of the electromagnetic waves' movement time through the power lines increases. The most widespread passive or wave methods are ones that use natural electromagnetic waves which appear in fault location [2]. As the experience in the operation of wave remote fault location shows, the accuracy of fault location without equipment cost's increasing and its operation is about ±300 meters (about one span of the line). But, in practice, the accuracy of the fault location

* Corresponding author: andrewyablokov@yandex.ru

 DOI: 10.1051/matecconf/201714101058

is significantly influenced by frequency characteristics of primary converters, accuracy in sensor signals, sampling accuracy, synchronization accuracy, accuracy in determination the wave parameters of the power lines and other factors [4, 5].

Also widespread methods of remote fault location are based on measurement of the emergency mode parameters (EMP). The advantages of these methods are simplicity of usage, absence of generating equipment, possibility of usage on every kind of lines. Also remote fault location, which based on measuring EMP has large (about 10-20%) accuracy caused by accuracy of current and voltage sensors [5, 6].

So, the development of new remote fault location algorithms based on signals from digital current and voltage transformers and research of accuracy in digital transformers are practical interesting.

2 Non-traditional current and voltage sensors

There were produced digital current and voltage transformers, combined transformers for voltage 6 (10), 35, 110 and 220 kV (fig. 1) at the Department of electric power systems automatic control, ISPEU. Nowadays the developed digital transformers are entered in the register measure instruments.

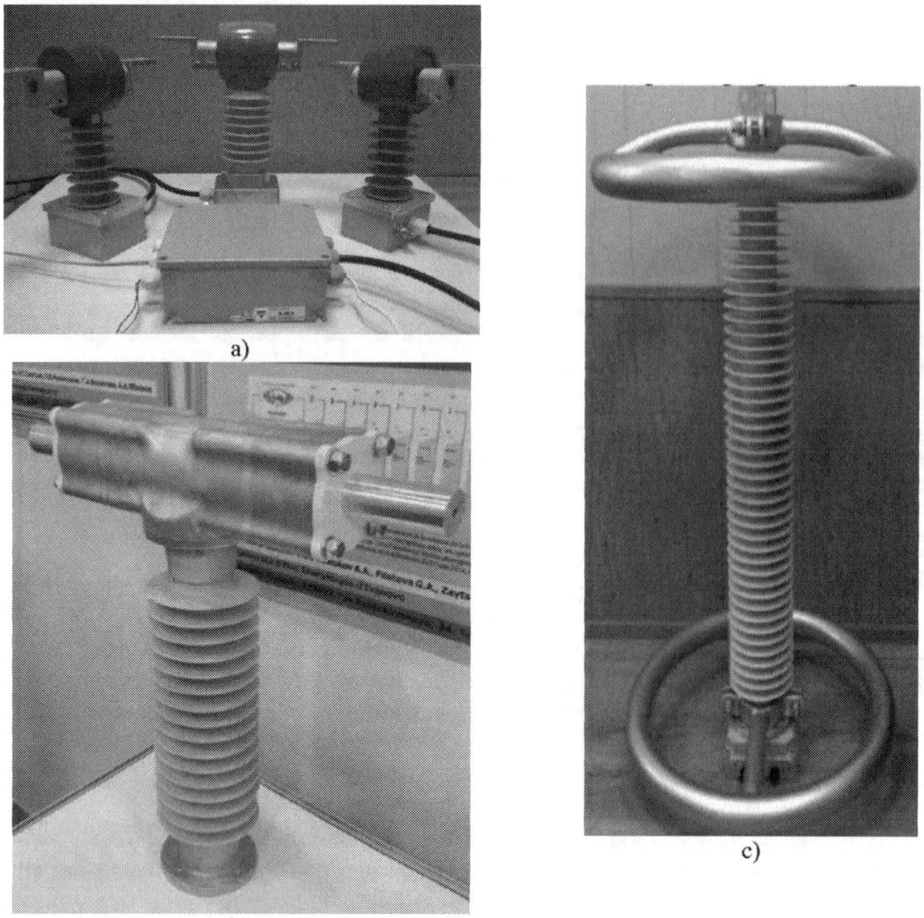

a) b) c)

Fig. 1. Experimental samples of digital current and voltage transformers: a) 6 (10) kV; b) 35 kV; c) 110 kV.

Primary converters of digital transformers contain several non-traditional current sensors based on different physical principles, and non-inductive voltage dividers. The tests on the laboratory equipment of the ISPEU and the accredited testing center for measuring instruments ("ICRM") showed that it's promising to use digital transformers for the fault location problems because of their better frequency characteristics and smaller accuracy (fig. 2-5).

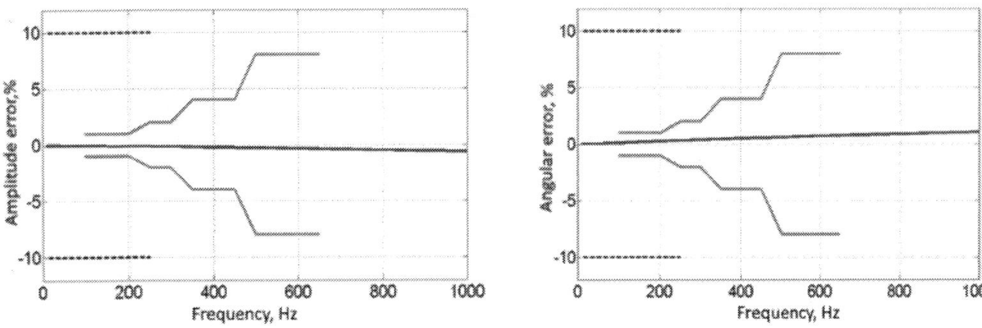

a) b)

Fig.2. Amplitude-frequency (a) and phase-frequency (b) characteristics of a small-sized current transformer with an accuracy margin for the calculation of electricity and protection for traditional usage in accordance with the Standard IEC 60044-8-2010.

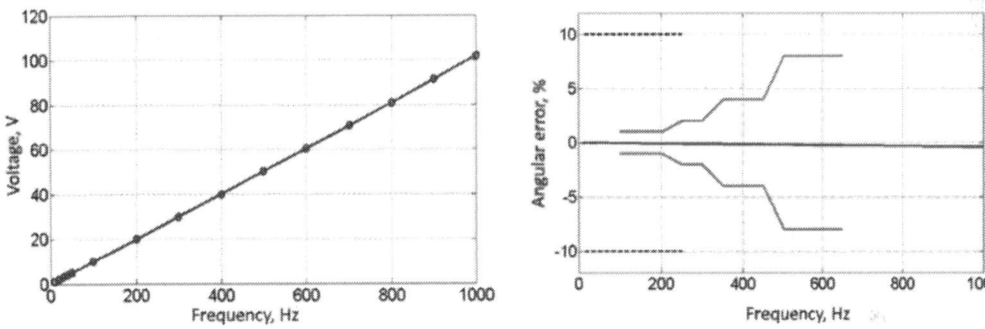

a) b)

Fig. 3. Dependence of the voltage on the Rogowski belt on frequency (a) and the phase-frequency characteristic (b) with a designation of accuracy limits for the calculation of electricity and protection for traditional usage in accordance with the Standard IEC 60044-8-2010.

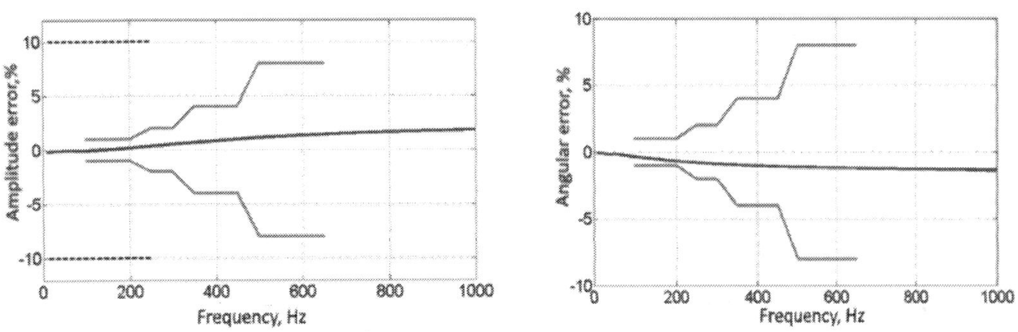

a) b)

Fig. 4. Amplitude-frequency (a) and phase-frequency (b) characteristics of the direct current sensor with an indication of the accuracy limits for the calculation of electricity and protection for traditional usage in accordance with the Standard IEC 60044-8-2010.

DOI: 10.1051/matecconf/201714101058

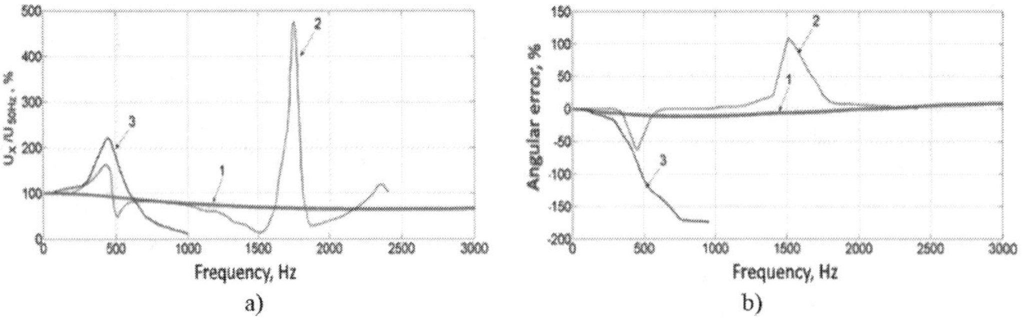

Fig.5. Amplitude-frequency and phase-frequency characteristics: a) primary voltage converter of digital voltage transformer (110 kV) 2) inductive voltage transformer (420 kV) with oil-and-paper insulation 3) capacitive voltage transformer.

Primary current converters, according to the test results, have a high accuracy in the wide range of currents from 1 to 120%In and with a reserve are placed within the limits of the accuracy class 0.2s with a three-fold overcurrent capability. It has the same results in a wide frequency range. The use of produced digital transformers with improved metrological characteristics in a wide frequency range will increase the accuracy of remote fault location based both on wave methods and on the measurement of EMP.

At the moment one-sided metering algorithms using as input signals currents and voltages obtained with the digital combined transformers are being developed. The algorithm is based on remote metering using the voltage measured from the Rogowski belt, proportional to the derivative of the primary current. Measurements occur at the moments of the transient current crosses zero values measured by a high-precision shunt. Measurement of voltage on the damaged phase is made by a resistive divider. The use of voltage from the Rogowski belt leads to an increase in the accuracy of the measurement compared with the use of the mathematically calculated derivative (fig. 6).

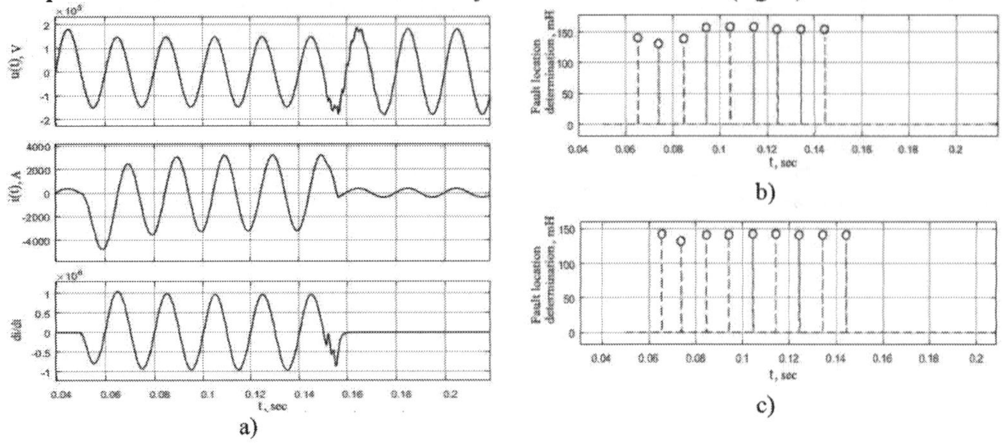

Fig.6. Electrical quantities during short circuit (a) and measured fault location – using mathematically calculated derivative (b) and voltage measured from the Rogowski belt (c), where 150 mH is true short circuit inductivity.

The estimation of the distorting factors (transient resistance in the fault location, load current, capacitive conductivity of the line, accuracy of specifying the specific parameters of the transmission line) is carried out on the models of electric networks in Matlab.

The research was made thanks to the grant from Russian Science Foundation (project No. 17-79-10455) at the Ivanovo State Power Engineering University.

References

1. V.D. Lebedev, A.A. Yablokov, IOP Conf. Series: Mater. Sci. Eng. **177**, 1 (2017), DOI: 10.1088/1757-899X/177/1/012099
2. G. Filatova, V. Lebedev, V. Shuin, ICIEAM, 1 (2016), DOI: 10.1109/ICIEAM.2016.7911011
3. Standard 56947007-29.240.55.224-2016. Studies of fault location determination on overhead power lines voltage 110 kV and more. The standard of company. Introduction date: 17.08.2016 [in Russian]
4. S. Marx, B. K. Johnson, A. Guzman, V. Skendzic, M. V. Mynam, Int. Conf. Power Syst. Transients, 15 (2015)
5. Yu.V. Bychkov, V.N. Koslov, K.I. Ermakov, J. Relay Prot. Electric Power Syst. Automatics, **1**, 42 (2016)
6. *Otchet po naturnym ispytaniyam ustrojstv OMP VLEHP* (Report on tests fault location devices). Report of FGC UES, 04.12.2015. Stream: http://aura-e.ru/files/ОТЧЕТ_ОМП.pdf [in Russian]

MATEC Web of Conferences **141**, 01059 (2017)
Smart Grids 2017
DOI: 10.1051/matecconf/201714101059

The analysis of the external factors influence on the efficiency of the absorption heat pumps inclusion in the scheme of a two-stage line installation of a STP

Dmitriy S. Luzhkovoy[1], *Olga Yu.* Romashova[1,*], *Marina A.* Vagner[1], *Aleksandr A.* Tubolev[1], *Juri V.* Kobenko[1], and *Elena S.* Riabova[2]

[1]National Research Tomsk Polytechnic University, 634050 Tomsk, Russia
[2]Samara University, 443086 Samara, Russia

Abstract. The article deals with a comparative analysis of the efficiency of a two-stage line installation in a heating turbine before and after the inclusion of absorption heat pumps into its scheme with a decrease in the outside air temperature. The research shows the dependence of the efficiency of the line installation on its heat load while using AHP in its scheme, as well as on the heat conversion factor of the absorption heat pumps.

1 Introduction

Two-stage heating of the delivery water at the CHP in comparison with the one-stage heating allows to increase the generation of electricity by heating consumption. Modern heating turbines exclude the possibility of maintaining an equal heating distribution of the delivery water within the entire heating period, which leads to a decrease of the efficiency of the heat-producing power generation [1-3].

It is known that the heating distribution substantially depends on outside air temperature [4]. When turbines are operating in the heating mode, the water heating in the lower line heater with a decrease in the outside air temperature increases with a corresponding decrease in the upper heater. Other things being equal, unevenness of the line heating increases with a decrease in the flow of the delivery water and an increase in its temperature in the return line [5].

The calculations of the operating modes of a two-stage line installation in a turbine of the T-110-130 type in Tomsk climatic conditions (the minimum design outside air temperature for heating $t_{out.tem.}^{d.}$ = -40 ° C) for the heating system graph 150/70 and the extraction factor α_{CHP} = 0.6 show that the heating proportion in the lower stage during the turbine operation in the heating mode with the nominal heat load varies from a value, that is close to 50% at the temperature of switching the peak water-heating boilers, to 80% at $t_{out.tem.}^{d.}$ [4].

* Corresponding author: roma@tpu.ru

This research shows the calculation analysis of the use of an absorption heat pump (AHP) to equalize the stage heating of the delivery water with a decrease in the outside air temperature. The calculations were carried out in two stages. At the first stage, the efficiency of the line installation with AHP was analyzed with a change in the heat load $q = Q_{l.inst.}/Q_{l.inst.}^{nom.}$ and the unchanged temperature graph of the heating system. At the second stage, the dependence of the line installation efficiency on the AHP heat conversion factor at the nominal heat load was analyzed.

The initial data for the research are presented in [6].

The design scheme of a two-stage line installation with AHP is shown in Figure 1.

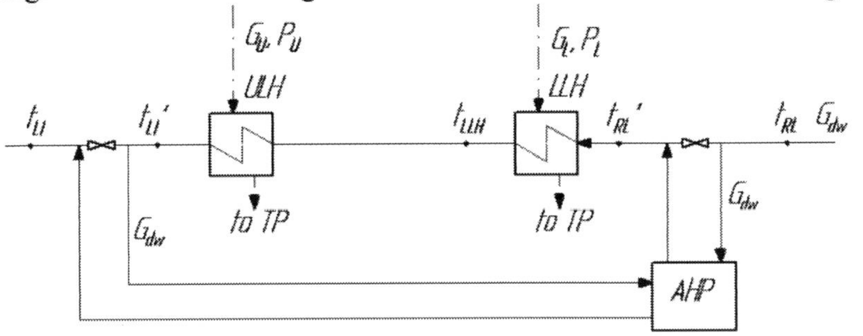

Fig. 1. The design scheme of a two-stage line installation with the AHP included in the scheme (ULH – upper line heater; LLH – lower line heater; TP – turbine plant; G_{dw} – delivery water discharge; G_U – vapor feed rate of selection on upper line heater; P_U – vapor pressure of selection on upper line heater; G_L – vapor feed rate of selection on lower line heater; P_L – vapor pressure of selection on lower line heater; t_{RL} – return line temperature; t_{LLH} – delivery water temperature after the lower line heater; t_{LI} – delivery water temperature after the line installation).

2 Calculation results

2.1 The efficiency of a line installation using AHP with a change in the heat load

The calculation results of the efficiency of a two-stage line installation with an AHP in comparison with the initial scheme at a minimum (design) outdoor air temperature and various heat loads are shown in Figures 2, 3.

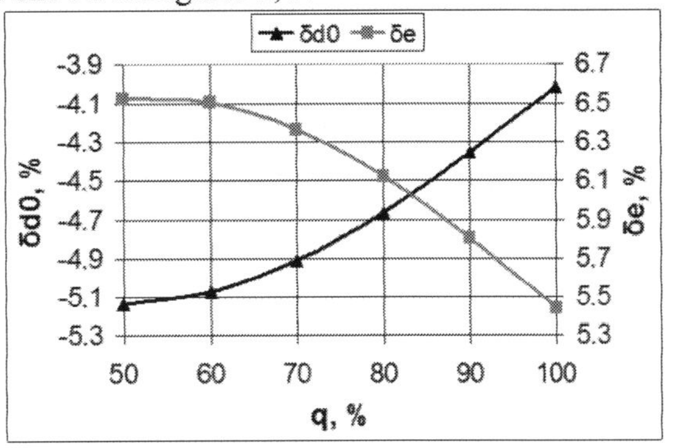

Fig. 2. The dependence of the relative change in the specific steam consumption for the turbine δd_0 and the relative change in the specific electricity generation at heating consumption δe from the heat load of the line installation with the inclusion of the absorption heat pump into the two-stage heating of the delivery water at $t_{out.tem.}^{d.}$

 DOI: 10.1051/matecconf/201714101059

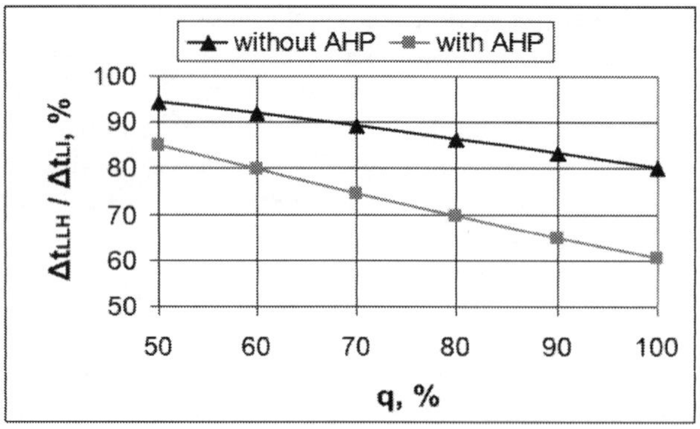

Fig. 3. The dependence of the proportion of the delivery water heating in the lower line heater from the heat load of the line installation in the original version and with the absorption heat pump in the scheme of the two-stage delivery water heating at $t_{out.tem.}^{d.}$

2.2 The efficiency of a line installation with an AHP with a change in the heat conversion factor

The results of calculating the efficiency of a two-stage line installation with an AHP in comparison with a line installation without AHP at a minimum (design) outdoor air temperature and various AHP heat conversion factors are presented in Figures 4, 5.

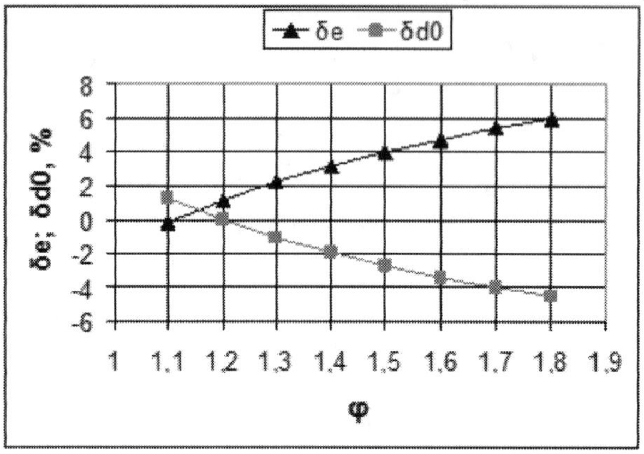

Fig. 4. The dependence of the relative change in the specific electricity generation at heating consumption δe and the specific steam consumption for the turbine δd_0 on the AHP heat conversion factor when it is included in the two-stage delivery water heating scheme at $t_{out.tem.}^{d.}$.

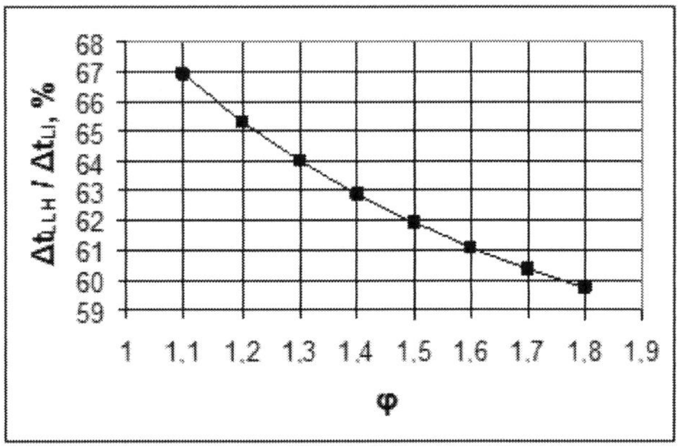

Fig. 5. The dependence of the delivery water heating proportion in the lower network heater on the heat conversion factor of the heat pump when it is included in the scheme of two-stage delivery water heating and the operation of the line installation according to the temperature schedule of the minimum design temperature of the outside air $t_{out.tem.}^{d.}$

Conclusions

1. The use of AHP for equalizing the delivery water heating in the two-stage line installation of the extraction turbine while working in the heating mode leads to the maximum increase in the indicators of heating efficiency in the operating mode at the minimum design temperature of the outside air.
2. The efficiency of the absorption heat pump inclusion in the scheme of a two-stage line installation of a STP for equalizing the delivery water heating at the minimum design temperature of the outside air depends significantly on the heat load of the heat exchangers and the AHP heat conversion factor.
3. The economic effect of using AHP increases with a decrease in the heat load of the line installation (the parameters of the heating system do not change). With a heat decrease from the nominal value of up to 50% for the turbine T-110-130, the decrease in the specific steam consumption for the turbine varies from 4% to 5.1%, and the increase in the specific heat production by heating consumption varies from 5.4 to 6.5% compared with the corresponding operating modes of the turbine without AHP.
4. The analysis of the influence of the AHP heat conversion factor φ on the efficiency of step heating at the nominal heat load of the extractions and the minimum design temperature of the outside air shows that an increase in the heating efficiency of the T-110-130 turbine unit is achieved at values greater than 1.15. The use of AHP with values φ = 1,8 allows to increase the electricity generation by heating consumption to 6% and to reduce the steam consumption for the turbine to 4.5% in comparison with the initial scheme.
5. The effect of reducing the specific steam consumption for the turbine and increasing the specific heat production by thermal absorption when introducing absorption heat pumps into the two-stage line installation is achieved by equalizing the heating of the delivery water in the line heaters. When the heat load of the line installation is reduced from 100 to 50%, the decrease in the proportion of heating of the delivery water in the lower line heater, due to the input of the AHP with the heat conversion factor φ = 1.7, changes from 19 to 9%. At the nominal heat load of the line installation, the increase in the AHP heat conversion factor from 1.1 to 1.8 leads to a change in the proportion of the delivery water heating in the lower line heater from 67 to 60%.

DOI: 10.1051/matecconf/201714101059

References

1. D.P. Elizarov, E.I. Tazhiyev, Electric stations, **2** (1994) [in Russian]
2. A. L. Buran, V. E. Kozlov & V. E. Filyukov, MATEC Web Conf. **110** (2017)
3. A. Gabdullina, N. Galashov, S. Tsibulskiy, I. Asanov, A. Kiselev, MATEC Web Conf. **91**, 01004 (2016)
4. O.Yu.Romashova, O.R. Gabidullin, Energetics: Ecology, reliability, safety, **9**, 239 (2003) [in Russian]
5. O.Yu.Romashova, O.R. Gabidullin, Energetics: Ecology, reliability, safety, **9**, 222 (2004) [in Russian]
6. O.Yu. Romashova, D.S. Luzhkovoj, A.A. Tubolev, M.A. Vagner, MATEC Web Conf. **110** (2017)

Simulation of an autonomous power supply system based on lithium-iron-phosphate ($LiFePO_4$)

Mikhail Popov[1], and *Oleg* Maniv[1,*]

[1]National Research Tomsk Polytechnic University, 634050 Tomsk, Russia

Abstract. Optimization of equipment composition and operation mode of power supply systems with renewable energy sources is an important and urgent task. As a result of the study, graphs of generation and consumption of electrical energy, charge-discharge characteristics of electric energy storage devices were constructed, the type and capacity of storage batteries were selected, the solar power plant scheme was developed and the main equipment was selected in accordance with the scheme, the economic efficiency of the solar power plant project was estimated and the payback period was determined.

1 Introduction

The most serious technical problem of the PV power plant (PVP) operation is a strong dependence of power generation on the plant's geographical location and climate conditions. Therefore, a reliable prediction of the electrical power generation is crucial for the PVP design to enable the appropriate choice of the plant equipment that ensures the energy and economic efficiency of the PVP.

Increase the capacity utilization factor of solar cells in autonomous power supply systems is possible through the use of energy storage devices, first of all it is batteries.

Different types of batteries are not only of different cost, but differ in the main parameters: the number of recharge cycles, the maximum shelf life, the yield capacity, internal resistance, size, volume of the electrolyte, the temperature range of work, the possibilities of accelerated charging. The discharge characteristics of the batteries depend on the discharge current. Battery life also depends on operational characteristics, frequency and depth of charge-discharge, charge and discharge current.

Different types of batteries have not only a different price, but differ in basic parameters: the number of charge cycles, the maximum period of storage, give capacity, internal resistance, the size, the electrolyte, operating temperature range, fast charging capabilities. The discharge characteristics of the batteries depend on the discharge current. Battery life also depends on operational characteristics, frequency and depth of charge-discharge, charge and discharge current.

* Corresponding author: mix_mix13@tpu.ru

DOI: 10.1051/matecconf/201714101060

When designing an autonomous power supply system based on renewable energy sources, we will consider lithium-iron-phosphate batteries. This is the most perfect, to date, type of battery. In addition, these types of batteries are low-maintenance, hence the operating costs associated with their use will be significantly lower. In this regard, the use of $LiFePO_4$ batteries is most convenient and profitable for a solar power plant. The main parameters that are usually used when choosing electric batteries are capacity, service life and cost.

The research aims to develop an improved method to predict the plant power output under actual operating conditions.

2 The object of the study and the method requirements

The object of the study is a stand-alone PVP, its typical scheme is shown in Fig. 1.

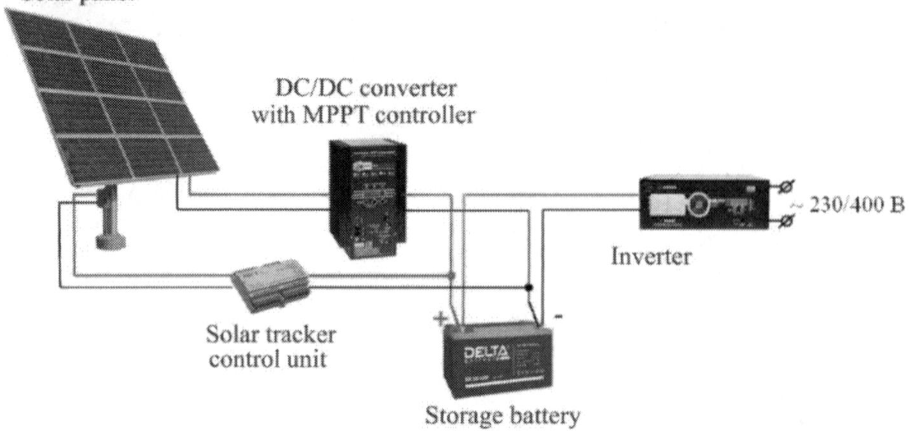

Fig. 1. Scheme of the PV power plant.

Solar energy comes from batteries on the battery and accumulates there a charge. Recharged completely, the battery begins to function as an uninterrupted power source during the night hours and hours of shortage of solar insolation. In hours of excessive insolation, the batteries act as energy storage devices, the system works directly on solar panels, consuming power for the work of power receiving devices [1-5].

For the power supply object considered in this work for power receivers in the city of Sochi in the Krasnodar Territory, the values of solar radiation are shown in Table 1.

Table 1. Solar radiation in Sochi.

Solar radiation Summ, kW·h/m²											
Jan	Feb	Mar	Apr	May	Jun	Jul	Aug	Sep	Okt	Nov	Dec
138.7	172.1	198.3	209.7	233.7	236.9	255.6	246.6	223.6	188.7	139.8	104.13

The average value of solar radiation for the year is 195.68 kW·h / m², the total generation of electricity for the year: 71421.87 kW·h, which is one of the best indicators in Russia. [6]

Within the studies performed is selected accessory to reserve the object with maximum power connected power devices - 102 kW [7, 8]. Below is a real daily load graph of the object under study (Fig. 2).

DOI: 10.1051/matecconf/201714101060

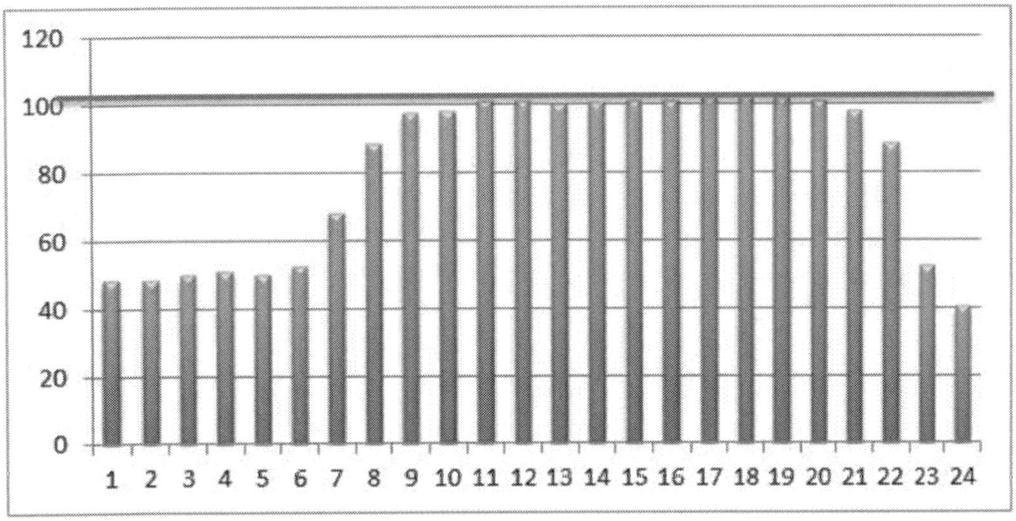

Fig. 2. Daily load graph.

2 Mathematical simulation of the PVP operating modes

The primary solar energy value available for conversion depends on the total solar radiation intensity at the installation site, geographical location of the plant, the solar panel spatial orientation, and the external meteorological factors: cloudiness level and surface reflection coefficient.

For effective use of batteries, it is necessary to accurately select the required capacity, taking into account the modes of operation of the consumer, the source and characteristics of the batteries. To do this, we will perform a mathematical simulation of the PVP operation. The simulation results are shown in the graphs below. On the figure 3 represents simulation results with insufficient battery capacity in the sistem. Figure 4 represents it closely.

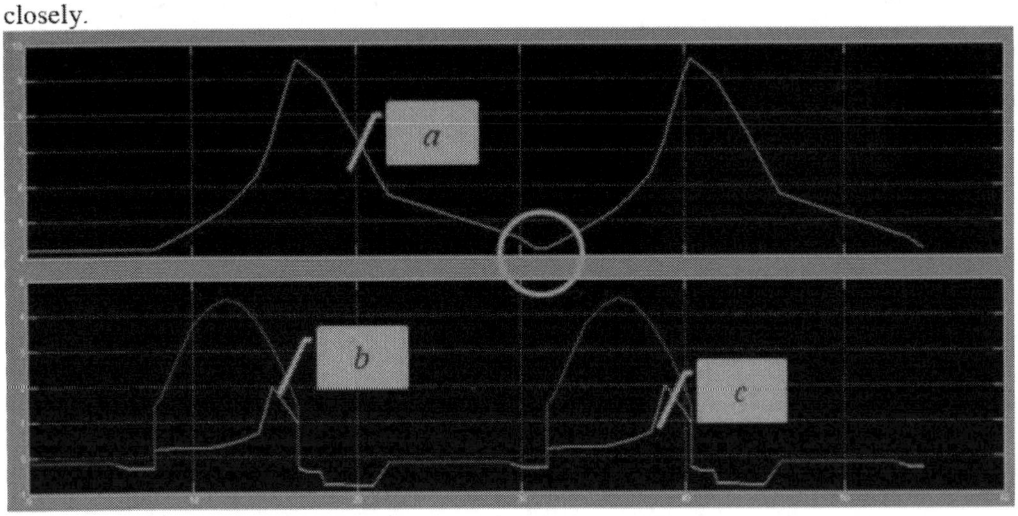

Fig. 3. *a* – Battery charge/discharge diagram, *b* – Electricity generation by solar panels, *c* – load.

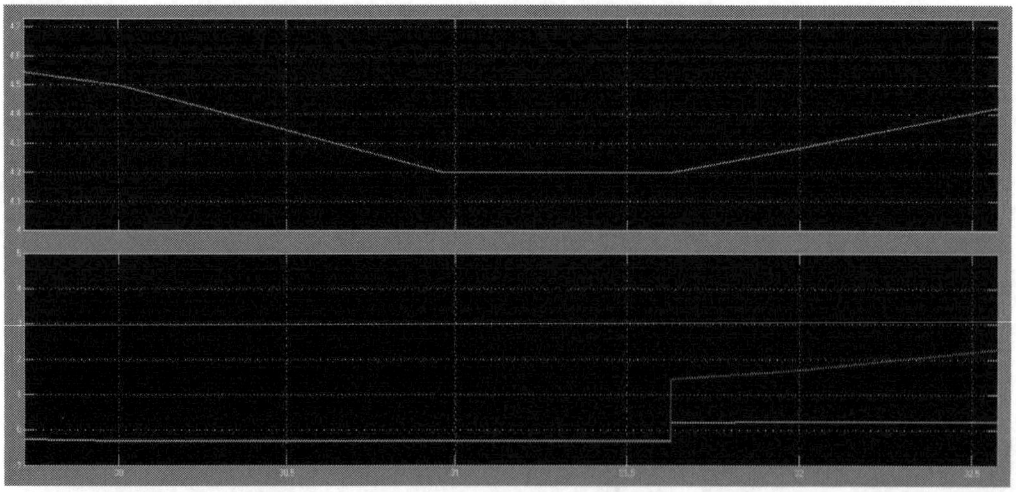

Fig. 4. Battery discharge depth.

When the battery capacity have coosen whrite, there is no unused battery capacity and there is no insufficient battery capacity, it is shown on the figure 5.

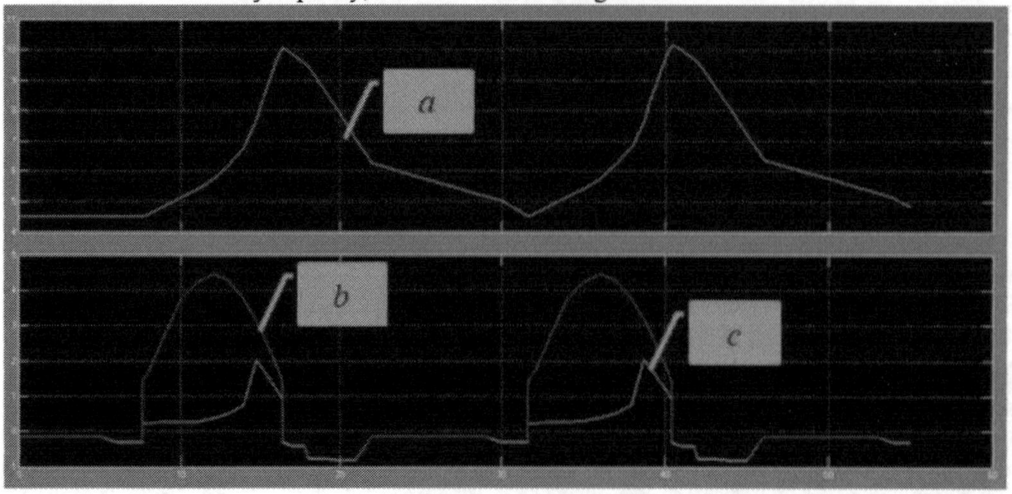

Fig. 5. *a* – Battery charge/discharge diagram, *b* – Electricity generation by solar panels, *c* – load.

3 Conclusion

An autonomous power supply system based on SES and battery was created. Two processes of the system were simulated for a given load schedule at different times, during the winter and summer solstice on the basis of simulation results, it was established that the system is fully operational and autonomous. An additional power supply is not required.

References

1. S. Obuhov, I. Plotnikov, M. Surkov, L. Sumarokova, K. Savkin, MATEC Web Conf. **91**, 01036 (2017)
2. B. Tsybikov, E. Beyerleyn, P. Tyuteva, MATEC Web Conf. **91**, 01034 (2017)

3. D.S.Akhmetbayev, D.A. Aubakir, Y.Z. Sarsikeyev, B.A. Bainiyazov, M.A. Surkov, V.I. Rozhkov, G.N. Ansabekova, A.S. Yerbolova, A.T. Suleimenov, M.S. Tokasheva, Results in Physics, **7**, 1644 (2017)
4. Y.Z. Sarsikeyev, B.V. Lukutin, D.Y. Lyapunov, M.A. Surkov, S.G. Obuhov, Adv. Mater. Res. **953-954**, 529 (2014)
5. Typical schemes of connection of RES, URL: http://www.suntechnology.ru
6. Average monthly electricity generation by regions, URL: http://www.helios-house.ru/
7. Technical conditions for connection to the electric grids of JSC "Kubanenergo". Sochi, (2014), URL:https://http://kubanenergo.ru/potrebitelyam/tekhnologicheskoe-prisoedinenie/obschaya-informatsiya-o-tekhnologicheskom-prisoedinenii/
8. Sochi is a resort city, URL:https://ru.wikipedia.org

MATEC Web of Conferences **141**, 01061 (2017) DOI: 10.1051/matecconf/201714101061
Smart Grids 2017

Solar energy for self-contained power supply

Maxim Zholobov [1], *Margarita* Gubaeva [1], *Nikolai* Kachalov [1,*], and *Olga* Kachalova [2]

[1]National Research Tomsk Polytechnic University, 634050 Tomsk, Russia
[2]The Russian Presidential Academy of National Economy and Public Administration, 634050, Tomsk Russia

Abstract. Solar energy relevance in self-contained utility system as well as economic feasibility for each class of consumers considered. The article will outline utilising features of self-contained photovoltaic stations in Middle East and Northern Africa.

1 Introduction

The world is increasingly interested in solar energy for technologies development and cutting of production costs together with conventional energy sources dearness, which have remained so for the last few years.

Middle East and Northern Africa (MENA) bear enormous potential for solar power development. After numerous theoretical studies and successful experiments conducted, the idea to harness and transform solar energy is gradually being implemented in these countries.

The challenge of responding remote regions utility-scale, industrial-scale and agricultural demand for electrical and thermal power, as well as the issues of sustainable growth and lowering polluting effect of energy production necessitates renewable energy development. By now there are national plans of renewable energy development in the number of countries. Mostly they focus on wind, solar and biomass energy.

Wide practical application of the first two is limited by well-understood disadvantages like low energy flux density and intermittence which result in significant costs on energy collection, storage and converting facilities. With regard to geographical position and climate pattern of MENA, using solar energy for power generation appears to be the most prospective method.

Relevance of solar energy as the power generation source is investigated in terms of MENA. Geographical location of these countries caused a big amount of isolated settlements long-distanced from well-developed grid. According to economic factors such as underpopulation and because of low energy demand these countries cannot be powered on to the grid. Self-contained utility system based on alternative energy can solve this problem.

* Corresponding author: mtfl@mail.ru

2 Solar irradiation characteristics

Solar power is an alternative energy, based on conversion of sunlight energy into any form. It is a renewable energy and is 'environmentally friendly', i.e. it is non-polluting during the active phase of utilising.

Solar energy comes from nuclear fusion [1] of hydrogen in the Sun core which is under high-pressure. Solar emission attenuated while direct solar radiation permeates the atmosphere and reaches the surface of Earth as direct and diffused light which behave differently: direct irradiation is manageable by various optical systems, while diffused light defy accumulation and control.

A unitary standard of solar energy measurements due to European Commission and International Electrotechnical Commission recommendations accepted normal atmosphere pressure [2]. Subsequently an additional decision was taken [3].

3 Solar power potential

Solar energy application efficiency mostly depends on the accuracy of designs in patterns and exact data on solar radiation arrival at the intended operation facilities. Creating minimum cost competitive solar units efficiently meeting the demand linked with the rationale for solar receivers optimum area, batteries rated power, type and capacity. Due to relatively low solar radiation flux density near Earth surface, the area of the prime mover receivers can be large so the receivers contribute to the whole facility cost.

Daily, seasonal, and weather intermittence of primary energy necessitates appliance of batteries and their capacity determines facility operating receivers size as well as the capacity of energy backup and, consequently, technical-and-economic indexes of the facility as a whole [4].

In general case information on solar irradiation flow and total incident energy can be obtained in the following ways:

• analytical, when the necessary parameters for a specific geographic point are determined by calculations;

• direct (usually short-term) measurements on site;

• unified methodology multi-year measurements at weather stations, the results of which are accumulated in climatic reference guides and databases.

The main source of the initial actinometric information is climatic reference guides and databases created by specialized organizations on the basis of long-term meteorological observations at weather stations results processing [5, 6].

We will consider solar power efficiency in terms of MENA.

MENA power development is influenced mainly by factors of own energy sources supply. But oil usage expansion magnifies environmental problems, moreover, recent political events proved that oil-fields cannot guarantee reliable energy supply and the orientation only on them put country's security at risk.

Solar power growth involves difficulties like the fact that the most suitable areas for its development are located far from water sources, and the most populated water adjacent areas do not have such favorable conditions for solar energy production.

Solar energy application contains two main issues: converting it into electricity and heat. By type of application solar power is divided into grid and self-contained utilities. The former includes solar power plants as a part of the grid or centralized system of the existing modern electric power industry, including transportation and distribution. In this case, at solar power plants even with a small relative density (capacity) among the traditional generating capacitance, no significant batteries are required for power fluctuations are damped by the grid [7, 8].

Self-contained solar power plants provide individual consumers, both connected and not connected with the grid. In the first case significant storage is not required as well for the grid supplies if solar energy deficiency happens, and intakes the energy surplus [9, 10].

According to the existing principle of solar energy conversion, solar power plants are divided into photovoltaic, implementing the method of direct (no-machine) conversion of solar energy into electrical energy, and thermodynamic in which radiant energy firstly converted into thermal, then in thermodynamic cycle of thermal machine the heat is converted into mechanical energy, and then in generating unit it's converted into electrical energy.

For these purposes various circuitry implemented. Management system tasks are:

• maximum efficiency of renewable energy utilising;

• generated and consumed electricity balancing, in most cases requiring batteries in the energy system;

• power converters operation control, generated electricity parameters regulation [11].

Photovoltaic method is preferable for self-contained low power solar plants (hundreds of Ws to tens of kWs) because of its constructive simplicity, which makes it more convenient to operate.

For the mentioned reasons, the fastest growth is observed in the field of photovoltaic plants. Beyond that, close attention is paid to the development of self-contained photovoltaic power plants for certain categories of consumers, for whom powering on to the grid is economically inefficient. This applies primarily to such consumers as pipelines cathodic protection stations, mobile networks current supply stations and remote agricultural consumers [12].

4 Self-contained photovoltaic stations

Photovoltaic power plants are self-contained supply systems based on renewable solar energy. This product is a source of energy to meet the needs of consumers not having a permanent connection to electric grids in the area. A significant benefit of these stations is the possibility to expand capacities with minimum costs for mechanical and start-up works, as well as to decrease power output while practical energy saving using modular solutions [13].

Mobile networks current supply photovoltaic stations. The issue of alternative energy supplying cellular network objects is being actively addressed. The focus is on the developing countries of MENA, where local energy networks often unreliable, and alternative energy sources available.

Microwave network in MENA consists of retransmission stations. The total number of photovoltaic stations in communication is currently well over 180 units with total capacity by the end of 2015.

Photovoltaic power plants in cathode protection. There is a wide network of pipelines in MENA. According to climatic conditions hybrid power plants are usually located at each cathode protection station. Implementation efficiency directly connected to the features of the region where consumers live. Presence of more than one energy source and overhead transmission lines absence along the pipeline routes significantly increase the certainty of power supply [14, 15].

Sometimes, to provide synchronous efficient operation of several cathode protection stations as well as other low-power consumers gas engine generator or gas turbine/ by-pass accumulator combination is used. Battery usage allows equating the graphics of power output and load, due to it all units repair interval increases and fuel consumption reduces.

For most countries the main obstacle to electrification of residential areas or regions remote from electrical grid is financial. High-voltage lines extension through the desert for electricity transmission to several hundred inhabitants is too cost-consuming. In many underpopulated countries electricity is available only in cities, and there is no onsite grid for feeding the rural areas.

5 Conclusion

1. Solar batteries application as well as self-contained solar power plants potential is undoubtable. Modern trends in this field show that solar energy large-scale implementation is unrelated to the problem development layer, but primarily to the economic feasibility for each class of consumers.
2. Based on climatic conditions, population density and remoteness of consumers from the grid, in MENA the front most is self-contained photovoltaic plants.

References

1. P. Mints, Refocus., **7**, 34 (2006), DOI: 10.1016/S1471-0846(06)70657-6
2. H. M. Brandhorst, Terrestrial solar cell calibration and measurement procedures. Proceedings of the Inter. Photovoltaic Solar Energy Conf., Luxemburg, 26–29 May 1977 (Reidel Publ. Co., Dordrecht-Boston, 1978)
3. *Terrestrial Photovoltaic Measurement Procedures* (Technical Memorandum 73702, NASA, Cleveland, Ohio, 1977)
4. R. E. H. Sims, Solar Energy, **76**, 9 (2004), DOI: 10.1016/S0038-092X(03)00101-4
5. A. S. Alam, L. S. Dooley, A. S. Poulton, Energy efficient relay-assisted cellular network model using base station switching, IEEE Global Telecommunications (GLOBECOM 2012): 2nd International Workshop on Multicell Cooperation, 3–7 December 2012 (Anaheim (California), USA, 2012)
6. A. F. Bosch, *Dynamic base station energy saving with relays: Research/Master Thesis* (UPC – Tsinghua University, 2011) URL: http://upcommons.upc.edu/pfc/bitstream/2099.1/10991/1/PFC.pdf
7. T. Lambert, P. Gilman, P. Lilienthal, *Micropower system modeling with homer* (John Wiley and Sons, 2005), DOI: 10.1002/0471755621.ch15
8. A. Balastov, N. Kachalov, E. Senkiv, O. Kachalova, MATEC Web of Conferences, **110**, 01008 (2017), DOI: 10.1051/matecconf/201711001008
9. J. Lorincz, T. Garma, G. Petrovic, Sensors, **12**, 4281 (2012), DOI: 10.3390/s120404281
10. D. Lee, S. Zhou, Z. Niu, GLOBECOM – IEEE Global Telecommunications Conference, 6134083 (2011), DOI: 10.1109/GLOCOM.2011.6134083
11. S. V. Glushkov, N. A. Kachalov, E. Senkiv, V. Glushkova, Matec Web of Conferences, **92**, 01072 (2016), DOI: 10.1051/matecconf/20179201072
12. C. Lubritto, *Telecommunication Power System: Energy Saving, Renewable Sources and Environmental Monitoring in Trends in Telecommunications Technologies, ed. by C.J. Bouras* (InTech, 2010), DOI: 10.5772 / 8493
13. O. Arnold, F. Richter, G. Fettweis, O. Blume, Future Network and Mobile Summit, 5722444 (2010)
14. S. N. Roy, International Telecommunications Energy Conference (Proceedings), 4664025 (2008), DOI: 10.1109/INTLEC.2008.4664025
15. J. Eyer, G. Corey, *Energy Storage for the Electricity Grid: Benefits and Market Potential Assessment Guide* (Nova Science Publishers, Inc., USA, 2011)

AUTHOR INDEX

AUTHOR INDEX